10th International Conference on Energy Materials and Environment Engineering (ICEMEE 2024)

E3S Web of Conferences Volume 560 (2024)

Kuala Lumpur, Malaysia
14-16 June 2024

Editors:

Khalida Binti Muda
Mohd Razif Harun
Goh Pei Sean
Norhaniza Binti Yusof

ISBN: 979-8-3313-0109-5

Printed from e-media with permission by:

Curran Associates, Inc.
57 Morehouse Lane
Red Hook, NY 12571

Some format issues inherent in the e-media version may also appear in this print version.

For additional information, please contact EDP Sciences – Web of Conferences
at the address below.

EDP Sciences – Web of Conferences
17, Avenue du Hoggar
Parc d'Activité de Courtabœuf
BP 112
F-91944 Les Ulis Cedex A
France

Phone: +33 (0) 1 69 18 75 75
Fax: +33 (0) 1 69 28 84 91

contact-edps@webofconferences.org

Additional copies of this publication are available from:

Curran Associates, Inc.
57 Morehouse Lane
Red Hook, NY 12571 USA
Phone: 845-758-0400
Fax: 845-758-2633
Email: curran@proceedings.com
Web: www.proceedings.com

TABLE OF CONTENTS

Preface

Nowadays, with global energy transition and environmental governance increasingly becoming the themes of the times, scientific research and technological innovation in the fields of energy materials and environmental engineering are particularly important. In order to further promote academic exchanges and cooperation in related fields, the 2024 10th International Conference on Energy Materials and Environment Engineering (ICEMEE 2024) was held via virtual form in Kuala Lumpur, Malaysia from June 14th to 16th, 2024.

The conference attracts about 80 experts and scholars, engineers and technologists from all over the world. The conference committee consists of many famous scholars, with the international one accounting for more than 60%, who have deep academic attainments and rich practical experience in their respective fields, providing a solid guarantee for the smooth progress of the conference.

The call for papers of ICEMEE 2024 covers a wide range of aspects of energy and environmental engineering, including but not limited to Renewable Energy Systems, Energy Efficiency, Biomaterials and Biological Devices, New Energy Materials, Environmental Management, Environmental Remote Sensing Applications, etc. These topics are the current research hotspots and cutting-edge issues in the field of energy and environmental engineering, attracting the attention and contributions of many scholars at home and abroad. After rigorous peer review, a number of high-quality papers were finally selected, providing the conference with rich academic content and in-depth discussion topics.

The main agenda of the conference is centered on energy and environmental engineering, aiming to build a high-level academic exchange platform and promote the collision of ideas and sharing of results between domestic and foreign experts and scholars. A number of well-known scholars worldwide gave dramatic keynote speeches and shared their latest research results and practical experience in related domains, providing participants with valuable academic insights and reference. Prof. Karamanis Dimitrios (University of Patras, Greece) presented and discussed the SERAS concept (sufficiency, efficiency, renewables and sharing) in BIPV deployment, and the prioritized integration and emerging material upgrades towards carbon neutral cities. Prof. Hossein Ganjidoust (Tarbiat Modares University, Iran) discussed the importance of obtaining biogas from municipal solid waste. It is worth mentioning that this conference also pays special attention to the industrialization cooperation of academic achievements. The participating experts and scholars not only conducted in-depth academic exchanges, but also actively discussed and cooperated on how to promote the industrialization in the field of energy and environmental engineering.

In conclusion, the 2024 10th International Conference on Energy Materials and Environment Engineering is a high-level academic event, which provides a valuable communication platform for experts and scholars engaged in the research of energy and environmental engineering. Our sincere gratitude and acknowledgement go to all the people involved in the pre-, mid- and post-session preparation of ICEMEE 2024. It is believed that in the near future, with the continuous development of energy and environmental engineering, this field will usher in a broader development prospect and richer academic achievements.

The Committee of ICEMEE 2024

Committee Member

Statement of Peer review

In submitting conference proceedings to *Web of Conferences*, the editors of the proceedings certify to the Publisher that

1. They adhere to its **Policy on Publishing Integrity** in order to safeguard good scientific practice in publishing.
2. All articles have been subjected to peer review administered by the proceedings editors.
3. Reviews have been conducted by expert referees, who have been requested to provide unbiased and constructive comments aimed, whenever possible, at improving the work.
4. Proceedings editors have taken all reasonable steps to ensure the quality of the materials they publish and their decision to accept or reject a paper for publication has been based only on the merits of the work and the relevance to the journal.

Title, date and place of the conference

2024 10th International Conference on Energy Materials and Environment Engineering (ICEMEE 2024)
June 14-16, 2024
Kuala Lumpur, Malaysia

Proceedings editor(s):

Prof. Khalida Binti Muda, Universiti Teknologi Malaysia, Malaysia
Assoc. Prof. Mohd Razif Harun, University Putra Malaysia, Malaysia
Assoc. Prof. Goh Pei Sean, Universiti Teknologi Malaysia, Malaysia
Assoc. Prof. Norhaniza binti Yusof, Universiti Teknologi Malaysia, Malaysia

Date and editor's signature

Khalida Binti Muda
Mohd Razif Harun
Goh Pei Sean
Norhaniza binti Yusof

17, avenue du Hoggar - PA de Courtabœuf – BP 112 - 91944 Les Ulis Cedex A (France)
Tél. : 33 (0)1 69 18 75 75 - Fax : 33(0)1 69 07 45 17 – www.edpsciences.org

A Review on Heat Transfer Performances of Thermosiphons Using Nanofluids

Huiying Ma[1,2,a*], Jiaokun Cao[1,2,b], and Qiang Sheng[1,2,c*]

[1] Technology and Engineering Center for Space Utilization, Chinese Academy of Sciences, Beijing, China
[2] University of Chinese Academy of Sciences, Beijing, China

Abstract: Thermosiphon transfers heat by phase-change process. Nanofluid differs from traditional working medium. Nanofluids in thermosiphon can enhance heat transfer efficiency. Different kinds of nanofluids have different effects when used in thermosiphons. Thermosiphon with nanofluid as a working medium can be used in heat collection and cooling systems and has good working performance.

1 Introduction

Thermosiphon is a highly efficient heat transfer equipment based on fluid phase change. Based on varying configurations, thermosiphons can be categorized into two types: two-phase closed thermosiphons (TPCTs) and loop thermosiphons, as depicted in Fig. 1. After entering the evaporator, the liquid absorbs heat and then evaporates to vapor, which flows through the adiabatic section or riser and finally releases heat in the condenser. The liquid created from condensation flows into the evaporator again. This process realizes heat transfer from evaporator to condenser. Gravity is the driven force, which means that heat transfer management can be achieved without any additional electricity power in the ground environment. Thus, thermosiphon costs less energy than pump-driven heat transfer devices.

Nanofluids are new heat transfer working mediums with high heat transfer performance. Nanofluids are engineered liquids that consist of nanoparticles, with diameters typically under 100 nanometers, homogeneously and persistently dispersed within traditional base fluids [1]. Fig. 2. illustrates the process of nanofluid preparation. Nanofluids' thermal conductivity differs from traditional liquids, which means that nanofluids can improve heat transfer properties. This paper reviewed several kinds of nanofluids' influence on the working properties of thermosiphons and gave relevant application scenarios. It was conducive to showing the working characteristics of thermosiphons with nanofluids as working mediums, reflecting heat transfer enhancement effect of nanofluids in thermosiphons, and improving the ideas for their applications.

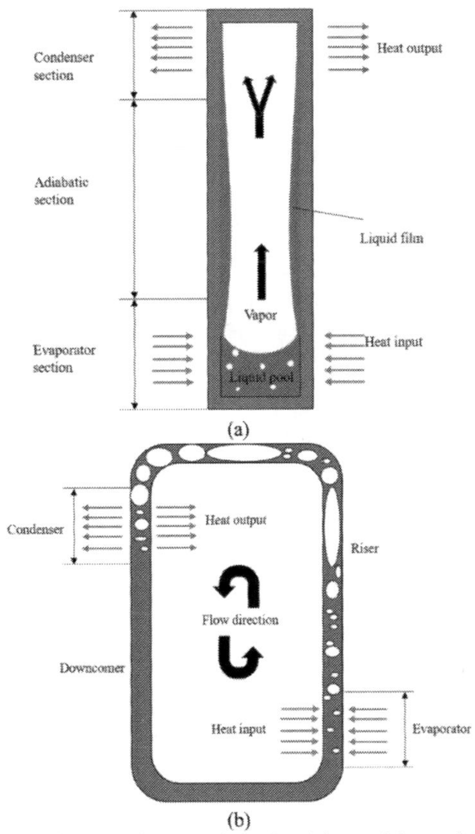

Fig.1. Diagrams of (a) two-phase closed thermosiphon and (b) loop thermosiphon

* Corresponding author: [a]huiyingma@163.com, [c]shengqiang@csu.ac.cn
[b]jiaokuncao@csu.ac.cn

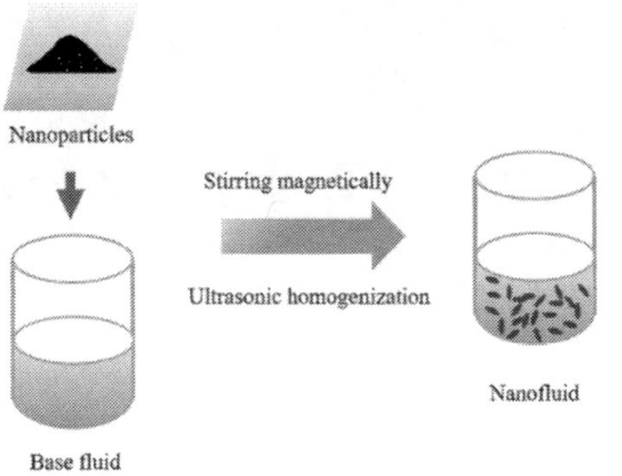

Fig. 2. Preparation of nanofluid

2 Nanofluids used in thermosiphons

2.1 Metal and metal oxide nanofluids

The use of nanofluids can lead to an improvement in thermal performance. Buschmann et al. [2] studied on titanium dioxide and gold nanofluid thermosiphon. The application of nanofluids led to a decrease in thermal resistance, with the maximum reduction observed at 24%, particularly at volume concentrations of 0.2% and 0.3%. In another study [3], the working performance of TPCT was studied under different conditions as shown in Fig. 3. and Fig. 4. Nanofluids could reduce temperature and thermal resistance.

Stability is an essential property of nanofluids. Cacua et al. [4] carried research on Al_2O_3-water nanofluids. Using nanofluids was beneficial to the working performance as shown in Fig. 5. They found compared with water, both stable and unstable nanofluids reduced the thermal resistance of TPCT by 24%. During evaporation, the nanofluids formed a precipitate and deposited on the wall. It is controversial that whether the vapor carries any nanoparticles. Buschmann et al. [2] found that when the nanofluid worked in the thermosiphon, the nanoparticles would not flow with the vapor, forming a precipitate in the condenser. Grab et al. [5] found some nanoparticles were transported with vapor. The deposited nanoparticles affected the working characteristic of the thermosiphon.

The liquid filling rate, and tilt angle affect the working process. In their research, Sarafraz and colleagues [6] discovered that the thermosiphon's maximum heat transfer efficiency was attained at a tilt of 65 degrees and a 60% fill rate.

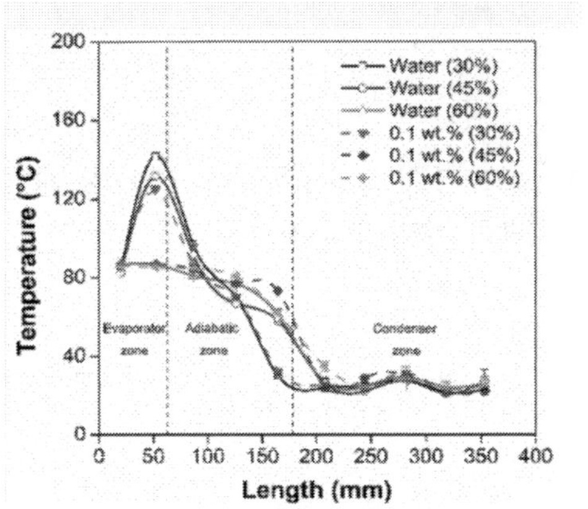

(a)

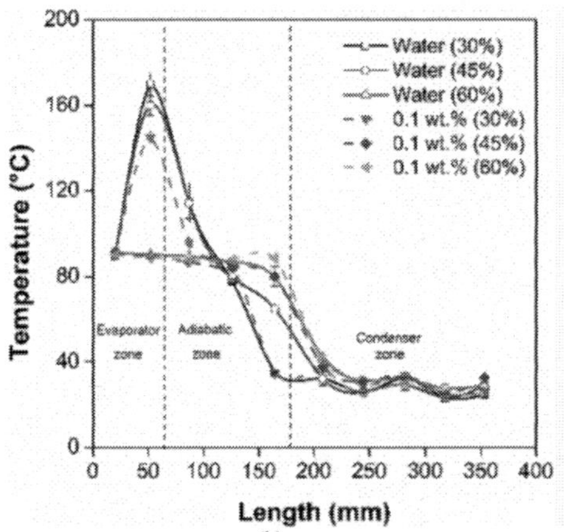

(b)

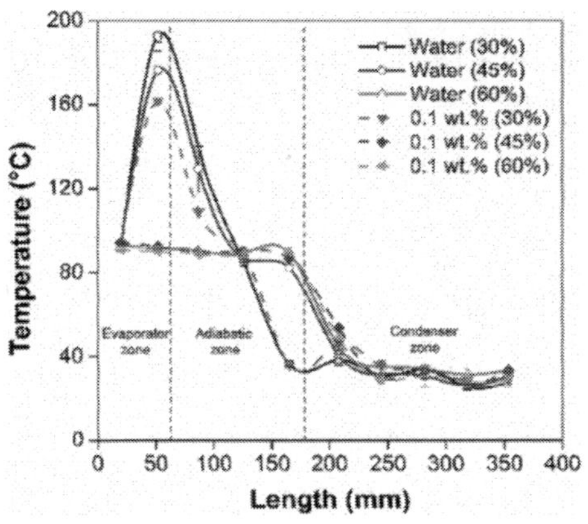

(c)

Fig.3. Temperature profiles of (a) 40 W and (b) 60 W and (c) 80 W [3]

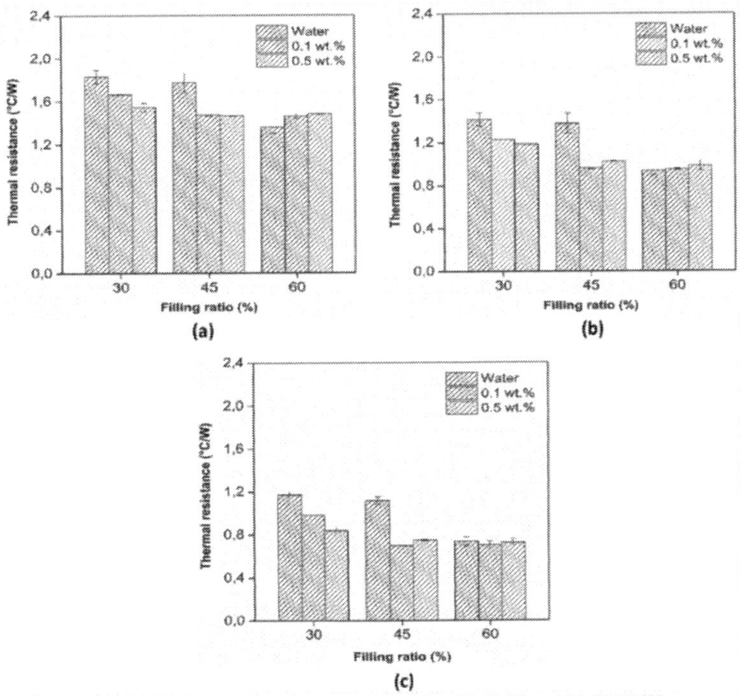

Fig. 4. TPCT thermal resistance at (a) 40 W, (b) 60 W and (c) 80 W [3]

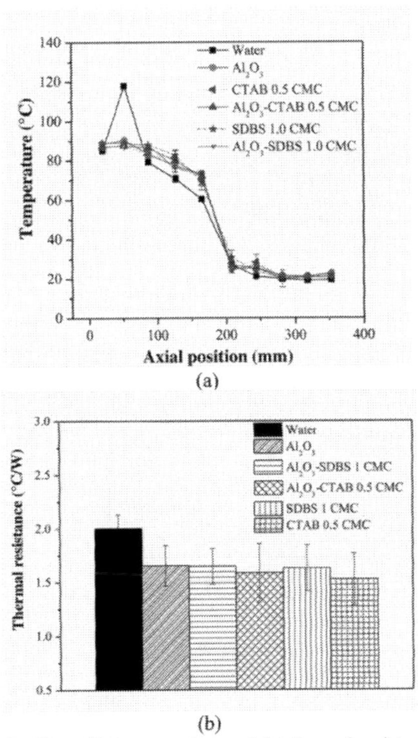

Fig.5. Profiles of (a) temperature and (b) thermal resistance [4]

2.2 Carbon-based nanofluids

Carbon-based nanofluids include graphene nanofluids, CNT nanofluids, and diamond nanofluids. The elevated thermal conductivity observed in carbon-based nanofluids surpasses that of their metal and nonmetal counterparts, attributed to the distinctive architecture of the nanoparticles involved. However, the stability is poor and the price is high, which restrict the usage of carbon-based nanofluids.

Graphene is a commonly used nanoparticle. The working performance of thermosiphon using graphene nanofluid as a working medium was studied at different thermal loads, temperatures, and tilt angles [7]. The findings indicated that the thermal conductivity increased, as illustrated in Fig. 6. In another study [8], the effects of different concentrations (0.3wt %, 0.5wt %, 1wt %) of graphene nanofluid and liquid filling rates on the performance of thermosiphon were studied through visual experiments. It was observed that at high concentrations (1wt %), many nanoparticles were carried into the condenser, causing blockage. Moreover, it caused the thermal resistance to rise.

Graphene oxide is an oxide of graphene. It can exist stably in aqueous solutions and polar solvents, so graphene oxide nanofluids have good stability. Afsari et al. [9] conducted a study examining the influences on a thermosyphon's thermal resistance and efficiency, which included factors such as graphene oxide (GO) nanofluid, the level of heat load, the presence of surfactants, and the concentration of the nanofluid. The thermal resistance experienced a 12% reduction in comparison to the scenario where water served as the working fluid. When the mass fraction was 0.3wt % and the heat load was 200W, the thermal efficiency of the thermosiphon with GO nanoparticles and SDS surfactant increased by 18%.

Carbon nanotube nanoparticles' thermal conductivity is higher than that of metal oxide nanoparticles. However, carbon nanotube nanoparticles are easy to precipitate, and the stability of the prepared nanoparticles is poor, which can be improved by functionalizing the nanotubes. Sardarabadi and colleagues [10] conducted research on the behavior of multi-walled carbon nanotubes in heat pipes that operate under the influence of gravity. Heat pipes with functionalized nanofluids had higher thermal efficiency as shown in Fig.7. Heat pipes using nanofluids operated at higher thermal loads (up to 90 W) than water.

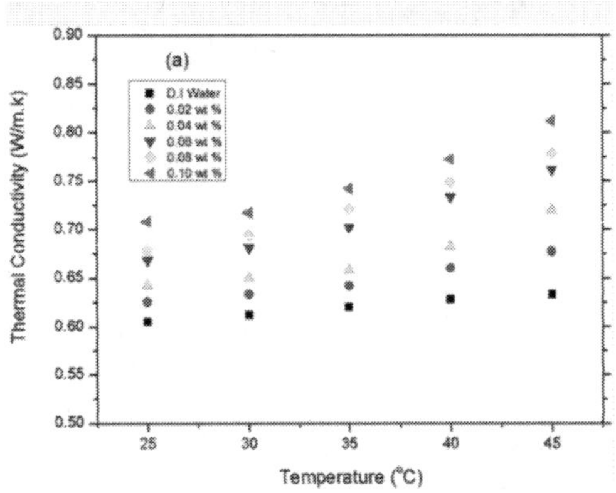

Fig.6. Profile of thermal conductivity [7]

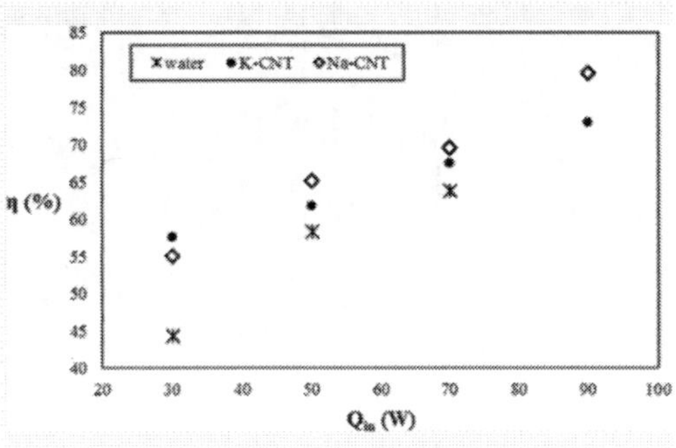

Fig.7. Profile of thermal efficiency [10]

2.3 Nonmetallic oxide nanofluids

Whether the effect of SiO_2 nanofluids when used in thermosiphon is positive or negative is controversial. Chen et al. [11] researched on loop thermosiphon with SiO_2 nanofluid as depicted in Fig. 8. Using nanofluids in the research paper had an adverse effect on the loop thermosyphon. Conversely, in a separate investigation [12] focusing on the impact of SiO2 nanofluid in an L-shaped heat pipe, there was a notable 24% decrease in thermal resistance observed. The nanofluids reported to be acting in thermosiphons has been mentioned in Table 1.

Table 1. Summary of nanofluids used in thermosiphons

Research	Nanofluid	Performance
[2]	Titanium dioxide and gold nanofluids	Thermal resistance reduction of 24%
[3]	Al_2O_3 nanofluid	With low concentration of nanoparticles or filling rate, the thermal resistance decreased.
[4]	Al_2O_3 nanofluid	Thermal resistance decrement of 24% in both stable and unstable nanofluids situations.
[5]	Titanium dioxide and gold nanofluids	The deposited nanoparticles influenced the performance.
[6]	Zirconia/ Acetone nanofluid	Heat transfer increment coefficient of 36.3%.
[7]	Graphene nanofluid	Thermal conductivity increment of 29%.
[8]	Graphene nanofluid	At high concentrations (1wt %), nanoparticle deposited in condenser, leading to rise of the thermal resistance.
[9]	Graphen oxide (GO) nanofluid	Thermal resistance reduction of 12%. Thermal efficiency of increment of 18%.
[10]	Functionalized multi-walled carbon nanotubes nanofluid	Higher operating thermal loads.
[11]	Water-based SiO_2 functionalized nanofluid	The heat transfer coefficient decreased.
[12]	Water/ SiO_2 nanofluid	Thermal resistance decrement of 24%.

3 Application of nanofluid thermosiphons

3.1 Heat collector

The heat collector collects heat so that the heat can be reused. Compared to pure water, nanofluids can enhance the heat collection effect. Alklaibi and colleagues [13] studied the operational efficiency of a solar collector utilizing nanodiamond-based nanofluid. Their findings indicated that the collector's peak thermal efficiency improved by 12.7% when using the nanofluid as opposed to using pure water. In another study [14], aqueous copper oxide nanofluid was used in a heat pipe which worked as an absorber of a solar collector. The influence of nanoparticle concentration and size, liquid filling rate were studied. It was found that solar collectors' efficiency was improved after using nanofluids.

https://doi.org/10.1051/e3sconf/202456001001

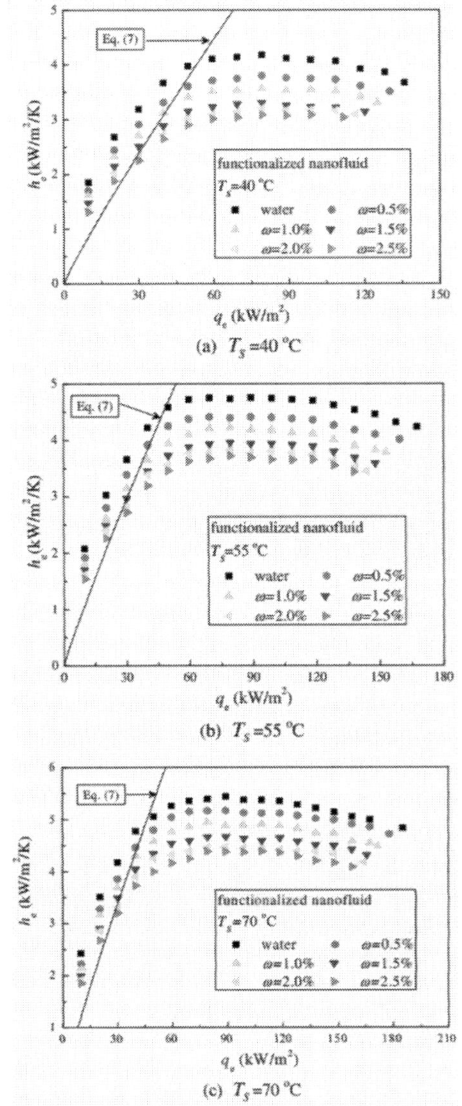

Fig.8. Heat transfer coefficient of (a) 40°C, (b) 55°C and (c) 70°C [11]

3.2 Photovoltaic/thermal (PV/T) system

A PV/T system is established by harnessing the heat from the solar collector to produce both thermal and electrical energy. Using methanol and Al_2O_3/methanol nanofluid, the properties at different liquid filling rates and nanoparticle concentrations were experimentally studied by Moradgholi et al. [15]. Fig. 9. shows the layout diagram of the system. The findings demonstrated that when compared to a standard photovoltaic (PV) panel, the temperature which incorporated nanofluid, was lowered by 14.52 °C, and it produced an additional 1.42 W of electrical energy.

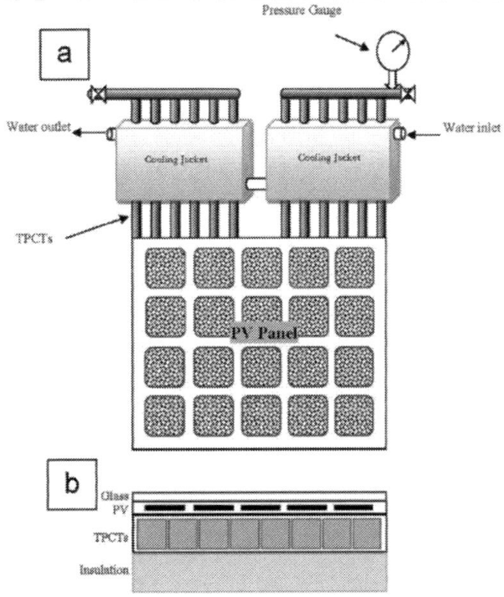

Fig. 9. The PV/T diagrams (a) front view (b) top view [15]

3.3 Chip cooling

During the operation of the chip, due to the small space, the heat flux generated is large, so the heating phenomenon can quickly burn the chip or reduce the service life, so the heat needs to be discharged in time. A micro-scale thermosiphon heat pipe uses phase conversion heat to cool the chip quickly. A micro-scale thermosiphon used for chip cooling using nanofluids (R113+surfactant+CuO nanoparticles) as the working medium was studied by Zhang et al. [16]. The structure was shown in Fig.10. The micro-channel design within the chip was employed to serve as the evaporative component of the thermosiphon for heat transfer. The experimental outcomes revealed that the implementation of nanofluids notably elevated the peak heat flux during the boiling phase of evaporation in the thermosiphon, suggesting a promising future for its application in cooling semiconductor chips.

In summary, nanofluid thermosiphons can be used to collect or release heat for the purpose of saving energy and maintaining ideal performance of equipment. Owing to the augmented heat transfer capabilities of nanofluids, there is potential for the large-scale industrial deployment of nanofluid-based thermosiphons. Additionally, these systems could be instrumental in achieving thermal regulation for lunar spacecraft operations.

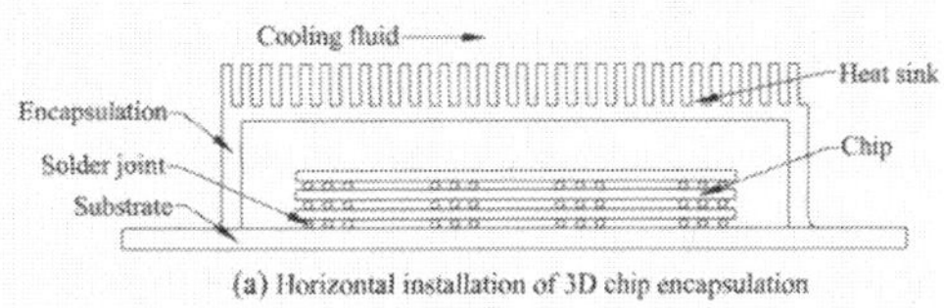

(a)

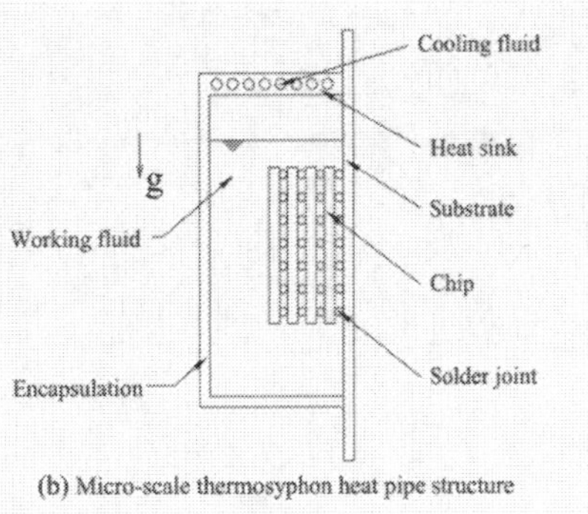

(b)

Fig.10. Diagram of (a) installation and (b) structure [16]

4 Conclusion

The application of nanofluids in thermosiphon technology was reviewed in this paper. Compared with traditional fluids, the thermophysical parameters of nanofluids are changed, which causes performance differences when working in a thermosiphon. Furthermore, the improved heat transfer efficiency and decreased thermal resistance in the thermosiphon due to the nanofluid are connected to the nanoparticles' accumulation on the evaporator's interior walls. Nonetheless, suboptimal design specifications could result in a decline of heat transfer effectiveness. Therefore, prudent design and computation are essential. The nanofluid thermosiphons can be used in heat collectors, solar photovoltaic systems, and chip cooling. But unclear effect of nanofluids on thermosiphons and unknown working life of nanofluid thermosiphons restrict the widespread of nanofluid thermosiphons. Therefore, there should be more investigations on how design parameters of nanofluid influence the thermosiphon's performance, the heat transfer enhancement mechanisms of nanofluid thermosiphons and operational reliability assessments of nanofluid thermosiphons.

References

1. Hajian, R., Layeghi, M., Abbaspour Sani, K. (2012) Experimental study of nanofluid effects on the thermal performance with response time of heat pipe. Energ. Convers. Manage., 56, 63-68. https://doi.org/10.1016/j.enconman.2011.11.010
2. Buschmann, M. H., Franzke, U. (2014) Improvement of thermosyphon performance by employing nanofluid. Int. J. Refrig., 40, 416-428. https://doi.org/10.1016/j.ijrefrig.2013.11.022
3. Gallego, A., Herrera, B., Buitrago-Sierra, R., Zapata, C., Cacua, K. (2020) Influence of filling ratio on the thermal performance and efficiency of a thermosyphon operating with Al_2O_3-water based nanofluids. Nano-Struct. Nano-Objects, 22, 100448. https://doi.org/10.1016/j.nanoso.2020.100448
4. Cacua, K., Buitrago-Sierra, R., Pabón, E., Gallego, A., Zapata, C., Herrera, B. (2020) Nanofluids stability effect on a thermosyphon thermal performance. Int. J. Therm. Sci., 153, 106347. https://doi.org/10.1016/j.ijthermalsci.2020.106347
5. Grab, T., Gross, U., Franzke, U., Buschmann, M. H. (2014) Operation performance of thermosyphons employing titania and gold nanofluids. Int. J. Heat Mass Tran., 86, 352-364. https://doi.org/10.1016/j.ijthermalsci.2014.06.019
6. Sarafraz, M. M., Pourmehran, O., Yang, B., Arjomandi, M. (2019) Assessment of the thermal performance of a thermosyphon heat pipe using zirconia-acetone nanofluids. Renew. Energ., 136, 884-895. https://doi.org/10.1016/j.renene.2019.01.035.
7. Das, S., Giri, A., Samanta, S., Kanagaraj, S. (2019) Role of graphene nanofluids on heat transfer enhancement in thermosyphon. J. Sci.: Adv. Mater. Devices, 4(1), 163-169. https://doi.org/10.1016/j.jsamd.2019.01.005
8. Fulpagare, Y., Tsai, D.-Y., Wang, C.-C. (2022) Performance of two-phase loop thermosiphon with graphene nanofluid. Appl. Therm. Eng., 200, 117714. https://doi.org/10.1016/j.applthermaleng.2021.117714
9. Afsari, K., Sarmasti Emami, M. R., Zahmatkesh, S., Jaromír Klemeš, J., Bokhari, A. (2023) Optimizing the thermal performance of the thermosyphon heat pipe for energy saving with graphene oxide nanofluid. Energy, 274: 127422. https://doi.org/10.1016/j.energy.2023.127422
10. Sardarabadi, H., Zeinali Heris, S., Ahmadpour, A., Passandideh-Fard, M. (2019) Experimental investigation of a novel type of two-phase closed thermosyphon filled with functionalized carbon nanotubes/water nanofluids for electronic cooling application. Energ. Convers. Manage., 188, 321-332. https://doi.org/10.1016/j.enconman.2019.03.070
11. Chen, Y.-J., Wang, P.-Y., Liu, Z.-H. (2013) Application of water-based SiO_2 functionalized nanofluid in a loop thermosyphon. Int. J. Heat Mass Tran., 56(1), 59-68. https://doi.org/10.1016/j.ijheatmasstransfer.2012.09.048
12. Khajehpour, E., Noghrehabadi, A. R., Nasab, A. E., Nabavi, S. M. H. (2020) Experimental investigation of the effect of nanofluids on the thermal resistance of a thermosiphon L-shape heat pipe at different angles. Int. Commun. Heat Mass, 113, 104549. https://doi.org/10.1016/j.icheatmasstransfer.2020.104549
13. Alklaibi, A. M., Sundar, L. S., Sousa, A. C. M. (2021) Experimental analysis of exergy efficiency and entropy generation of diamond/water nanofluids flow in a thermosyphon flat plate solar collector. Int. Commun. Heat Mass, 120, 105057. https://doi.org/10.1016/j.icheatmasstransfer.2020.105057
14. Chilbule, P. V., Dhole, L. P., Chavhan, G. R. (2023) Optimization of heat pipe charged with CuO

nanofluid using Taguchi technique. Mater. Today: Proc.. https://doi.org/10.1016/j.matpr.2023.08.343

15. Moradgholi, M., Mostafa Nowee, S., Farzaneh, A. (2018) Experimental study of using Al_2O_3/methanol nanofluid in a two phase closed thermosyphon (TPCT) array as a novel photovoltaic/thermal system. Sol. Energy, 164, 243-250. https://doi.org/10.1016/j.solener.2018.02.055
16. Zhang, K.-l., Liu, Z.-H., Zheng, B.-c. (2016) A new 3D chip cooling technology using micro-channels thermosyphon with super-moist fluids and nanofluids. Energ. Convers. Manage., 128, 44-56. https://doi.org/10.1016/j.enconman.2016.09.065

A Study of the Effect of CO_2-Water-Rock Reaction on Caprock Mechanical Sealing Ability

Bin Liu [1,2,a*], and Tianxiang Hu [1,b]

[1]School of Electrical Engineering and Information, Northeast Petroleum University, Daqing 163318, China.
[2]Bohai Energy Research Institute, Northeast Petroleum University, Qinhuangdao 066004, China.

Abstract: In this paper, the Johansen formation in the Norwegian CO_2 geologic sequestration project is taken as the research object, and a multi-field coupled chemical-seepage-stress model is established to analyze the effects of CO_2-water-rock reaction on the mechanical structural parameters, such as the pore pressure and the effective stress. Tensile safety factor and shear safety factor of the caprock are selected as evaluation indexes to describe the evolution of the caprock mechanical sealing ability over time dynamically. The simulation results show that the mechanical parameters change a lot during 0-50 years, and the change of mechanical parameters near the injection wells is larger than that far away from the injection wells, indicating that mechanical damage is more likely to occur near the injection wells. However, in 200 years, the tensile safety factor and shear safety factor of the caprock are always greater than 0, indicating that no tensile damage or shear damage occurs in the caprock, and the caprock sealing ability is good.

1. Introduction

With the rapid development of society, the excessive use of fossil fuels such as coal, oil and natural gas in industrial production and people's daily lives has led to a sharp increase in CO_2 emissions[1]. There is a global consensus that if nothing is done to curb the problem, increasing CO_2 concentrations will disrupt the Earth's climate, contribute to sea level rise and damage ecosystems. In order to reduce atmospheric CO_2 emissions, CO_2 geologic sequestration (CGS) has become one of the most promising approaches. One of the key conditions for the implementation of CO_2 sequestration project is to check the caprock sealing ability after CO_2 injection into the formation, and whether it can effectively prevent the leakage of CO_2[2].

At the present stage, the research on this problem can be divided into two categories, the first category is to study the geochemical reactions during the storage process alone, and the second category is to study the change of the microporous structure of the formation caused by the change of the formation's stress state alone[3]. However, for CO_2 storage, the three processes exist simultaneously and interact with each other. In this paper, the Johansen Formation in the Norwegian North Sea is taken as the research object, and a multi-field coupled chemical-seepage-stress model is established to reveal the mechanism of caprock sealing ability failure, to clarify the conditions of caprock mechanical damage, and to establish a method for evaluating the mechanical sealing ability of the caprock by taking the tensile safety factor and the shear safety factor as evaluation indexes.

* Corresponding author: [a]liubinnepu@163.com
[b]htx18812039070@163.com

2. Basic theory

2.1. Coupling mechanism

The basic equation for fluid conservation in deformable porous media can be expressed as follows:

$$\frac{\partial}{\partial t}\left[\phi\rho_f(1-\varepsilon_v)\right]-\nabla\left[\rho_f\frac{k}{\mu}(\nabla p-\rho_f b)\right]=Q_f \tag{1}$$

Where b is the volumetric force per unit mass of fluid, k is the permeability, p is the pressure at time t, Q_f is the fluid velocity, ε_v is the volumetric strain, ϕ is the porosity, μ is the fluid viscosity, ρ_f is the fluid density. Changes in porosity and volumetric strain are caused not only by mechanical changes but also by geochemical effects such as dissolution and precipitation of minerals.

The constitutive equation for the stress-strain relationship is shown below:

$$\sigma_{ij}=2G\varepsilon_{ij}+\left(K-\frac{2G}{3}\right)\varepsilon_{kk}\delta_{ij}+\alpha p\delta_{ij} \tag{2}$$

Where σ_{ij} is stress tensor, G is shear modulus, K is bulk modulus, p is the fluid pressure, α is Biot coefficient ($0\leq\alpha\leq1$), δ_{ij} is Kronecker function, ε_{ij} is strain tensor, ε_{kk} is volumetric strain ($=\varepsilon_v$).

The effective stress in a porous medium can be expressed as:

$$\sigma_{ij}'=\sigma_{ij}-\alpha p\delta_{ij} \tag{3}$$

Where σ_{ij}' is the effective stress, σ_{ij} is total stress.

2.2. Methods for evaluating caprock mechanical sealing ability

2.2.1. Tensile damage

The tensile safety factor is used as an evaluation index to evaluate whether the caprock is tensile damaged or not, the tensile safety factor can reflect the ability of the caprock to resist tensile damage under the combined effect of chemical reaction and pore pressure:

$$\xi = \frac{\sigma_3^{'}}{\sigma_3} \tag{4}$$

Where $\sigma_3^{'}$ is the minimum effective stress, and σ_3 is the minimum effective principal stress. When the value is less than 0, the rock undergoes tensile damage; when the value of the tensile safety factor is greater than 0, the rock will not be tensile damaged. The smaller the value, the greater the risk of tensile damage.

2.2.2. Shear damage

Shear damage can be expressed as the shear safety factor, which can reflect the ability of caprock to resist shear damage under the combined effect of chemical reaction and pore pressure. From the derivation of Mohr-Coulomb principal, the maximum shear stress to which the rock is subjected after CO_2 injection is:

$$\tau_m = \frac{(\sigma_1^{'} - \sigma_3^{'})}{2} \tag{5}$$

The maximum shear stress that a rock can withstand is:

$$\tau_m^{*} = \left[C \cot\varphi + (\sigma_1^{'} - \sigma_3^{'})/2 \right] \sin\varphi \tag{6}$$

Then the shear safety factor can be expressed as:

$$\chi = 1 - \frac{\tau_m}{\tau_m^{*}} = \frac{(\sigma_1^{'} - \sigma_3^{'})/2}{c\cos\varphi + (\sigma_1^{'} + \sigma_3^{'})\sin\varphi/2} \tag{7}$$

Where, c is the cohesion, φ is the angle of internal friction, $\sigma_1^{'}$ is the maximum effective principal stress, $\sigma_3^{'}$ is the minimum effective principal stress. If $\chi > 0$, the caprock will not undergo shear damage. If $\chi \leq 0$, the caprock undergoes shear damage.

3. Model description

The Johansen Formation is located in the southeastern part of the central block of the Norwegian North Sea, offshore from the municipality of Bergen on the western coast of Norway, with approximate coordinates between 60° and 61° N and 3° and 4° E. The Johansen Formation consists of three main units, including the Hegre Formation of the Upper Triassic, the Cook Formation of the Lower Cretaceous, and the Frigg Formation of the Lower-Middle Cretaceous[4].

In this study, a coupled chemical-seepage-stress model is developed to investigate the effect of geochemistry on the mechanical properties of the caprock Figure 1 shows a schematic of a CO_2 storage site modeled by a dense mudstone caprock and a sandstone reservoir. Three points at the top of the caprock are selected as monitoring points for analysis, with coordinates a(46,48,1), b(39,48,1), and c(32,48,1), respectively.

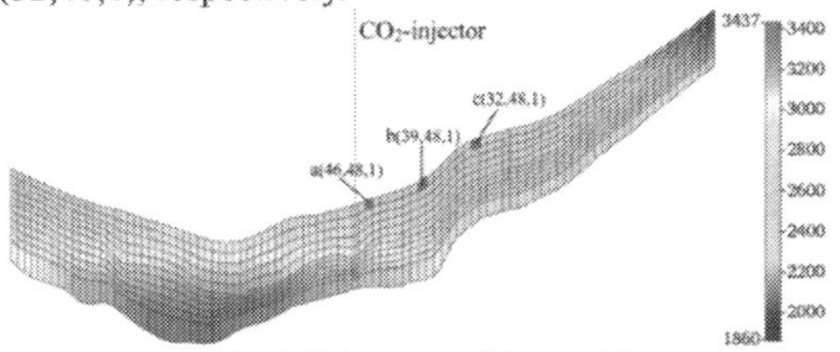

Fig. 1. Schematic of the model

The basic parameters of the model are shown in Table 1, the depth of the model is 1860-3437 m, the vertical thickness is 10-133 m. It is divided into 100×100×11 grid blocks, and the injection well is perforated at (48,48,9), (48,48,10), and (48,48,11). CO_2 is injected at a rate of 1.4×10^4 m^3 per day for 25 years, and the simulation lasted for 200 years[5].

Table 1. Parameters of the model

Parameters	Caprock	Reservoir
Porosity	0.01	0.22
Permeability (mD)	0.001	225
Average temperature (℃)	94	
Reference pressure (kPa)	31310kPa at 3100m	
Injection rate (m³/day)	1.4×10^4	
Rock compressibility (1/kPa)	4.35×10^{-7} (at 100kPa)	

Injection of CO_2 into saline aquifers results in chemical reactions with brine and minerals. Table 2 shows the mineral volume fractions used to initialize the model, and Table 3 lists the initial viscosity of each ion in the brine. The mechanical parameters of the caprock and reservoir entered in this model are shown in Table 4[6].

Table 2. Initial volume fraction of minerals (%)

Mineral	Quartz	Calcite	Illite	Kaolinite	Chamosite	Dolomite
Caprock	44.48	0.70	8.46	18.40	23.69	0.49
Reservoir	70.31	0.09	3.54	1.35	8.60	0.53

Table 3. Initial concentration of brine ions (mol/kg H_2O)

Ionic	Viscosity	Ionic	Viscosity
Fe^{2+}	6.45×10^{-4}	Al^{3+}	2.32×10^{-11}
Ca^{2+}	2.22×10^{-2}	SiO_2(aq)	5.63×10^{-3}
Mg^{2+}	2.18×10^{-3}	Na^{+}	3.30×10^{-1}
K^{+}	2.89×10^{-3}	Cl^{-}	3.77×10^{-1}

Table 4. Mechanical parameters of caprock and reservoir

Parameters	Caprock	Reservoir
Young's modulus (GPa)	7.1	19.9
Poisson's ratio	0.369	0.227
Cohesion (MPa)	21.6	12.0
Internal friction angle (°)	24.16	40.77
Initial stress field (MPa)	(20, 20, 20)	

4. Simulation results and analysis

4.1. Pore pressure

CO_2 injection destroys the original pressure balance of the formation, leading to significant changes in the pore pressure distribution of the formation, and the change in pore pressure is also the main reason for the changes in other mechanical parameters. Figure 2 shows the pore pressure change curves at the three monitoring points, from which it can be seen that the pore pressures at all three monitoring points show a trend of first increasing, then decreasing, and finally leveling off. The pore pressure at monitoring point a doesn't change during the first 5 years of injection because the vertical transportation distance of CO_2 hasn't affected the top of the caprock. It gradually increases from 28.8 MPa to 41.2 MPa at the end of the injection period (t=25 year), and then doesn't decrease immediately after the end of the injection period because CO_2 continued to diffuse upward. Between 25 and 40 years, it continues to increase to 43.6 MPa, and after 40 years, it begins to gradually decrease to 43.0 MPa, and finally stabilizes, with a maximum change of about 51.4%. The maximum change in pore pressure at monitoring point b is about 50.7%. The maximum change in pore pressure at monitoring point c is about 50.6%.

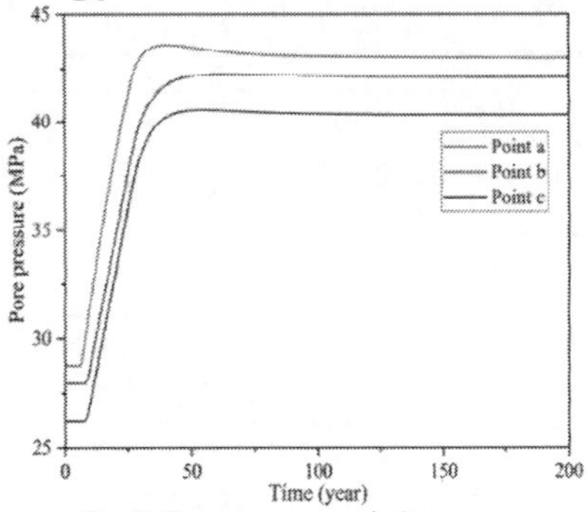

Fig. 2. Pore pressure variation curve

4.2. Effective stress

According to the principle of effective stress in porous media mechanics, the effective stress is equal to the total stress minus the pore pressure, the decrease of the effective stress due to the injection of CO_2 is equal to the increase of the pore pressure, so that the effective stress as a whole exhibits the opposite trend of change with the pore pressure.

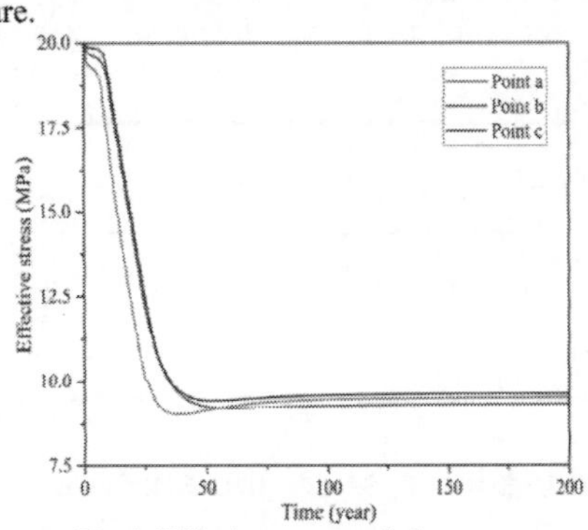

Fig. 3. Effective stress variation curve

Figure 3 shows the change curves of the effective stress at the three monitoring points, from which it can be seen that the effective stress shows an overall trend of first decreasing, then slightly increasing, and finally leveling off. The effective stress at monitoring point a decreases from 20 MPa at the beginning to 10 MPa during the injection period, continues to decrease to 9 MPa from 25 to 40 years after the end of the injection, and then begins to gradually increase to 9.5 MPa, and finally tends to level off, with a maximum change of about 55%. The maximum change in effective stress at monitoring point b is about 54%. The maximum change in effective stress at monitoring point c is about 53%. It can be seen that the effective stress variation is greater near the injection wells.

4.3. Tensile safety factor

Figure 4 shows the variation curves of the tensile safety factor at the three monitoring points, and the tensile safety factor shows a trend of decreasing, then increasing, and finally leveling off. The tensile safety factor at monitoring point a decreases from 0.41 to 0.152 at the end of the injection (t=25 year), and then continues to decrease from 25 to 40 years, decreasing to 0.106 at 40 years, and then slowly recovers after 40 years to 0.117 at 100 years, after which it gradually stabilized. The tensile safety factor at monitoring point b decreases from 0.417 to 0.213 at the end of injection, and then decreases to 0.119 at the end of 60 years, and then starts to recover slowly to 0.123 at the end of 200 years. At monitoring point c, the tensile safety factor decreases from 0.433 to 0.217 at the end of the injection, and then decreases to 0.122 at the end of 60 years, and then begins to recover slowly to 0.128 at the end of 200 years. The change in the tensile safety factor at monitoring point a is greater than that at the other two monitoring points, suggesting that there is a greater risk of tensile damage in the vicinity of the injection wells.

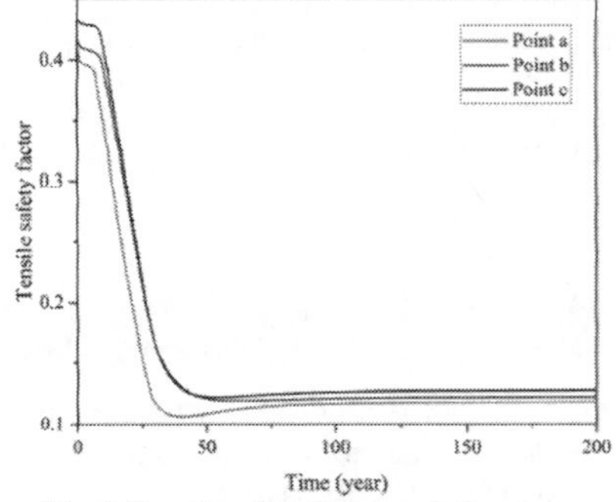

Fig. 4. Tensile safety factor variation curve

The reason for the above changes is due to the interaction between CO_2-water-rock during the CO_2 injection stage, under the influence of this action, the tensile strength of the rock shows a decreasing trend, and at the same time, due to the injection of CO_2, the minimum effective stress gradually decreases and the minimum principal stress gradually increases, and the tensile safety factor of the rock decreases at a faster rate under the combined influence of these two factors. After stopping the CO_2 injection, the rock tensile safety factor will continue to decrease in a short time due to the diffusion and transportation of CO_2 in the vertical direction. After that, the rock tensile safety factor slightly recovered because both the minimum principal stress and the minimum effective stress were restored to the initial state, but the tensile safety factor finally did not change significantly as the minimum principal stress and the minimum effective stress gradually stabilized. In this process, the tensile safety factor is always greater than zero, indicating that the caprock will not undergo tensile damage.

4.4. Shear safety factor

Figure 5 shows the variation curves of the shear safety factor at the three monitoring points, and the shear safety factor shows a similar trend to the tensile safety factor, which first decreases, then undergoes a slow recovery and finally stabilizes. The shear safety factor at monitoring point a decreases from the initial 1.000 to 0.912 at the end of injection (t=25 year), then continues to decrease from 25 to 40 years due to diffusive transport of CO_2 to 0.872 at 40 years, and then slowly recovers and levels off after 40 years to 0.877 at 200 years. The shear safety factor at monitoring point b decreases from 1.000 to 0.935 at the end of injection, to 0.884 at 60 years, and then starts to slowly recover and level off, recovering to 0.885 at 200 years. The tensile factor of safety at monitoring point c decreases from 1.000 to 0.921 at the end of injection, to 0.876 at 60 years, and then starts to slowly recover and level off, recovering to 0.877 at 200 years. The change in the shear safety factor at monitoring point a is greater than at the other two monitoring points, indicating that there is a greater risk of shear damage in the vicinity of the injection well.

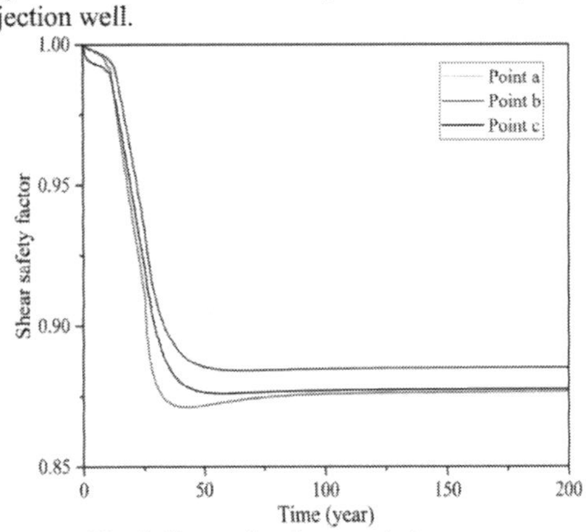

Fig. 5. Shear safety factor variation curve

The reason for the above changes is that during the injection stage, as the pore pressure increases, the effective stress of the formation decreases, causing the Mohr stress circle to move closer to the damage envelope, which results in a decrease in the rock shear safety factor. After stopping CO_2 injection, the shear safety factor of the rock continues to decrease in a short period of time due to the diffusion and transportation of CO_2. With the pore pressure of the formation slightly reduces, the effective stress of the formation also slightly rebounds, so that the Mohr circle is slightly away from the damage envelope, so the rock shear safety factor will be slightly rebounds, but always smaller than the initial state. However, it can be seen from Figure 5 that the shear safety factor of the rock is always in the state of greater than 0, indicating that the caprock will not occur shear damage.

5. Conclusions

In this paper, a chemical-seepage-stress three-field coupling model is established to simulate the changes of mechanical structural parameters, such as pore pressure and effective stress. The simulation results show that the pore pressure shows a trend of increasing and then decreasing, and finally tends to level off, and the maximum change of pore pressure is 51.4%. The effective stress shows a trend of decreasing and then increasing, and then tends to level off, and the maximum change of the effective stress is about 55%. The tensile safety factor and shear safety factor of the caprock show a trend of decreasing and then increasing, and then tend to level off. However, the tensile safety factor and shear safety factor of the caprock are always greater than 0, indicating that no tensile or shear damage occurs in the caprock, and the caprock sealing ability is good.

Acknowledgment

This research is funded by the Natural Science Foundation of Heilongjiang Province of China, grant number LH2023D004; National Key Research and Development Program of China, grant number 2022YFE0206800; CNPC Innovation Fund, grant number 2021DQ02-1103.

References

1. Masson-Delmotie V, Zhai P, Portner H O, et al. (2022) Global Warming of 1.5°C: IPCC special report on impacts of global warming of 1.5°C above pre-industrial levels in context of strengthening response to climate change, sustainable development, and efforts to eradicate poverty. Cambridge University Press, England.
2. Deng Y R, Wang Y H, Zhao Y J, et al. (2023) Carbon dioxide storage in China: Current status, main challenges, and future outlooks. Earth Science Frontiers, 30: 429-439.
3. Han X, Feng F, Cong Z. (2020) Dynamic evaluation method of caprock microscopic sealing in CO_2 sequestration project. Geofluids, 2020: 1-12.
4. Eigestad G T, Dahle H K, Hellewang B, et al. (2009) Geological modeling and simulation of CO_2 injection in the Johansen formation. Computational Geosciences, 13: 435-450.
5. Adu-Gyamfi B, Ampomah W, Tu J, et al. (2022) Assessment of chemo-mechanical impacts of CO_2 sequestration on the caprock formation in Farnsworth oil field, Texas. Scientific Reports, 12: 13023.
6. Sundal A, Hellevang H. (2019) Using reservoir geology and petrographic observations to improve CO_2 mineralization estimates: examples from the Johansen formation, North Sea, Norway. Minerals, 9: 671-683.

Simulation and analysis of radiant intensity for critical

Shaoze Li[1,a*],Fuqi Si[2,b]

[1]College of Biofood, Environment and Biology, Hefei University, Economic Development Zone, Hefei, China
[2]Key Laboratory of Environmental Optics and Technology, Anhui Institute of Optics and Fine Mechanics, Hefei Academy of Physical Sciences, Chinese Academy of Sciences, Shushan District, Hefei, China

Abstract: Atmospheric oxygen A-band (759-767nm) airglow has strong radiation intensity in the visible band, and the global atmospheric temperature field can be obtained based on the oxygen A-spectrum line. In order to realize the marginal observation of oxygen A-airglow, the intensity distribution and change at different heights need to be simulated. Firstly, the body emissivity at different heights and solar zenith angles is obtained based on the photochemical model of oxygen A-band airglow. Secondly, the oxygen airglow radiation transmission process is established based on the observation geometry of the edge. Then, the airglow radiation intensity is simulated based on the detection geometry path of the edge to obtain the airglow spectrum lines at different observation heights of the edge.

1 Preface

The near-space atmosphere (60~110km) will affect the accuracy of weather forecasting and the safety assessment of spacecraft launch and orbit[1][2], and the observation and study of near-space atmosphere has important military value and scientific signific-ance[3][4]. Oxygen A-band gas glow is one of the brightest spectra in the visible spectrum, and the distribution and change of gas glow spectral lines can be used to study the problems of atmospheric composition and temperature distribution. The wavelength region of oxygen A-band gas glow spectrum is 759~767nm, and it has strong self-absorption below 70 km[5].Spaceborne detection can achieve global observation in a short time, provide global temperature distribution data, and have a large spatial coverage. The Space Physics Laboratory of the United States uses the high-resolution Doppler imager (HRDI) to measure the temperature of the emission spectral lines in the oxygen A-band[6].The optimal estimation method based on the perturbation theory was used to invert the atmosphere at an altitude of 80~100km, and the temperature profile with an error of about 7K was obtained by using the coordinated inversion of three spectral lines. The Odin satellite is a Swedish and Canadian satellite that carries OSIRIS instruments, which include spectrometers and infrared imaging systems. These instruments are designed to make observations of atmospheric composition and structure. OSIRIS obtains information near a certain altitude in the atmosphere by observing the emission spectrum of oxygen in the A-band, which is the radiation of oxygen molecules in the A-band. The accuracy of the temperature inversion is about ±2K for the MLT region, but about ± 6K for the higher-level atmospheric temperature inversion, which indicates that the accuracy is relatively low in the higher-level temperature measurement. This discrepancy can stem from a number of factors, including instrument resolution, observation techniques, inversion algorithms, data processing methods, and different understandings of atmospheric properties.

In this paper, a radiative transmission model for the critical observation of oxygen A-band gas glow is established, and the emission spectrum intensities at different tangent point heights are obtained, which are used for the selection of detection schemes and inversion algorithms for the detection scheme and inversion algorithm of oxygen glow in the marginal atmosphere. The first part introduces three mechanisms of photochemical reaction mechanism of oxygen A-band gas glow. The second part is the calculation of bulk emissivity at different altitudes and at different solar zenith angles according to three mechanisms. In the third part, the critical observation gas light transport model is established, and the gas light radiation intensity at different heights is obtained.

2 Photochemical model of atmospheric oxygen A gas

A photochemical model is a mathematical model used to describe photochemical processes in the atmosphere, which combines the theory of photochemistry, the principle of photochemical reactions, and the theory of atmospheric dynamics. Its main purpose is to reveal how the concentration of atmospheric components affects the radiative properties of the atmosphere, especially the emissivity. Photochemical models typically employ differential equations to describe the evolution of

* Corresponding author: [a]994482658@qq.com
[b]sifuqi@aiofm.ac.cn

photochemical reactions and component concentrations in the atmosphere.

In the photochemical model, oxygen A-band gas glow refers to the spectrum produced by the radiation of oxygen molecules in the excited state, which usually includes spectral lines with a wavelength of 763.84 nm. This gas glow is mainly related to the excited state of oxygen molecules $O_2(b^1\Sigma_g^+)$. This excited state is usually formed through the following three processes:

1. Resonance scattering: When the energy of the photon is close to the energy of the excited state of the oxygen molecule, the photon and the oxygen molecule are resonantly scattered, so that the oxygen molecule enters the excited state.

2. Collision annihilation of ground state oxygen and atomic oxygen: Ground state oxygen $O(^3P)$ collides $O(^1D)$ with atomic oxygen produced by photolysis, This causes the annihilation of excited oxygen. This process occurs mainly in the higher atmosphere, where the concentration of ground state oxygen and atomic oxygen is higher.

3. Barth chemical reaction[7][8]: This reaction is a chemical reaction between an oxygen atom and an oxygen molecule. It is caused by the transfer of energy when an oxygen molecule encounters other oxygen molecules in the atmosphere, causing some oxygen molecules to enter an excited state.

These mechanisms work together to cause oxygen molecules to radiate out of the A-band spectrum when excited. As shown in Figure 1[9]in the photochemical model, J, k, and g represent the reaction coefficients of these generating mechanisms, which are used to describe the rates of various reactions.

By establishing photochemical models, combined with atmospheric dynamics theory and known reaction mechanisms, it is possible to simulate the phenomenon of atmospheric glow in order to study the changes in atmospheric chemical and optical processes and their effects on the atmospheric radiation balance and the environment.

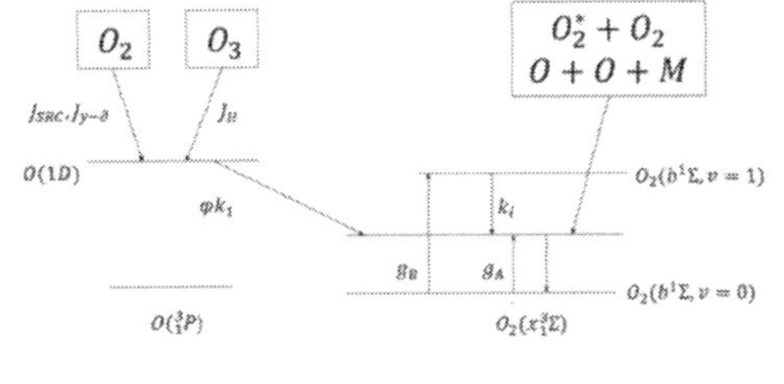

Fig. 1. Processes that oxygen A-band airglow production.

Oxygen molecules are resonantly excited by photons that absorb solar radiation at a wavelength of 763.84 nm, transitioning from the ground state to the excited state $O_2(b^1\Sigma_g^+)$. This process involves a resonant matching between the energy of the photon and the energy level of the excited state of oxygen. The reaction equation is as follows:

$$O_2 + hv \rightarrow O_2(b^1\Sigma_g^+)g \qquad (1)$$

where g represents the transition of oxygen molecules to the excited state, and hv represents the photons of solar radiation, representing the $O_2(b^1\Sigma_g^+)$ oxygen molecules in the excited state.

Collision annihilation is a photochemical reaction process in which the excited state of oxygen molecules $O_2(b^1\Sigma_g^+)$ undergoes a quenching reaction with the excited state of oxygen atoms through collision $O(^1D)$ to form non-excited state oxygen molecules $O(^3P)$. This process can usually be represented by the following reaction equation:

$$O(^1D) + O_2 \rightarrow O_2(b^1\Sigma_g^+) + O(^3P)k_{D,O_2} \qquad (2)$$

where k_{D,O_2} is the response rate.

The Barth chemical reaction is a two-step mechanism in photochemistry involving the generation of gas glows in the oxygen A-band (763.84 nm). The following is a description of the two steps:

The first step describes the collision of an oxygen atom and a quenched molecule, resulting in the production of an excited state of an oxygen molecule O_2^* :

$$O + O + M \xrightarrow{k_{14}} O_2^* + V \qquad (3)$$

$$V = O_{2,}N_{2,}k_{14} = 4.7\times10^{-33}(300/T)^2 \partial cm^6 s^{-1} .$$

The second step describes the O_2^* collision annihilation reaction between the excited state oxygen molecule and the oxygen molecule, and finally produces the excited state $O_2(b^1\Sigma_g^+)$:

$$O_2^* + O_2 \rightarrow O_2(b^1\Sigma_g^+) + O_2 \quad k_{15} \qquad (4)$$

Efficiency factor in the $k_{15} = 5\times10^{-13}\beta cm^{-1}s^{-1}$, quenching rate $\alpha\beta = 0.8$ reaction.

In the two-step Barth reaction, an intermediate product is obtained, and the intermediate product O_2^* is quenched with the oxygen atom to produce a non-excited state of the oxygen molecule and the oxygen atom The reaction equation is as follows:

$$O_2^* + O \rightarrow O_2 + O \quad k_{16} \qquad (5)$$

Quenching rate $k_{16} = 3\times10^{-11} cm^3 s^{-1}$.

The 763.84 nm radiation emitted by the reaction of the three mechanisms is the A-band gas glow, which A_Σ is the spontaneous emissivity coefficient and F_C the Frank-Condon factor of the A-band[10] The A-band volume emissivity is:

$$\eta_{O_2(b^1\Sigma_g^+)}(z) = F_C A_\Sigma n_{O_2(b^1\Sigma_g^+)}(z) \qquad (6)$$

3. Bulk emissivity simulation calculations

As can be seen from Figure 2, the bulk emissivity of the solar glow is stronger than that of other altitude ranges in the altitude range of 40-100 km. In this altitude range, the bulk emissivity of the air glow is affected by a variety of factors. First of all, resonance scattering is an important mechanism, contributing to bulk emissivity in the altitude range of 30-120km. Resonance scattering refers to the resonance scattering of photons and oxygen molecules when the energy of photons is close to the energy of the excited state of oxygen, so that the oxygen molecules enter the excited state. This process works in the higher layers of the atmosphere, bringing the range of bulk emissivity to 30-120 km. Secondly, the photolysis of oxygen and ozone also plays an important role in the bulk emissivity of solar light. The photolysis of oxygen mainly occurs at an altitude of about 100 km, whereas the photolysis of ozone mainly occurs at an altitude of about 40 km. Taken together, the photolysis process of oxygen and ozone plays a key role in the generation of solar glow, by excitation of gas molecules and resulting in the emission of light at specific wavelengths. These processes also reflect the chemical and photophysical properties of the atmosphere at different altitudes. The photons produced by these photolysis excite oxygen molecules, resulting in the bulk emission of solar light. In addition, the figure shows that the contribution of the Barth effect to solar energy is relatively small. The Barth effect refers to a chemical reaction between oxygen molecules, which causes some oxygen molecules to enter an excited state. However, in the overall emissivity, the contribution of the Barth effect is relatively low, and it has not become the main mechanism for generating solar glow.

The bulk emissivity of the low-altitude atmosphere may be affected by the optical thickness, which reduces the intensity of solar radiation. In addition, the low solar photolysis coefficient at low altitude may lead to a decrease in body emissivity at low altitudes. At altitudes above 120 km, the optical thickness is smaller in this altitude range, but the bulk emissivity may also decrease due to a sharp decrease in reactant concentration. This is because bulk emissivity is related to the concentration of reactants, and within this altitude range, a decrease in reactant concentration can be a major factor. The optical thickness and sunlight photolysis coefficient at low altitude are larger, which may lead to a decrease in body emissivity in the low altitude range. At altitudes over 120 km, the optical thickness is small, but a decrease in reactant concentration may also lead to a decrease in bulk emissivity. Together, these factors influence atmospheric optical and chemical processes in different altitude ranges.

In general, resonance scattering and photolysis of oxygen and ozone are the main generating mechanisms, while the contribution of the Barth effect is small. Together, these mechanisms form the bulk emissivity intensity of solar glow in the altitude range of 30-100 km. Atmospheric conditions and chemical processes in this altitude range make solar glow radiation relatively strong. The optical thickness and sunlight photolysis coefficient at low altitude are larger, which may lead to a decrease in body emissivity in the low altitude range. At altitudes over 120 km, the optical thickness is small, but a decrease in reactant concentration may also lead to a decrease in bulk emissivity.

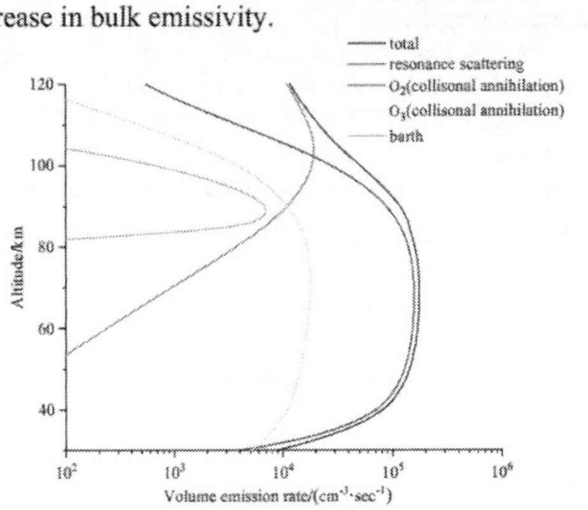

Fig.2. Simulated profile of the emissivity of oxygen in the A-band of solar gas glow.

Changes in the sun's zenith angle have a direct impact on atmospheric radiation, especially in the emission of solar globes. Changes in the sun's zenith angle directly affect the intensity of sunlight hitting the atmosphere. When the sun's zenith angle is smaller, the gaseous glow of sunlight directly into the atmosphere undergoes a more intense photochemical reaction, resulting in a relatively high bulk emissivity. This is because a smaller solar zenith angle means that the path of sunlight passes through a longer layer of the atmosphere and interacts more with atmospheric molecules, resulting in more photochemical reactions. As the zenith angle of the sun increases, the sunlight path is shorter and the radiation angle is more sloping, and the gas glow of sunlight into the atmosphere is less, so the bulk emissivity decreases. At an altitude of around 90 km, the change in the solar zenith angle has the least effect on the bulk emissivity, because the solar glow in this altitude range is mainly produced by the Barth effect and has less to do with solar radiation. When the solar zenith angle is 0° , that is, when the solar height is at its highest at noon, the gaseous glow directly from the sunlight into the atmosphere undergoes the most intense photochemical reaction, resulting in a stronger bulk emissivity relative to other altitudes. This phenomenon can be observed in Figure 3.

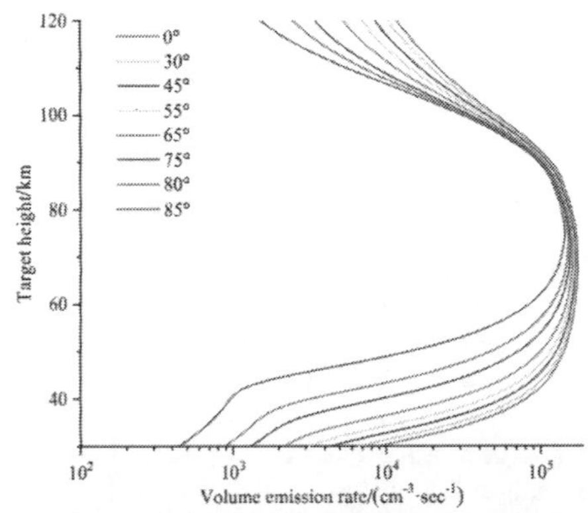

Fig.3. Changes in the emissivity profile of solar airglow at different angles.

4 Simulation results of critical radiant intensity

4.1 Radiative Transfer Mode

When describing the emission of photons in the oxygen A band, the radiant intensity of a particular rotational line is usually taken into account. This can be represented by the following relationship:

$$S_u(T_0)=\frac{T_S S_u(T_S)}{T_0}\times\exp\left(\frac{1.439E^{"}(T_0-T_S)}{T_0 T_S}\right) \quad (7)$$

where the $T_S = 296K$ intensity of the spectral line at the reference temperature is denoted as $S_u(T_0)$ and the vibrational energy level of the $O_2(b^1\Sigma_g^+)$ excited oxygen molecule $E^{"}$ is denoted as . The formula [(8)] is a v_1 Doppler line with the center of the spectral line:

$$D(V)=\frac{1}{\partial_D\sqrt{\pi}}\exp\left(-\left(\frac{v-v_1}{\partial_D}\right)^2\right) \quad (8)$$

$$\partial_D=\frac{V_0}{C}\sqrt{\frac{2KT}{m}} \quad (9)$$

where C is the speed of light, Boltzmann's constant is expressed as K, and m is the mass of the oxygen molecule.

Because there is self-absorption in the oxygen A-band, it is necessary to consider the calculation of transmittance. The expression can be written as:

$$\tau(v,s)=\exp\left[-\sec\theta\int_z n_{o_2}(z)\sum_J S_J(T)D(V,T)dz\right] \quad (10)$$

In the above equation, the line-of-sight path of the satellite scan is expressed as s, the $n_{o_2}(z)$ density of oxygen number at a certain height z, and σ the v absorption cross section at the wavenumber. The formula for the absorption section is:

$$\sigma=\sigma_{ref}\frac{T_{ref}}{T}\exp\left(\frac{1.439E"(T-T_S)}{TT_S}\right) \quad (11)$$

In the above equation is $E"$ the vibrational energy level of the $O_2(b^1\Sigma_g^+)$ excited oxygen molecule, σ_{ref} which is the absorption cross section at the reference temperature. The absorption section is sensitive to temperature changes in the line-of-sight path, so changes in temperature across the integration path need to be considered.

The oxygen number density is calculated from the oxygen number density at temperature and reference height using the following formula:

$$n(z)=\frac{n(z_0)T(z_0)}{T(z)}\exp\left[-\int_{z_0}^{\infty}\frac{mg}{kT(z')}dz'\right] \quad (12)$$

where g is the gravitational force and the integral term represents the z_0 hydrostatic effects in the range of heights from the reference height to infinity.

In summary, the integral term takes into account the effects of oxygen absorption over the entire altitude range from the current gaze path position to infinity. The entire formula can be expressed as:

$$B(v,z_t)=\int_{-\infty}^{+\infty}\eta(s)S(s)D(v,s)\exp\left[-\int_s^{\infty}n(s')\sigma(s')D(v,s)ds'\right]ds \quad (13)$$

Among them, $S(s)$ is the intensity of the rotating spectral line in the line-of-sight path, the $n(s')$ oxygen number density in the line-of-sight path, and $\sigma(s')$ the absorption cross-section. Figure 4 is a critical line-of-sight path model.

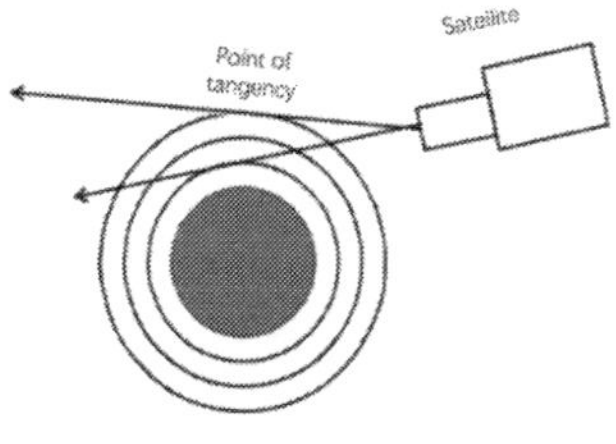

Fig.4. Limb line of sight path model.

4.2 Radiation intensity simulation results

Obtaining the optical thickness requires the calculation of the absorption coefficient, and in the atmosphere, the widening around the transition wavenumber v_{ij}, the expansion of which is expressed by the normalized linear function $f_L(v,v_{ij},p,T)$. In the lower atmosphere, the pressure broadening of the spectral lines predominates, and the Lorentz distribution can be assumed:

$$f_L(v,v_{ij},p,T)=\frac{1}{\pi}\frac{\gamma(p,T)}{\gamma(p,T)^2+\left[v-\left(v_{ij}+\delta(p_{ref})p\right)\right]^2} \quad (14)$$

In the low-pressure environment of the upper atmosphere, the Doppler broadening dominates in the linear and can be assumed to be a Gaussian distribution:

$$f_G(v,v_{ij},T)=\sqrt{\frac{\ln 2}{\pi\alpha_D^2}}\exp\left(-\frac{(v-v_{ij})^2\ln 2}{\alpha_D^2}\right) \quad (15)$$

Absorption coefficient $k_{ij}(v,T,p)=S_{ij}(T)f_L(v,v_{ij},p,T)$

Figure 5 shows the normalized absorption coefficient of the A-band of oxygen from 30 to 80 km. Table 1 shows the data of the MISI database data for different heights of data density, temperature, and pressure.

 https://doi.org/10.1051/e3sconf/202456001003

Table 1. Number density, temperature, and pressure at 30-80km.

height(km)	Oxygen number density (g/cm3)	Temperature（K）	pressure (Pa)
30	6.727E+16	214.3	0.199
40	1.35E+16	244.8	0.0456
50	3.43E+15	255.1	0.012
60	9.40E+14	239.9	0.0031
70	2.37E+14	223.7	0.0007
80	5.34E+13	215.3	0.0001

The HITRAN database sample line strengths selected at different heights are shown in the figure below.

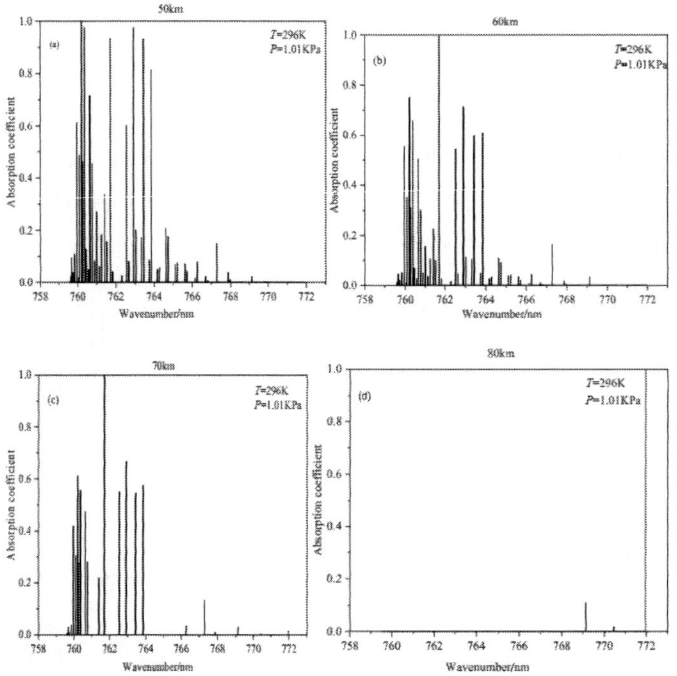

Fig.5. Schematic diagram of absorption coefficient spectral lines in the A-band of oxygen.

According to the schematic diagram of the absorption coefficient spectral line of oxygen A-band, it can be seen that the absorption coefficient is negligible after reaching 80 km, and the influence of the absorption coefficient needs to be considered when calculating the optical thickness of oxygen below 80 km, and the influence of the absorption coefficient can be ignored above 80 km.

Based on the calculated bulk emissivity of the oxygen A band, the HITRAN database, and the established radiative transfer model, the oxygen A emission lines at different heights were obtained, as shown in Figure 6.

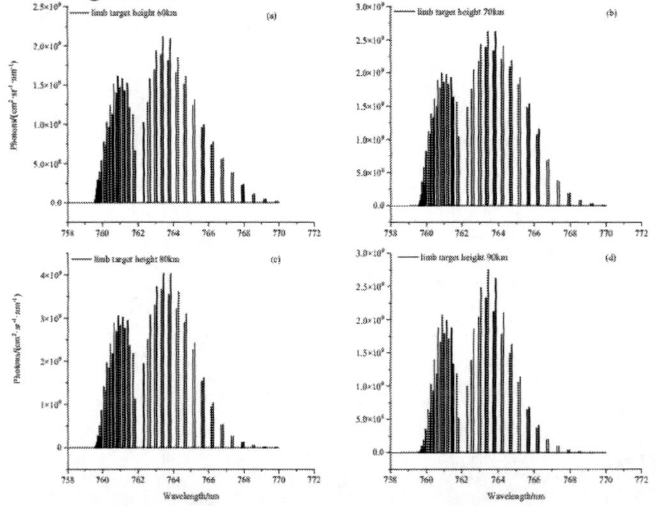

Fig.6. Emission Spectral Intensity of Oxygen A-band.

Figure 6 is the emission line under the simulated high resolution, in the actual observation, the spectral resolution of different instruments is different, the spectral response function of the instrument and the high-resolution spectrum line are convoluted, and the emission spectrum line at different resolutions can be obtained:

$$C = \int_{-\infty}^{+\infty} I(\lambda) B(\lambda) d(\lambda) \tag{16}$$

For the grating spectrometer, the spectral response function is usually of the Gaussian type, and the half-height and width of the cm^{-1} Gaussian function are selected to obtain the simulated values of the critical radiant intensity at different tangent point heights, as shown in Figure 7.

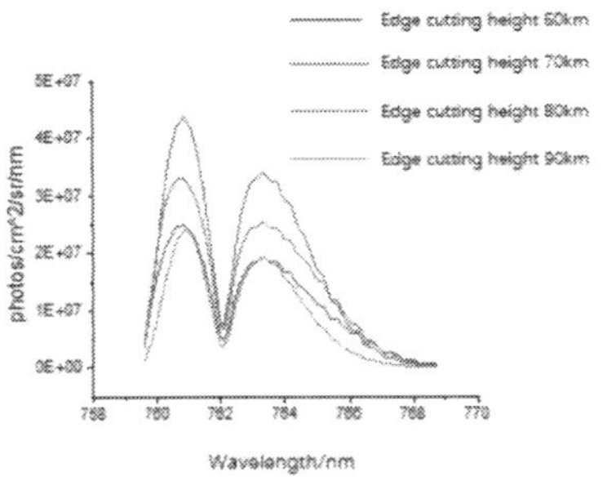

Fig.7. Convolutional Results of Oxygen A Spectral Lines.

In the figure, the radiation intensity of gas glow is different at different altitudes, the radiation intensity is stronger at 80 km, and the radiation intensity is weaker below 70 km, mainly because the intensity of the gas glow spectral line decreases due to the strong oxygen self-absorption, and the weak radiation intensity of the gas glow above 90 km is mainly due to the thin atmosphere and the low concentration of participating in the photochemical reaction. At the same time, the influence of temperature on the shape of the spectral lines can be seen, for example, the lower temperature at 90 km weakens the radiation intensity on both sides of the spectral line, which indicates that different temperatures determine different spectral radiation shapes, because with the increase of temperature, the contribution of higher rotational energy levels to the spectrum increases, affecting the shape of the spectral bands. Therefore, by detecting the radiation intensity and distribution of oxygen A gas glow lines, the temperature field of the global atmosphere can be detected.

5. Conclusion

In this paper, the simulation method of atmospheric oxygen A emission line is studied. Firstly, the photochemical model of the gas glow of oxygen A band is established $O_2(b^1\Sigma_g^+)$, the bulk emissivity of the gas glow of the A band is calculated according to the photochemical model, and then the radiation transfer model of the gas glow of oxygen A band is established $O_2(b^1\Sigma_g^+)$ The radiative transmission model is carried out based on the critical edge detection geometric path to simulate the radiant intensity of the gas glow, and the factors affecting the emissivity and radiant intensity of the gas glow in the oxygen A-band are analyzed, and the following factors are verified, such as the solar zenith angle, when the solar zenith angle is 0°, that is, when the sun shines directly into the atmosphere, the bulk emissivity is the strongest. This is because the direct exposure of sunlight into the atmosphere causes strong photochemical reactions, including the photolysis of oxygen and ozone, resulting in a higher bulk emissivity, and the effect of temperature changes on the emission spectrum of oxygen in the A-band is symmetrical, with less radiation in the center and less radiation in both sides.

Acknowledgments

Natural Science Foundation of Anhui Province (1808085QD114)

References

1. Wang Ying Jian, SCIENCE IN CHINA(Scries A), 2000.30 (S1) 17.
2. LIU Zhenxing.Space Physics[M].Harbin:Harbin Institute of Techology Press,2005
3. CHRISTENSEN A B,YES J H,BISHOP R L,et al.Observations of molecular oxygen atmospheric band emission in the thermosphere using the near infraredspectrometer.on.the.ISS/RAIDS.experiment[J].j.Geophys.Res.SpacePhys,2012,117(A4):DOI:10.1029/2011ja016838
4. Hays P.B,Abreu V.J, Dobbs M.E, et al. High Resolution Doppler Imager on Upper Atmosphere Research Satellite, Geophysical Research: Atmospheres, 1993,98 (D6) : 10713-10723
5. Zarboo A,Bender S,Burrows J.P, et al. The volume emissivity of $O_2(1\Sigma)$ and $O_2(1\Delta)$ in the mesosphere and low thermosphere was retrieved using SCIAMACHY MLT edgewise scanning.Atmospheric Measurement Technology.2018, 11:473-487
6. Skinner W.R.,Hays P.B.,Grassl H.J.,et al. High resolution Doppler imager on the Upper Atmosphere Research Satellite. International Society of Optics and Photonics,1994
7. Bucholtz A, Skinner W. R, Abreu V.J., et al. O_2 Daily airglow of atmospheric band system. Planetary and Space Science, 1986,34 (11) : 1031-1035
8. WallaceL, HuntenD.M. Daylight in the oxygen A zone. J. Geophysical Research, 1968,73 (15) :4813-4834
9. Llewellyn E. J., Lloyd N.D., Degenstein D.A., et al. The OSIRIS instrument on the Odin spacecraft. Canadian Journal of Physics, 2004,82 (6) : 411-422

10. Nicholls R. W.Franck-Condon decomposed the high vibrational quantum number v-o for the O2 band system. National Bureau of Standards Research: Part A - Physics and Chemistry, 1965, 69A:369-373

Research on detection technology of photolysis rate of atmospheric trace gases

Zhihong Liu [1,2], Fuqi Si [2*], Haoran Li [2,3], Runze Song [2,3]

[1] College of Biology, Food and Environment, Hefei University, Hefei, Anhui 230601, China
[2] Key Laboratory of Environmental Optics and Technology, Anhui Institute of Optics and Precision Machinery, Chinese Academy of Sciences, Hefei, Anhui 230031, China
[3] University of Science and Technology of China, Hefei, Anhui 230036, China

Abstract. A detection system for measuring the optical velocity of trace gases was built based on a micro-spectrometer , including an optical receiver, a radiation calibration system, a spectrometer system, and a host computer software . Using Python language and PyQt5 framework, the system's host computer software was designed and developed, including the data collection and spectrometer control module storage module, and the real-time data display module for online continuous monitoring of the photolysis rate. Through field experiments, the photolysis rates of NO_2 , O_3 , H_2O_2, HONO, and HCHO were measured and inverted. The photolysis rate measurement results were compared and verified with a standard instrument (UF-CCD photolysis spectrometer produced by Metcon, Germany) . The results show that the photolysis rate measurement system and the standard instrument results are linearly fitted, and R^2 is both higher than 0.90. At the same time, the diurnal variation patterns of NO_2 , HONO, and H_2O_2 were analyzed . The results showed that the pattern of gas photolysis rate was consistent with changes in solar radiation intensity. The gas photolysis rate gradually increased in the morning, reached a peak at noon, and then reached a peak in the afternoon. gradually decreases. By analyzing the uncertainty of the system, the total system error is approximately 5.07 %. The feasibility and measurement accuracy of the developed photolysis rate measurement system were verified.

1 Preface

The impact of environmental changes and atmospheric pollution on ecosystems and human health has become an important issue of global concern. In recent years, with the changes in human production and lifestyle such as energy, industry, transportation, etc., the type of air pollution in China has gradually changed from the original single bituminous coal pollution to the composite pollution of coal smoke, automobile exhaust, and industrial waste gas[1][2]. The absorption of ultraviolet and visible light by pollutants that cause complex atmospheric pollution involves a variety of complex photochemical reactions. Among them, there are three main types of photochemical reactions that have a greater impact on the atmospheric environment and have been studied more: photolysis reactions, reactions of excited molecules, and photochemical reactions. catalytic reaction[3]. The principle of photolysis reaction is that under solar radiation, higher concentrations of nitrogen oxides (NOx), volatile organic compounds (VOCs), etc. in the atmosphere that are emitted from human life and naturally emit photochemical reactions, that is, atmospheric molecules absorb photon energy. Finally, the covalent bonds are broken and dissociated to produce highly reactive components (atoms and free radicals). These substances will further react under certain conditions to generate new chemical substances. These substances will further participate in the reaction to generate harmful substances to human health and ecology. Environmental stability has an important impact. Accurate measurement of photolysis rates is therefore important for understanding these processes.

Since scientists realized the measurement of photolysis rate in 1975, with the continuous development of science and technology, the technologies currently available for gas photolysis rate mainly include chemical photometer measurement method, radiation measurement method, radiation transfer model, etc. Chemiphotometry is a method for direct detection of actinic flux using compounds of interest[4]. The gas of interest is exposed to solar radiation in a static or flowing gas system and its rate of photochemical conversion to products is measured. The conversion rate is then used to calculate the photolysis rate[5]. Radiometry is a general method of measuring radiation using optical instruments using photodetectors. In atmospheric chemistry, radiation measurement methods have been modified to derive photolysis rates from the measured intensity of incident radiation. Radiometry can be further divided into methods of measuring radiation that are either spectrally resolved (narrowband spectral radiometry) or spectrally integrated

* Corresponding author: sifuqi@aiofm.ac.cn

(broadband radiometry). Each method can be further divided into methods for measuring actinic flux or irradiance. As a result, various types of radiometers, such as filter radiometers[6], actinic flux radiometers[7], etc., determine the photolysis rate directly from the measured radiation intensity. The radiative transfer model is a method that uses a theoretical model to calculate the photolysis rate. By inputting relevant parameters (atmospheric composition, solar radiation intensity, temperature, humidity and other related parameters) into the model, the dynamic process of photochemical reactions in the atmosphere is simulated to determine the substance. Photolysis rate method. Relative to chemical methods, the most common method is actinic flux radiometer measurement. It measures the actinic flux of the radiation spectrum and can be used to determine the photolysis rate of various gases from the same measured radiation spectrum. It also has the advantages of high sensitivity and high responsiveness, enabling rapid detection of the photolysis rate.

This article builds a photolysis rate measurement system based on a micro-spectrometer (band range 290nm~408nm). By converting the collected solar spectrum data into actinic flux, it can meet the measurement of photolysis rates of various atmospheric trace gases. The system's host computer software was designed and developed based on Python language for online real-time monitoring of photolysis rate. Through the field comparison test, the gas photolysis rate measurement results are highly correlated, which verifies the feasibility and accuracy of the photolysis rate measurement system.

2 Experimental plan and device

2.1 Photolysis rate measurement system experimental device

The system as a whole is composed of software and hardware and can be subdivided into multiple subsystems. Including optical receiver, radiation calibration system, spectrometer system, and host computer software. The optical receiver is the core component of the system and consists of a quartz receiving head, a shadow ring, a quartz light guide rod , and a drying box. The quartz receiving head is made of an oval quartz ball cover, which is processed from a 1mm thick quartz material. The optical receiver collects solar radiation from all directions. In order to reduce light loss, a quartz light guide rod is added between the quartz receiving head and the quartz fiber inlet. The light radiation inside the dome passes through the quartz light guide rod and is completely reflected inside it. Uniform light is coupled to the quartz fiber and incident on the spectrometer (QE65000, Ocean optics). The light is then collected by the host computer software . Combined with the laboratory calibration parameters , the collected signal value is converted into actinic radiation flux, and then calculated The photolysis rate of atmospheric trace gases . The system structure diagram is shown in Figure 1.

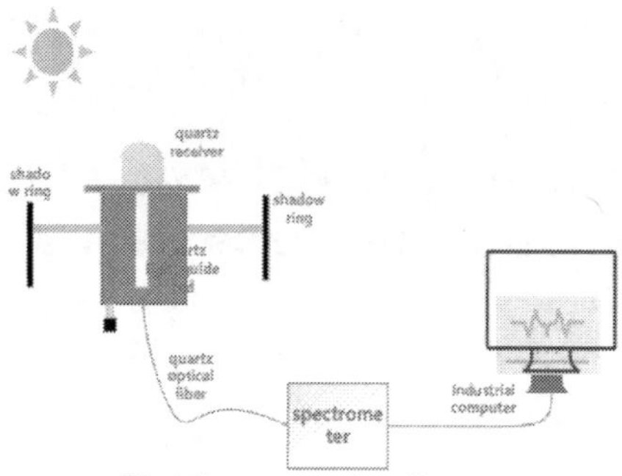

Fig.1 System structure diagram

2.2 Principle of photolysis rate measurement

The formation mechanism of atmospheric chemical components is complex. Without an accurate understanding of the kinetic rates of the photolysis process, it is impossible to quantitatively describe atmospheric chemistry. Assume that compound AB produces products A and B under solar radiation,

$$\mathrm{AB} + h\nu \rightarrow \mathrm{A} + \mathrm{B} \quad (1)$$

$h\nu$ represents solar radiation, and the following kinetic rate coefficient J can be defined:

$$J = -\frac{1}{[\mathrm{AB}]}\frac{\mathrm{d}[\mathrm{AB}]}{\mathrm{d}t} \quad (2)$$

In the formula, the square brackets represent the number density of each molecule in the gas phase (cm^{-3}), and J is the first-order rate coefficient (s -1) called the photolysis rate or J value . It describes the probability per second that a bulk compound decomposes into its products in a given radiation field. Note that J is not a constant value but depends on the local radiation field that changes in space and time. For any first-order decay reaction, the reciprocal of the rate coefficient is the average lifetime of the decomposing molecules:

$$\tau_{\text{photolysis}} = \frac{1}{J} \quad (3)$$

If the J value of the photolysis reaction is known, the chemical loss rate and productivity can be calculated. For general chemical reactions such as formula (1) , it is as follows:

$$-\frac{\mathrm{d}[\mathrm{AB}]}{\mathrm{d}t} = \frac{\mathrm{d}[\mathrm{A}]}{\mathrm{d}t} = \frac{\mathrm{d}[\mathrm{B}]}{\mathrm{d}t} = J[\mathrm{AB}] \quad (4)$$

In the formula, J[AB] is called the photolysis rate of AB. It is a measure of the number of AB molecules broken down per cubic centimeter and per second, producing corresponding numbers of light fragments A and B in the same volume and time. J is collectively called the photolysis rate.

The photolysis rate is the integral of the product of substance absorption cross section, quantum yield and actinic flux. Therefore, the photolysis rate of substance AB can be further expressed as:

$$J[\mathrm{AB}] = \int \sigma(\lambda,T)\Phi(\lambda,T)F_{\lambda}(\lambda)\mathrm{d}\lambda \quad (5)$$

where $\sigma(\lambda,T)$ is the absorption cross section of the substance, $\Phi(\lambda,T)$ is the quantum yield of the substance, and $F_\lambda(\lambda)$ is the photochemical flux.

The relationship between actinic flux and radiance can be expressed as:

$$F(\lambda) = \int\int_0^\pi L(\lambda, \theta, \varphi) \sin\theta d\theta d\varphi \quad (6)$$

$L(\lambda, \theta, \varphi)$ is the radiance, θ is the solar zenith angle, φ is the solar azimuth angle

3 Experimental results and discussion

3.1 Angular responsivity test

The details of the angular responsivity test are as follows. Place an angle meter at a specific distance from the xenon lamp on the experimental table. Place the optical receiving head on the angle meter so that the quartz receiving head is placed in the center of the angle meter. The center position is 0 degrees polar angle. Start the xenon lamp, preheat it for 10 minutes, and start measuring data after the light source is stable. The angle meter rotates 2 degrees each time, measures the angular response of the receiving head from 0 to 90 degrees, records and saves the data.

Calculate the standard deviation of the spectral data at each angle, normalize it with an angle of 0 degrees, and fit the data with a third-order polynomial according to the data distribution characteristics. The processed data results are shown in Figure 2. The angular responsivity value and The relationship between angles is:

$$Y = 0.99 + 0.07 * X - 3.23 \times 10^{-4} * X^2 + 1.97 \times 10^{-6} * X^3 \quad (7)$$

In the above formula, Y represents the normalized angular responsivity value, and X represents the angle.

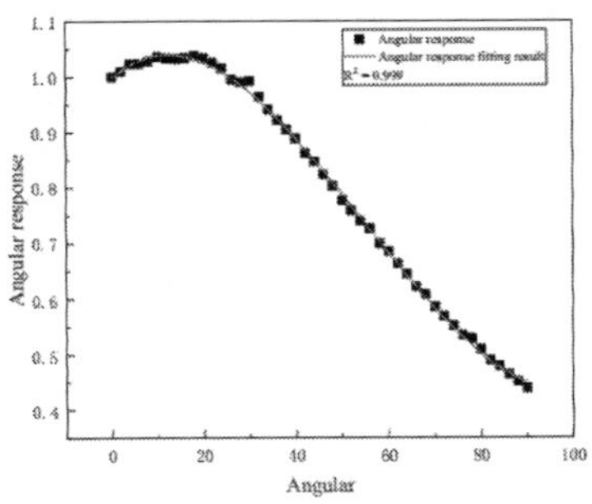

Fig. 2 Angular responsivity fitting result chart

Since the receiver is designed to achieve uniform angular responsivity in all directions, the ideal responsivity at each angle is 1 and the error does not exceed 5%. So we need to weight each angle algorithmically. Customize a weighting function. Taking the reciprocal of the function of the formula as the weighting factor, the correction formula for the angular responsivity of the instrument is as follows:

$$Z = 1 / Y \quad (8)$$

When performing measurement processing, use the solar zenith angle calculation tool, input the longitude, latitude and measurement time of the measurement location, calculate the solar zenith angle from 0 to 24 o'clock on the day, and then substitute the weighting factor into Formula 6 for calculation. The use of angular responsivity weighting algorithm can improve the reception of sunlight at all angles of the receiver and improve the accuracy of photolysis rate measurement. The comparison of angular responsivity after correction is shown in Figure 3.

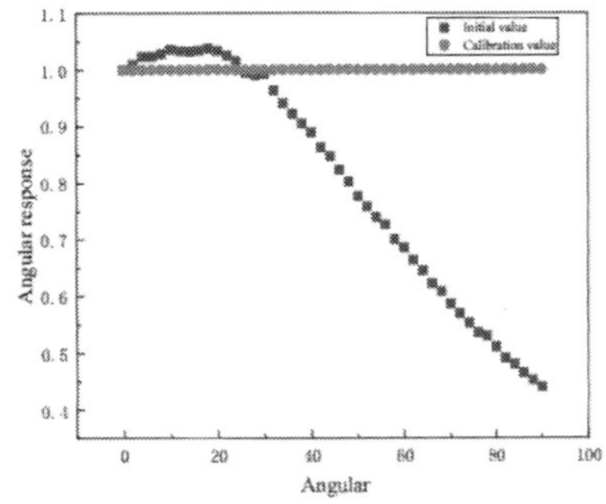

Fig.3 Comparison of angular responsivity before and after correction

3.2 Radiance calibration

The radiance calibration of the photolysis rate measurement system is the prerequisite for inverting the gas photolysis rate. The integrating sphere method is used for radiance calibration. Calculated as follows

$$\frac{L(\lambda)}{I_s(\lambda)} = \frac{DN(\lambda) - dark(\lambda)}{I_{lamp}(\lambda) - dark(\lambda)} \quad (9)$$

L(λ) represents the solar radiance, DN(λ) represents the spectral data of solar radiation; dark(λ) represents the background noise during the measurement process, which is subtracted during calculation; I_{lamp} represents the measured integrating sphere spectral data; I_s represents the calibration standard value corresponding to the integrating sphere.

In order to avoid the influence of stray light in the experimental environment, the integrating sphere calibration was performed in a completely dark environment. First, the solar radiation receiver is placed at the light outlet of the integrating sphere, and the other end is connected to the spectrometer and computer through an optical fiber. Then start the main control switch of the integrating sphere control system, and turn on the tungsten halogen lamp control power supply in order to preheat for 10-15 minutes. After the light source is stable, record the data received by the spectrometer. And save the measurement results to a file. The radiance results after calibration are shown in Figure 4.

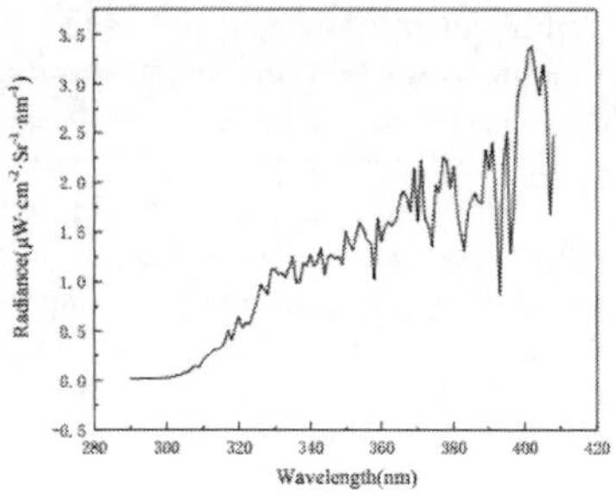

Fig. 4 Radiance calibration results

3.3 System software

In order to realize the online continuous measurement of the system, an automated photolysis rate measurement system host computer software was developed. The software mainly includes four modules: data acquisition module, spectrometer control module, real-time data display module and storage module. The main architecture of the software is shown in Figure 5

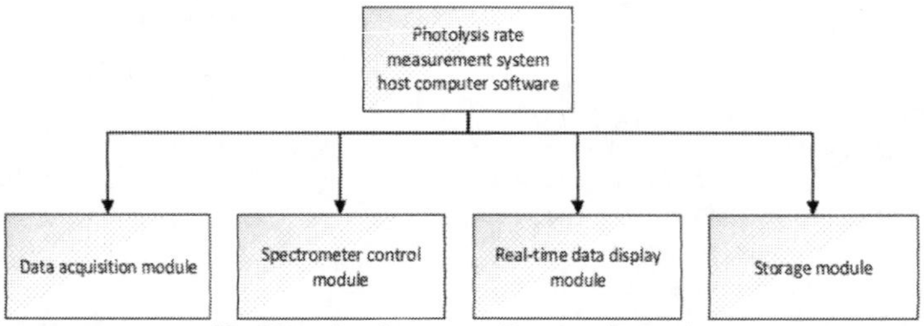

Fig.5 Functional structure diagram of PC software

The software is developed based on the Python programming language, and the system display interface uses Qt designer for rapid design. The data acquisition module needs to install the Python-Seabreeze library to communicate with the spectrometer. The control module can set and initialize various parameters of the spectrometer. Set the integration time, average times and data saving waiting time of the spectrometer according to the actual measurement requirements. The display module is based on the matplotlib library, which displays the photolysis rate data of each species calculated by the system in the form of a curve on the software interface in real time. Since the system requires long-term observation, it is necessary to facilitate storage and management. The storage module mainly adopts two solutions, local file saving and database saving. Local saving means that each photolysis rate data is saved to the system disk in the format of a text file. The database storage uses the relational database MySQL. In Python programming, the pymysql library needs to be used to interact with the database.

3.4 Field test

In order to verify the feasibility of the instrument to measure the photolysis rate, field experiments were carried out. The test site is on the roof of the comprehensive observation field of the Anguang Institute of the Chinese Academy of Sciences . The surrounding environment is open and the field of vision is wide, without the obstruction of high-rise buildings and tall trees. The optical receiver is fixed on a 2.5m-high triangular bracket. In order to prevent meteorological factors from affecting the system measurement results, the spectrometer is placed indoors and the receiver and spectrometer are connected through quartz optical fiber. The integration time of the spectrometer is set to 1 second. , the spectrum is averaged once, the data is saved every 3 minutes, and the measurement time is from 9:00 am to 18:00. The field test is shown in Figure 6 .

Fig.6 Field test diagram

The UF-CCD photolysis spectrometer product produced by German Metcon Company is used as the standard instrument. The instrument can measure the solar spectrum in the 280~650nm band, obtain the original spectrum data based on the measured spectral flux, and then use the spectrum detector and analysis software to obtain the photolysis rate value of important substances in the corresponding band, which can be used for ground observations. Atmospheric research in station and aerial survey modes . The measurement results of the system in this study were compared and verified with the measurement results of standard instruments. During the observation, the two instruments were placed at the same location so that the shadows of the instruments would not affect each other. Experimental observations were conducted for many days. Calculate and analyze the data measured by the two instruments, and provide the measurement results and correlation analysis of nitrogen dioxide (NO_2) and hydrogen peroxide (H_2O_2) gas photolysis rates on June 1, 2023, as shown in Figure 7 and Figure As shown in 8 , the results show that the trends and values of the measurement values of the two instruments have high correlation and good consistency, and the correlation coefficients of each gas result exceed 0.90.

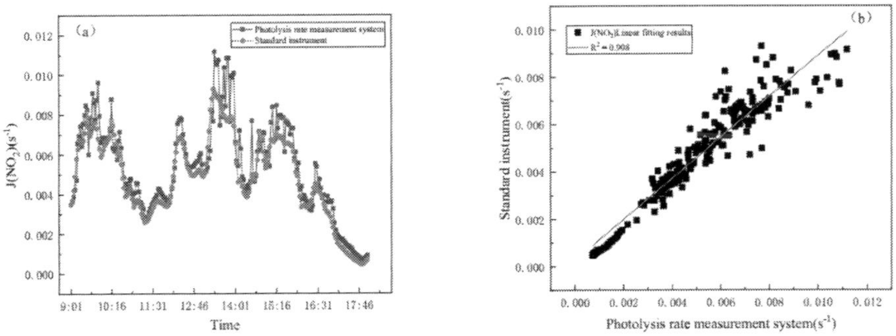

Fig.7 Photolysis rate comparison chart (a) and correlation analysis (b) of NO_2

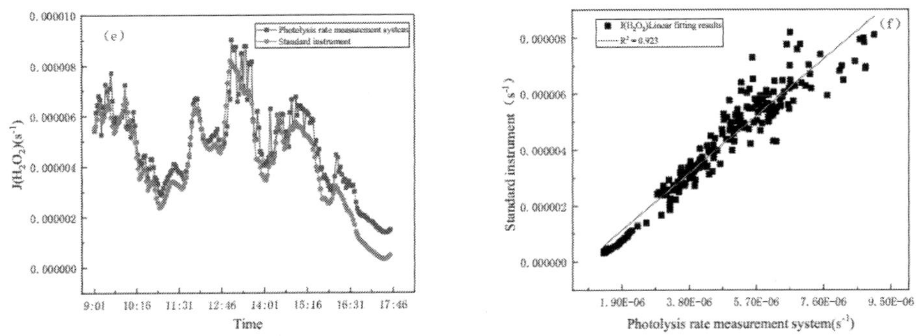

Fig.8 Photolysis rate comparison chart (e) and correlation analysis (f) of H_2O_2

By observing the photolysis rate comparison image, it can be found that the measurement values of the photolysis rate measurement system developed in this article are basically consistent in order of magnitude and curve trend compared with the measurement results of standard instruments. The observation results of the photolysis rate measurement system in this study are overall The trend is slightly higher than the standard instrument value. According to the analysis of the radiance calculation formula and the photolysis rate calculation formula, the possible reason is caused by the uncertainty of the head distance of the optical receiving head during the integrating sphere calibration. When the receiving head is too close to the integrating sphere mouth, the calibration The resulting radiance value will be too small. Since the photolysis rate is proportional to the radiance, the photolysis rate will be too small. When the distance is too far, the calculated result of the photolysis rate will be too large.

In order to further verify the accuracy of the instrument, measurements were carried out at the same location for three consecutive days on June 8 and June 10, 2023. In order to ensure the consistency of the measurement results, the measurement parameter settings and the initial measurement test of the observation time range were consistent. Comparing three days of observation data with standard instruments, the analysis results of each gas are shown in Table 1. The results show that the correlation of gases measured by the two instruments exceeds 0.9, and the average correlation of each gas for three consecutive days exceeds 0.91, with high accuracy. However, the correlation of ozone is lower than that of other gases. This is because the sensitivity of the spectrometer is low in the wavelength range where ozone photolysis reaction occurs. Therefore, follow-up work can consider adding a correction factor to optimize the measurement of ozone photolysis rate.

Table 1 Correlation of instrument measurement results

Correlation	NO_2	$O_3(O^1D)$	HONO	H_2O_2	HCHO-r(HCO)	HCHO-m(CO)
2023.6.8	0.917	0.901	0.922	0.935	0.938	0.932
2023.6.9	0.921	0.915	0.934	0.939	0.943	0.941
2023.6.10	0.945	0.926	0.956	0.947	0.968	0.965
average value	0.928	0.914	0.937	0.940	0.950	0.946

3.5 Instrument detection limit analysis

The detection limit is an important parameter for evaluating system performance. To evaluate the detection limit of a photolysis rate measurement system. Twenty days of field observation experiments were conducted on the roof of Villa No. 3 of the Anguang Institute of the Chinese Academy of Sciences. Use statistical analysis method to evaluate the performance of the photolysis rate measurement system, take twenty consecutive days of observation data, and calculate the detection limit of the photolysis rate of various gases. The amount of sample data for each gas is 4820, and the statistical calculation is The mean square error of each gas, according to analytical chemistry[8], the instrument detection limit can be defined by the following formula:

$$LOD = 2\sigma = 2\sqrt{\frac{\sum_{i=1}^{n}(X_i - \overline{X})^2}{n}} \tag{10}$$

Among them, σ is the system mean square error, which is generally used to represent the pure noise of the system. LOD=2σ represents the system detection limit. *Xi* represents the measurement value, represents the average value of the measurement results, $\overline{X}$ and N represents the total number of statistical samples. The final statistical analysis results of the photolysis rates of each gas are shown in Table 2

Table 2 Statistical analysis results of photolysis rate of each gas

Species	Number of samples	maximum value	minimum value	Mean square error (σ)	Detection limit 2σ (s^{-1})
$O_3(O^1D)$	4820	0.000121557	5.00376E^{-5}	1.52234E^{-5}	3.04468E^{-5}
NO_2	4820	0.009597879	0.000235655	0.002039554	0.004079108
HONO	4820	0.001540709	4.01893E^{-5}	0.000331283	0.000662566
H_2O_2	4820	7.52869E^{-6}	8.30706E^{-7}	1.50028E^{-6}	3.00056E^{-6}
HCHO-r(HCO)	4820	2.66596E^{-5}	2.22406E^{-6}	5.54103E^{-6}	1.10821E^{-5}
HCHO-m(CO)	4820	3.65E^{-5}	1.72662E^{-6}	7.8605E^{-6}	0.000015721

3.6 Daily variation pattern of photolysis rate

In order to further analyze the changing rules of gas photolysis rate. Using the above twenty-day data as a sample, the daily variation pattern of photolysis rate was analyzed. The method is to accumulate the values of each gas at each moment and average them. The analysis results are shown in Figure 9. The photolysis rate of nitrogen dioxide gradually increases in the morning, reaches a peak at noon, and then gradually decreases in the afternoon. This trend is consistent with changes in sunlight intensity, since the photolysis rate is directly affected by the intensity of solar radiation. As solar radiation increases in the morning, the photolysis rate of NO_2 increases; at noon, when the solar radiation is strongest, the photolysis rate of NO_2 reaches the maximum value of the day; then as the solar radiation weakens, its photolysis rate also decreases. The diurnal variation pattern of photolysis rates of HONO and H_2O_2 is very similar to that of NO_2. In general, the changing trends of the photolysis rates of these three substances over time indicate the diurnal cycle characteristics of atmospheric chemical reactions and reflect the direct impact of light conditions on the rate of atmospheric chemical reactions.

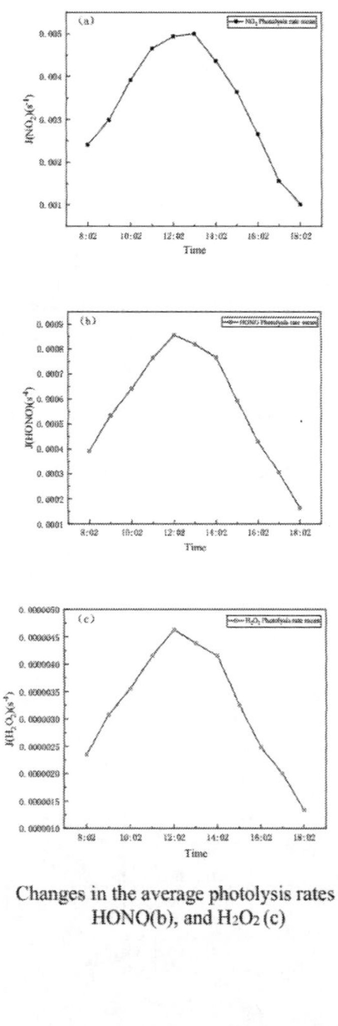

Fig 9 Changes in the average photolysis rates of NO_2(a), HONO(b), and H_2O_2 (c)

3.7 Uncertainty Analysis

The uncertainty in the measurement of the photolysis rate measurement system comes from many aspects. It mainly includes the accuracy of hardware (optical receiver , spectrometer), integrating sphere calibration error and algorithm processing , and errors caused by field tests . It is necessary to further evaluate and quantify each source of uncertainty, and further analyze the uncertainty of the overall system by combining verification tests and tests of the system and standard instruments.

1) Errors in the system hardware itself , including errors in the optical receiver [9]and errors in the spectrometer. The design of the optical receiver should theoretically make the relative angular response in all directions uniform, that is, normalized with a 0-degree polar angular responsivity, and the ideal value of each angular response is 1. This paper uses a weighted algorithm to make the receiving head of the system achieve uniform response in all directions. The error of the weighting factor corrected for non-ideal angular responsivity does not exceed 2%.

temperature drift and sensitivity problems caused by changes in the temperature of the spectrometer during the observation process , which affect the measurement results. Due to the off-the-shelf spectrometer with temperature control used in this study. In the measurement comparison test, the spectrometer was placed indoors, and the cooling was turned on throughout the process. The operating temperature was much lower than the indoor temperature, so the error of the spectrometer was negligible.

2) The error caused when using the integrating sphere for calibration. The uncertainty of the stability of the integrating sphere itself and the uniformity of the integrating sphere surface during the integrating sphere calibration can be obtained from the test report of the calibration system. It is about 2.8 % . The error caused by the distance between the calibration photolysis rate measurement system and the integrating sphere does not exceed 1% , and the total error in this part is about 3.8 % .

3) The integral error in calculating the photolysis rate value is numerical processing error[9]. In the algorithm design, the processing of discrete data is to convert the resolution into 1nm for integration. Research shows that the uncertainty caused by numerical integration does not exceed 1%. .

4) System stability error during field testing . For example, uncertainty arising from disconnecting and reconnecting optical fibers [10], uncertainty in actinic flux was determined to be less than 2% , and uncertainty arising from optical receiver adjustment levels was approximately 0.5 %.

Based on the various factors analyzed above, using the error transfer formula (11) , the system uncertainty is set to Δ, Δ_i indicating one of the errors, and the total error of the system is estimated to be about 5.07%.

$$\Delta = \sqrt{\sum_i \Delta_i} \qquad (11)$$

4 Conclusion

Field tests were conducted using the built photolysis rate measurement system, and compared with standard instruments. The results showed that the correlation between the two was high, with the correlation reaching more than 0.90, which verified the accuracy and reliability of the system's measurement of photolysis rate. Then the detection limit of the system was analyzed, and the daily variation pattern of the photolysis rate of NO_2 , HONO and $H_2O_2$was analyzed . The results showed that the pattern of gas photolysis rate was consistent with the change of solar radiation intensity. The gas photolysis rate gradually increased in the morning. increases, reaches a peak at noon, and then gradually decreases in the afternoon. Finally, the system uncertainty was analyzed from the hardware and software aspects, and the total system error was approximately 5.07 %.

Acknowledgments

This research was supported by the National Key R&D Program of China(2023YFC3705304).

References

1. Li Kangwei. Experimental study on the formation mechanism of secondary aerosols under different atmospheric composite pollution scenarios[D]. Zhejiang University, 2021.
2. Cao Lingyan. Research on local governance model of urban air pollution [D]. Lanzhou University, 2019.
3. Wang Anpu. Overview of air pollution chemistry research[J]. Environmental Chemistry, 1992(06):1-13.
4. Kuhn H J, Braslavsky S E, Schmidt R. Chemical actinometry (IUPAC technical report)[J]. Pure and Applied Chemistry, 2004, 76(12): 2105-2146.
5. Calvert J G, Pitts J N. jr., Photochemistry[J]. Wiley, New York, 1966, 8: 99.
6. Junkermann W , Platt U , Volz-Thomas A . A photoelectric detector for the measurement of photolysis frequencies of ozone and other atmospheric molecules[J]. Journal of Atmospheric Chemistry, 1989, 8(3):203-227.
7. Mccluney W R . Introduction to Radiometry and Photometry (1st ed 1994). 1994.
8. Thalman R, Baeza-Romero M T, Ball S M, et al. Instrument intercomparison of glyoxal, methyl glyoxal and NO 2 under simulated atmospheric conditions[J]. Atmospheric Measurement Techniques, 2015, 8(4): 1835-1862.
9. Hofzumahaus A , Kraus A , Müller, Martin. Solar actinic flux spectroradiometry: a technique for

measuring photolysis frequencies in the atmosphere[J]. Appl Opt, 1999, 38(21):4443-4460.

10. Eckstein E , Perner D , Ch. Brühl, et al. A new actinic flux 4π-spectroradiometer: Instrument design and application to clear sky and broken cloud conditions[J]. ATMOSPHERIC CHEMISTRY AND PHYSICS, 2003, 3(6).

Reactivated Spent Bleaching Earth as a New Path in Waste Resource Utilization for the Crude Palm Oil Bleaching Process

Renita Manurung*, *Siti* Maisarah, *Hamidah* Harahap, *Sri* Zahrani Dwi Mauliyah Parinduri, and *Andre* Pranata

Department of Chemical Engineering, Faculty of Engineering, Universitas Sumatera Utara, Medan, 20155, Indonesia

Abstract. The palm oil refining process usually uses adsorbents. The commonly used adsorbent is Bleaching Earth. However, this raises a new problem: hazardous waste in the refining process, namely Spent Bleaching Earth (SBE). For this reason, one of which is the utilization of SBE as an adsorbent through a reactivation process. This study aims to determine the characterization of RBE produced from the microwave reactivation process and acid solution as an adsorbent in the palm oil (CPO) bleaching process. The SBE reactivation process was carried out using microwaves and sulphate acid solution at concentrations 3N. RBE characteristics were analyzed using Scanning Electron Microscopy - energy dispersive X-ray spectroscopy (SEM-EDS) and Brunner Emmet Teller (BET) - Barrett-Joyner-Halenda (BJH). The resulting RBE characteristics include surface area, pore volume, and pore size of 436,820 m2/g, 0.544 cc/g, and 2.488 nm, respectively. Patricelli model is more appropriate to use to explain the mechanism of the SBE desorption process using the MAE method with the coefficient of determination value of 0.9931. The palm oil adsorbed using RBE has 10R/20Y/0.9W of color quality, 3.93% free fatty acid number, and 11.61 meq O2/kg peroxide number. In this study, SBE was successfully converted into RBE by microwave reactivation process, and the excellent performance of RBE in palm oil bleaching was verified. This research not only helps to solve the environmental problems in the process of palm oil refining but also provides a new way for waste resource utilization.

1 Introduction

Palm oil is one of important commodity for Indonesia on the domestic and international side. The most application of palm oil is for edible oil and biofuel. Indonesia is one of the world's leading palm oil producer, supplying approximately more than 80% of the commodity globally. In Indonesia, plantations have predominantly been developed in Sumatra and Kalimantan (Indonesian Borneo), with rapid expansion in recent decades. Palm plantation has the highest yield of oil per unit area cultivated and produces 10 times more oil per hectare than other oilseed crops such as sunflower, rapeseeds, cottonseed, peanut [1]. The palm oil is extracted from the Elaeis guineensis fruit mesocarp. The harvesting palm tree should be consider average 24–30 months after planting. The oil palm tree age and type of oil palm seeds are important thing that should consider for harvesting the oil palm fruits. 8 to 15 fresh fruit bunches (FFB) are produced from each oil palm tree producing per year. There are about 1000–1300 fruitlets consist in each fresh fruit bunch. This each fruitlet contain with fibrous mesocarp or flesh layer and the endocarp (shell) which contains the kernel [2]. Palm oil production from fresh fruit bunch are proceed by several step such as, sterilization, stripping, digesting, pressing, extraction and refining process. Palm oil contains large amount of mayor compound which is fatty acids. Otherwise, large amount of minor compounds in palm oil, are tocopherol, tocotrienol, squalene, phytosterol, phenolic compounds and carotenoids. Regarding carotenoids, the main type present in crude palm oil is ß-carotene (56%) and a-carotene (35%), both recognized for their pro-vitamin A activity.

The refining process of crude vegetable oil goes through several stages: degumming, neutralizing, bleaching, and deodorizing. Bleaching is the most important aspect of palm oil refining, as the colour of palm oil is usually the main attraction influencing consumer choice. In this stage, carotene compounds, peptoids, gums, and some impurities compounds are separated through adsorbent media or what is commonly called Bleaching Earth (BE) [3]. BE improves the final product's appearance, taste, and stability by adsorbing unwanted colouring matter and residues from the crude oil [4]. However, this bleaching process generates a large amount of solid waste, namely BE residue or Spent Bleaching Earth (SBE), which is usually disposed to the plant landfill without any treatment [3,5,6]. SBE contains 20-40% residual oil and other organic matter (carotene, peptoids, gum, and impurities compounds) [3]. The oil content contained in SBE may contribute to spontaneous combustion. In addition, there is the potential for fat leakage into water and soil, which can cause environmental damage and economic losses [7]. So, it is

* Corresponding author: renita.manurung@usu.ac.id

necessary to make efforts to overcome this, one of which is by reusing SBE.

Reusing SBE as an adsorbent requires a reactivation process first. The aim is to remove the oil content and impurities in the SBE surface, which can impact the efficiency of the SBE sorption power. Several studies have reported the reactivation of SBE using socket extraction [8], calcination [9], pyrolysis [10], and using acid solutions [11]. However, most methods used have drawbacks, including the need for significant energy consumption, long time, and high temperature. It adversely affects the ability of the SBE used because it can damage the structure of the SBE itself [12]. One of the efforts that can be made to overcome this is by using microwaves. Lee et al. (2018) reported using microwaves saves time and energy [13]. Therefore, the reactivation process of SBE in this study uses microwaves combined with an acid solution. This study aims to determine the performance of regenerated SBE produced from the microwave reactivation process and acid solution as an adsorbent in the palm oil (CPO) bleaching process.

2 Material and Method

2.1. Material

Spent Bleaching Earth (SBE) and palm oil (CPO) were obtained from one of the palm oil industries in Medan City. Hexane and sulphuric acid solvents used were pure analysts (Merck)

2.2 Spent bleaching earth reactivation

First, desorption of oil content contained in SBE using a microwave extraction. The extraction process was carried out using hexane solvent at a solvent ratio 1:4 with operating conditions of 450 watts of microwave power and variation time 10 - 60 minutes. After that, the SBE was dried in an oven at 110°C. The result called desorbed bleaching earth (DBE). Second, the DBE was activated using sulphuric acid solution 3 under 300 rpm, stirring at 70°C for 2 hours with a ratio of 1:10 (DBE: sulphuric acid solution). After the activation process was completed, it was filtered and then washed with hot water until the pH of the washing water was >3. Next, it was dried at 105 °C. The result called reactivated bleaching earth (RBE).

2.3 Characterization of reactivated bleaching earth

The surface morphology and chemical element analysis of RBEs were analyzed using Scanning Electron Microscopy - energy dispersive X-ray spectroscopy (SEM-EDS) at a magnification of 15,000 with a voltage acceleration of 10 kv. Brunneur Emmet Teller (BET) - Barrett-Joyner-Halenda (BJH) analysis was evaluated by nitrogen gas absorbance method using Quantachrome Novatouch LX-4.

2.4 Application: CPO Bleaching Process

First, 100 g of CPO was heated at 110°C under 300 rpm, stirring using a magnetic stirrer in an Erlenmeyer. Then, add RBE. The CPO bleaching process was carried out for 45 minutes at the ratio of RBE: CPO (0.8, 1.5 and 3%). After completion, RBE and bleached oil were separated using filter paper. Furthermore, bleached oil was analyzed for free fatty acid content, color, and peroxide number.

3 Results and Discussion

3.1. Characterization of reactivated bleaching earth

Table 1. SEM-EDS Analysis on ABE, SBE, DBE and RBE

Adsorbent	Morphology	Detail
ABE (Activated Bleaching Earth)	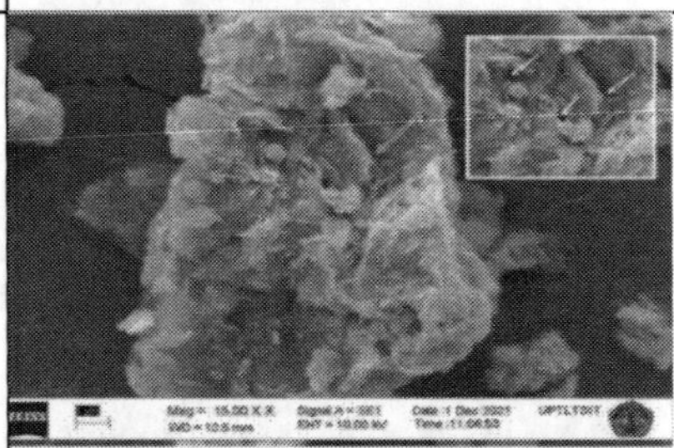	opened pores
SBE (Spent Bleaching Earth)	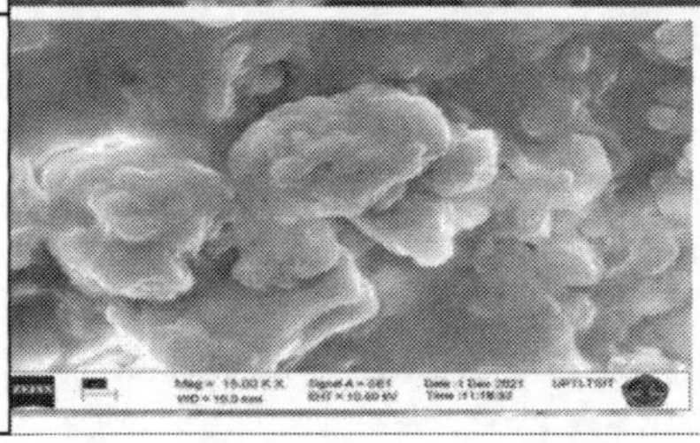	closed pores

DBE (Desorbed Bleaching Earth)	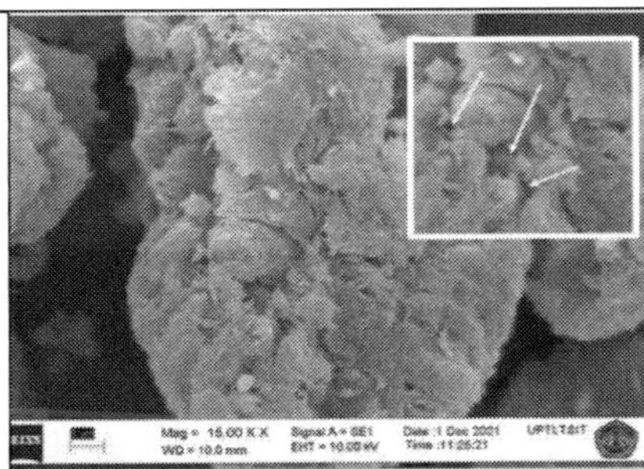	the pore slightly open
RBE (Reactivated Bleaching Earth)	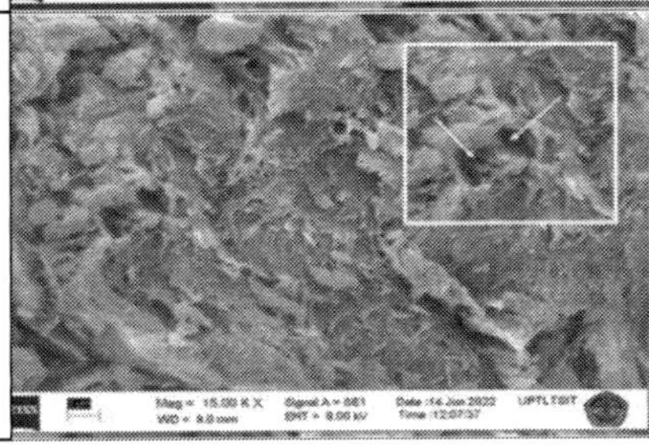	opened pore

3.1.1 Scanning Electron Microscopy - Energy Dispersive X-ray Spectroscopy (SEM- EDS)

The characteristics of ABE (Activated Bleaching Earth), SBE (Spent Bleaching Earth), DBE (Desorbed Bleaching Earth), and RBE (Reactivated Bleaching Earth) concerning morphology and elemental composition were analyzed using SEM-EDS, as shown in Table 1. Based on Table 1, the ABE sample has a flat surface and tends to be rough; the pores are visible, as indicated by the arrows. Meanwhile, morphology SBE sample has a surface like a mound of soil (rounded). After the desorption process using the microwave, the surface of the DBE sample slowly opens again. The shape of the DBE surface returns to a flat shape and tends to be rough. However, there are not as many visible pores as open pores in the ABE sample. Then, after being given an acid activation treatment, the RBE pore surface already resembles the shape of the ABE surface. It shows that the acid activation process is beneficial for compounds in improving the pore structure or particle surface; this certainly affects the ability of RBE to adsorb in its reusability.

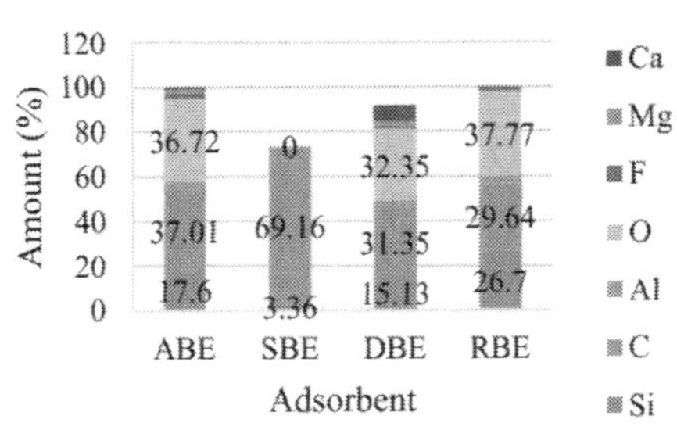

Fig. 1. Compound Composition of ABE, SBE, DBE and RBE

Figure 1 shows that the composition of compounds in ABE, SBE, DBE, and RBE samples is known. It is concluded that the Si content in ABE is 17.60, and in SBE, the Si content decreases to 3.36; then, after the desorption process, the Si content increases again to 15.13% and continues to increase in RBE up to 26.70%. Likewise, the composition of Al, Mg, and Ca compounds decreased in SBE and increased again during desorption. However, it is different with the C compound, wherein in the ABE sample, the amount is 37.01% and increased in the SBE sample by 69.16%, then decreased in the quantity of 31.35 and 29.64% in DBE and RBE. It shows that microwaves and sulphuric acid solution will remove the oil content and impurities found on the surface of SBE. The loss of impurities on RBE will prove that RBE has high efficiency as an adsorbent in the CPO oil bleaching process.

3.1.2 Brunauer-Emmet-Teller – Barret Joyner Halenda (BET- BJH)

Table 2. SEM-EDS Analysis on ABE, SBE, DBE and RBE

Properties	ABE	SBE	DBE	RBE
Surface area (m^2/g)	124.602	6.192	26.064	436.820
Pore volume (cc/g)	0.195	0.032	0.085	0.544
Pore size (nm)	3.128	10.529	6.499	2.488

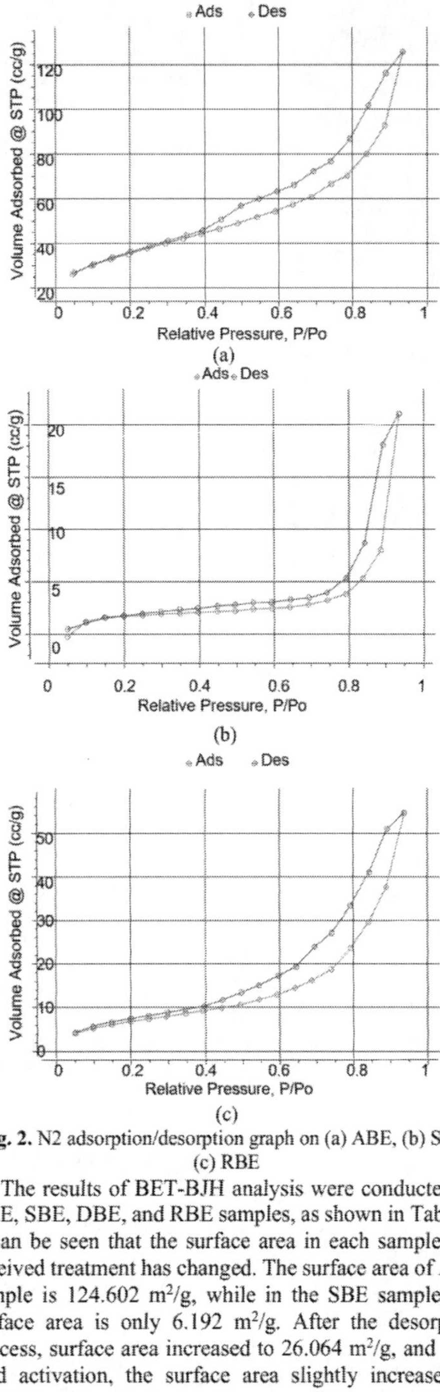

Fig. 2. N2 adsorption/desorption graph on (a) ABE, (b) SBE, (c) RBE

The results of BET-BJH analysis were conducted on ABE, SBE, DBE, and RBE samples, as shown in Table 2. It can be seen that the surface area in each sample that received treatment has changed. The surface area of ABE sample is 124.602 m^2/g, while in the SBE sample, the surface area is only 6.192 m^2/g. After the desorption process, surface area increased to 26.064 m^2/g, and after acid activation, the surface area slightly increased to 436.820 m2/g. Similarly, the pore volume size of each ABE, SBE, DBE, and RBE sample changed while it was increasing on DBE and RBE samples. It shows that microwave treatment and acid activation will remove impurities that cover SBE pores.

Adsorption and desorption with the N2 method was used to determine the porosity of the adsorbent. The graph of nitrogen adsorption desorption isotherms on ABE, SBE and RBE is shown in Fig. 2. The presence of hysteresis loops on the isotherm curve indicated the mesoporous structure, similar to the type II isotherm according to the International Union of Pure and Applied Chemistry (IUPAC) classification. Type II isotherms are S-shaped or sigmoid curves and are generally found in materials with larger pores than micropores. The graph shows that the hysteresis loops are closely spaced and the adsorption and desorption are almost parallel. It shows pores with a regular geometry. The isotherm graph shows capillary condensation at the initial stage (at a relative pressure of 0.5), with nitrogen condensation occurring in the internal and second mesopores at a higher partial pressure (P/Po 0.8), which is associated with the presence of porosity between the particles. This shows the ability or absorption to enter triglycerides into the pores of the catalyst in the transesterification reaction [14].

3.2. Reactivation Process: Effect Microwave Time

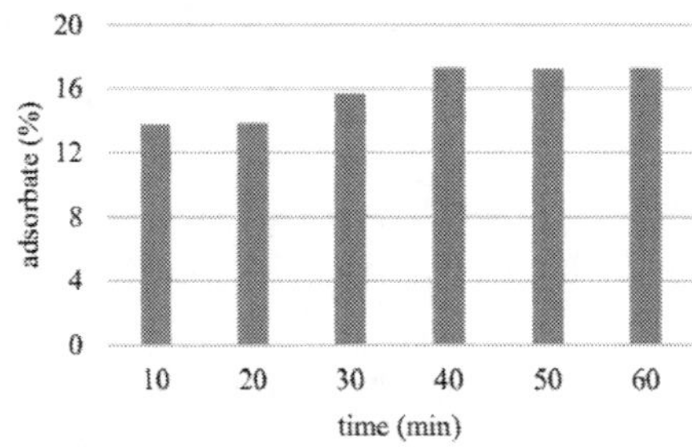

Fig. 3 Relationship between variations in microwave time and the amount of adsorbate released

In Figure 3, The graph shows the amount of adsorbate released at varying times of 10, 20, 30, 40, 50, and 60 minutes with a power of 450 watts. At 450 watts, the adsorbate released at a time variation of 10 minutes to 60 minutes was 13.77%, 13.87%, 15.69%, 17.33%, 17.22%, and 17.29%. The amount of adsorbate released from the adsorbent is greatly influenced by many factors, including contact time, temperature, wave power, solvent, and methodology. The adsorbate component released from the adsorbent consists of oil residue and other organic materials such as carotene, peptoids, gum, and impurity compounds [15]. It is shown that the longer the time, the greater the amount of adsorbate released.

At each time variation, temperature measurements are taken. There is a continuous increase as time increases. The temperature increased at 10, 20, 30, and 40 minutes, namely 79 C, 83 C, 90 C, and 120 C. However, for 50 minutes, the temperature decreased to 106 C and increased again at 60 minutes to 114 C. Based on research by Luo and Gross (2004), it is explained that increasing temperature will increase the desorption rate [16]. High temperatures will activate the substance molecules so that the collision process between molecules occurs more quickly and opens the pores of the substance so that the amount of impurity released increases [17]. The graph shows that the longer the time, the higher the temperature, which causes more adsorbate to be released.

3.3. Desorption Kinetic

This research conducted a kinetics study of the SBE desorption stage with the best operating conditions at 40 minutes and 450 watts of power. This operating condition provides the highest % released adsorbate, which is 17.33%. Table 3 shows the kinetic model parameters and coefficient of determination values.

Table 3. Kinetic Model Parameters and Determination Coefficient Values

Parameter	Model		
	1st Orde	**2nd Orde**	**Patricelli**
K (per min)	0.1457	0.01337	0.0248
R2	0.9724	0.9862	0.9931
YM (%)	16.6894	18.3356	11.7050
RMSE	0.1284	0.0602	0.0016

The experimental data obtained was input into the equation of each kinetic model using OriginLab software.

1st Order Kinetic Model Equation:

$$\ln CA = -kt + \ln CA0 \quad (1)$$

Patricelli's kinetic model equation:

$$\frac{1}{CA} - \frac{1}{CA0} = kt \quad (2)$$

The experimental data and kinetic model are plotted on a graph like Fig. 3 below to see the size of the model's suitability to the experimental data.

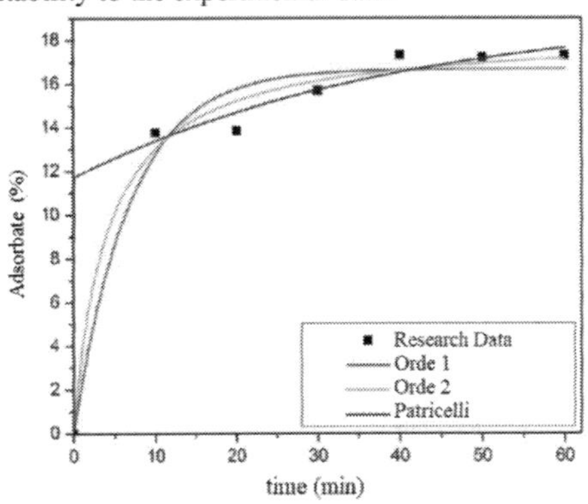

Fig. 4. Kinetic graph of experimental data based on Order 1, Order 2 Models and Patricelli Models

In Fig. 4, the kinetic graph of experimental data based on the kinetic models of Order 1, Order 2, and the Patricelli Model produces a significant increase in the amount of released adsorbate at the beginning of the desorption process (0-10 minutes). After 10 minutes until the end of the process, the increase in the amount of released adsorbate is not very significant.

In Table 4, each model's coefficient of determination R2 value provides a high value. The Patricelli model provides the highest coefficient of determination value of 0.9931 compared to the first-order model (0.9724) and the second-order model (0.9862); this shows that the Patricelli model is more appropriate to use to explain the mechanism of the SBE desorption process using the MAE method following research by Pihtili (2013) which states that the coefficient of determination value is a linearity test value which states the relationship between perfection and final concentration [18]. So, a higher coefficient of determination value indicates a higher level of conformity to the experimental data with the kinetic model.

Patricelli's kinetic model approach explains that SBE desorption occurs in 2 stages: the washing and diffusion stages. First is the washing stage, where the sample and solvent mix at a specific wave power. Otherwise, it will decrease mass transfer limitations and increase solvent penetration into the ruptured cell walls. The next stage is diffusion, where solutes from the internal cell walls diffuse into the solvent. Usually, the diffusion stage runs slower than the washing stage due to limited mass transfer [19].

3.4. Application: Bleached Palm Oil Quality

After palm oil bleaching process using ABE and RBE, the palm oil bleached was analysed the free fatty acid and peroxide value which can be seen in Table 4. Bleached palm oil quality are approach same using ABE and RBE. It's proven that SBE can be reactivated using microwave and sulphuric acid activation for palm oil bleaching process.

Table 4. Free Fatty Acid and Peroxide Value Bleached Palm Oil using ABE and RBE

Adsorbent	Unit	ABE	RBE
Free Fatty Acid	%	3.70	3.93
Peroxide Value	meq O_2/kg	16.13	11.61
Color Quality	-	10R/20Y/0.8W	10R/20Y/0.9W

4 Conclusion

The reactivation of spent bleaching earth using microwaves and sulphuric acid solution can restore optimal absorption in bleaching crude palm oil (CPO). The resulting RBE characteristics include surface area, pore volume, and pore size of 436,820 m2/g, 0.544 cc/g, and 2.488 nm, respectively. Patricelli's model is more appropriate for explaining the mechanism of the SBE desorption process using the MAE method with a coefficient of determination value of 0.9931. Bleached palm oil quality using 3% RBE is 10R/20Y/0.9W colour quality, 3.93% free fatty acid number, and 11.61 meq O2/kg peroxide number. And longer reactivation time using microwave will cause higher temperature, which affects more adsorbate to be released.

References

1. B. C. Steffens, B. N. Segala, E. H. Tanabe, C. A. Ballus, D. A. Bertuol. IChemE: Food and Bioproducts Processing **143,** 158–169 (2024)

2. M. H. M. Yasin, R. Mamat, G. Najafi, O. M. Ali, A. F. Yusop, M. H Ali. Renewable and Sustainable Energy Reviews **79**, 1034–1049 (2017)
3. B. Mu, & A. Wang. Handbook of Ecomaterials **5**, 3147–3167 (2019)
4. N. S. A. Wafti, C. K. Yoo, S. W. Lin, T. C. S. Yaw, & L. C. Abdullah. Journal of Oil Palm Research **23**, 999–1004 (2011)
5. A. Boukerroui, & M. S. Ouali. Journal of Chemical Technology and Biotechnology **75**, 773–776. (2000)
6. J. I. Chang, H. S. Tai, & T. H. Huang. Environmental Progress **25**, 373–378 (2006)
7. C. Dejkajorn, P. Suttiarporn, H. Sukkathanyawat, K. Wichianwat, S. Sriphan, & T. Srisurat. E3S Web of Conferences **302**, 01009 (2021)
8. R. A. Majid & C. R. C. Mat. Journal of Oil Palm Research **29**, 579–587 (2017)
9. L. Liu, Y. Li, B. A. Yoza, K. Hao, Q. X. Li, Y. Li, Q. Wang, S. Guo, & C. Chen. 699. (2020)
10. C. Su, L. Duan, F. Donat & E. J. Anthony. Applied Energy **210**, 117–126 (2018)
11. M. R. Sabour, M. Shahi & G. A. Dezvareh. GJESM **3**, 299–310 (2017)
12. A. A. Al-Zahrani & M. A. Daous. Process Safety and Environmental Protection **78**, 224–228 (2000)
13. H. G. Lee, Y. Jo, K. Ameer, & J. H. Kwon. Food Science and Biotechnology **27**, 1607–1617 (2018)
14. R. Manurung, S.Z.D.M. Parinduri, R. Hasibuan, B.H. Tarigan, and A.G.A. Siregar. Case Studies in Chemical and Environmental Engineering **7** (2023) 100345.
15. B. Mu, & A. Wang. Handbook of Ecomaterials **5**, 3147–3167 (2019)
16. W. Luo & K. J. Gross. Journal of Alloys and Compounds **385**, 224-231 (2004)
17. X. Zhao, D. Bradshaw & M. J. Rosseinsky. Science **306**, 1012-1015 (2004)
18. Pihtili, H. Tribology in Engineering. BoD–Books on Demand (2013)
19. A. N. Mustapa, A. Martin, J. R. Gallego, R. B. Mato & M. J. Cocero. Chemical Engineering and Processing: Process Intensification **97**, 66-74 (2015)

Research Progress of AlGaN Ultraviolet Light-Emitting Diodes

Yiming Zhao[1], Zijie Wang[2], Xiaolin Feng[3,*]

[1]School of Computer Science, South China Business College Guangdong University of Foreign Studies, Guangzhou, 510545, The People's Republic of China
[2]School of Internet Economics and Business, Fujian University of Technology, Fuzhou, 350118, The People's Republic of China
[3]Faculty of Engineering and Life Sciences, Universiti Selangor, Bestari jaya, Selangor 45600, Malaysia

Abstract: The research progress of AlGaN ultraviolet light-emitting diodes (UV LEDs) is explored in this paper. Firstly, the significance and application areas of AlGaN UV LEDs are introduced, emphasizing their wide-ranging applications in fields such as disinfection, medical devices, and sensors. Subsequently, an overview of the material science and manufacturing processes used in the production of AlGaN UV LEDs is provided, including crystal growth techniques and fabrication processes. Next, the design, structure, and operating principles of different types of AlGaN UV LEDs are described, with a focus on the devices optical performance, electrical characteristics, and thermal management effects. Finally, the latest research findings are summarized, and the challenges and future development directions facing current technology are discussed. This paper aims to provide readers with a comprehensive understanding of the research field of AlGaN UV LEDs and serve as a reference for future research and applications.

1. Introduction

In the realm of modern technological applications, ultraviolet light-emitting diodes (UV LEDs) are emerging as a significant area of research, particularly those made from AlGaN materials, due to their unique optical and electrical properties[1]. AlGaN-based UV LEDs possess the capability to emit light efficiently within the ultraviolet spectrum range (approximately 200 to 350 nanometers), demonstrating broad application potential in critical areas such as water and air disinfection, medical diagnostics, biochemical assays, and high-density data storage[2].With increasing global demands for environmental protection and energy efficiency, the use of AlGaN UV LEDs is also expanding in fields such as environmental monitoring and hazardous substance detection[3]. These devices offer advantages over traditional mercury lamps and other types of UV sources due to their durability, higher energy efficiency, and operation without heavy metals, making them safer and more environmentally friendly[4].

Despite the clear advantages of AlGaN UV LEDs, their development faces several challenges, including improving light output efficiency, reducing manufacturing costs, and enhancing device stability and lifespan[5]. In recent years, researchers have made significant advances in material synthesis, device structure optimization, and the development of innovative manufacturing technologies, providing potential solutions to these issues. This paper will review the research history, current technological status, and latest scientific advancements of AlGaN UV LEDs, aiming to provide a comprehensive reference resource for both the academic and industrial communities, thus fostering further exploration and technological innovation in this field.

2. The opportunities and challenges of AlGaN-based UV LEDs

Table 1 comprehensively details the main applications of three types of ultraviolet light: UVA, UVB, and UVC. UVA light spans a wavelength range of 315 to 400 nanometers. Its ability to trigger the curing reactions in photosensitive resins makes it particularly valuable in printing and nail art. It is widely used in fluorescence excitation and photopolymerization. UVB light, with wavelengths between 280 and 315 nanometers, penetrates deeper into the skin, influencing biological processes and promoting the synthesis of vitamin D, which is essential for bone health. Therefore, it is primarily utilized in medical treatments (such as psoriasis), vitamin D synthesis, and biological research. UVC light, possessing the shortest wavelengths ranging from 100 to 280 nanometers, can disrupt the DNA and RNA structures of microorganisms, effectively killing or inactivating bacteria and viruses. Its potent disinfecting ability is extensively applied in the sterilization of air and water, surface disinfection, and security verification.

Mercury lamps are currently common UV light sources; however, UV LEDs offer several advantages over traditional mercury lamps, making them increasingly attractive for replacement in various applications. Firstly,

* Corresponding author: 18345148026@163.com

UV LEDs are more energy-efficient and consume less electricity, which helps reduce long-term operational costs. Secondly, LEDs typically have a much longer lifespan than mercury lamps, often lasting tens of thousands of hours, thereby decreasing the frequency of replacements and maintenance costs. The environmental benefits of UV LEDs are particularly notable. Unlike mercury lamps, which contain toxic heavy metals, LEDs are free of hazardous substances and are more environmentally friendly. Additionally, LEDs have a rapid response time, enabling instant on/off capabilities, whereas mercury lamps require a warm-up period. UV LEDs also offer superior controllability, allowing for the adjustment of light intensity and wavelength as needed, unlike the fixed spectral output provided by mercury lamps. This flexibility makes UV LEDs especially important in precise scientific experiments and industrial applications, such as in photopolymerization and biomedical imaging.In accordance with the Minamata Convention on Mercury formulated by the United Nations Environmental Programme, of which China is a key signatory, the production and use of high-pressure mercury lamps and mercury-containing fluorescent lamps are gradually being phased out starting from 2020. Therefore, the future trend in the industry is to develop a high-efficiency, energy-saving, and environmentally friendly alternative to traditional mercury lamps[6].

Table 1. Applications of Different Ultraviolet Wavelengths

Type	Wavelength Range	Primary Applications
UVA	315-400 nm	Fluorescence excitation, photopolymerization,
UVB	280-315 nm	Medical treatments (e.g., psoriasis), vitamin D synthesis, biological research
UVC	100-280 nm	Air and water disinfection, surface sterilization, security verification

UV LED light sources offer several advantages over traditional mercury lamps, including adjustable emission wavelengths, non-toxicity, portability, low power consumption, long lifespan, and high integration. They are poised to replace traditional mercury lamp sources and play a significant role in fields such as food preservation, sterilization, and disinfection. However, the quantum efficiency of deep ultraviolet LED devices emitting at 280 nm is currently below 20%, with the majority hovering around 4%. Even for those deep ultraviolet LEDs that have entered mass production, their quantum efficiency is even lower, ranging from only 2% to 3%. Therefore, there is still significant room for improvement in the quantum efficiency of deep ultraviolet LEDs, making it a key focus of research in the field of ultraviolet LEDs[7]. The main factors influencing the quantum efficiency of deep ultraviolet LED devices are as follows:

a. High dislocation density within AlGaN-based materials: According to the theoretical predictions, when the dislocation density within the deep ultraviolet LED material exceeds 10^{10} cm^{-2}, the internal quantum efficiency is around 5%, whereas when the dislocation density is less than 10^{8} cm^{-2}, the internal quantum efficiency reaches approximately 60%. This indicates that the material density within the LED significantly affects its external quantum efficiency. High-quality LED epitaxial materials are essential for achieving efficient deep ultraviolet LED devices, which necessitates matching the lattice parameters and thermal expansion coefficients of the substrate with those of the epitaxial layers. However, due to the lack of AlGaN homogeneous epitaxial substrates, current deep ultraviolet LED structures are based on heterogeneous epitaxy on sapphire substrates. As shown in Figure 1, the lattice mismatch between the sapphire substrate and the bottom AlN layer in the LED can reach as high as 13.3%[8].

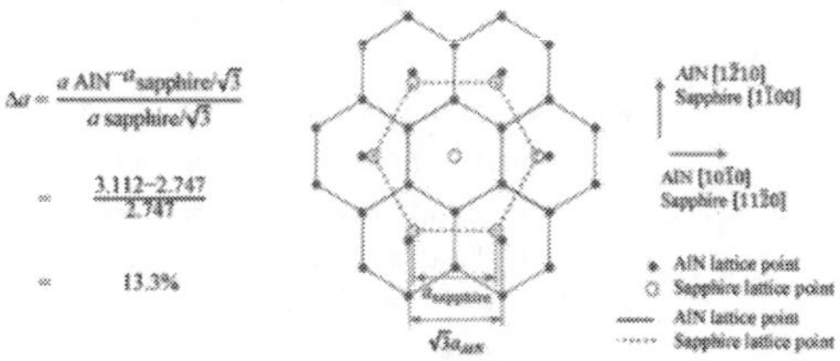

Fig. 1. Schematic diagram of the lattice arrangement of AlN film grown on a *c*-plane sapphire substrate

Consequently, as the epitaxial layer thickness increases during LED growth, internal stress gradually accumulates, ultimately leading to stress relaxation via dislocations or cracks. Furthermore, the slow migration rate of Al atoms makes it difficult for them to migrate to optimal nucleation positions, thereby increasing the dislocation density within the material. These dislocations penetrate into the quantum well active region, forming non-radiative recombination centers, thereby reducing the radiative recombination efficiency of the device. Additionally, there is a significant difference in thermal expansion coefficients between the sapphire substrate and the epitaxial layers, and since deep ultraviolet LED growth typically occurs at high temperatures, this can lead to increased material cracking, resulting in decreased yield of deep ultraviolet LEDs.

b. Difficulty in extracting transverse magnetic (TM) mode emission from high-Al-content AlGaN quantum wells: In gallium nitride (GaN) materials, the crystal field splitting energy Δ_{CF} is positive at 38 meV, resulting in heavy hole (HH) and light hole (LH) bands located above the crystal field splitting band (CH), with emission predominantly in the transverse electric (TE) mode (Figure 2(a)). However, in aluminum nitride (AlN) materials (Figure 2(b)), the crystal field splitting energy ΔCF is -219 meV, resulting in the crystal field splitting band (CH) located above the heavy hole (HH) and light hole (LH) bands, with emission predominantly in the transverse magnetic (TM) mode. In deep ultraviolet LEDs, AlGaN quantum wells are typically used as the active region, and as the LED wavelength decreases, the AlGaN composition of the quantum wells gradually increases, resulting in a transition from TE mode emission, characteristic of GaN materials, to predominantly TM mode emission, characteristic of AlN materials[9]. Figure 2 illustrates the polar plot of the emission direction of light generated in the active region during the transition from

TE to TM mode, showing that TM mode light is difficult to extract from the surface of the ultraviolet LED. Therefore, the transition in emission mode of the quantum wells is another major factor limiting the improvement of the quantum efficiency of deep ultraviolet LEDs.

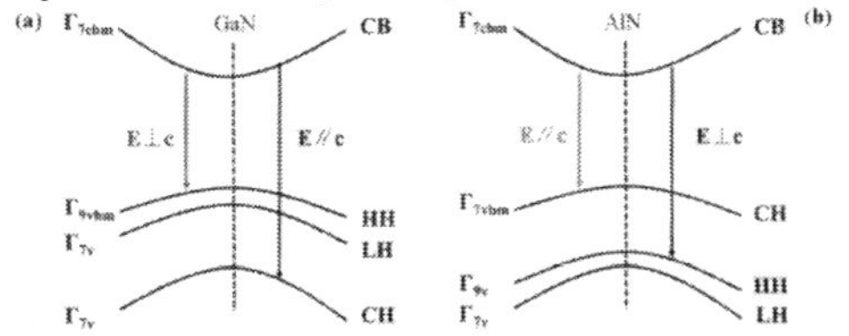

Fig. 2. Energy band structure of GaN(a) and AlN(b) films

c. Low carrier injection efficiency of deep ultraviolet LEDs: In addition to the challenges of high dislocation density and difficulty in TM mode light extraction, the carrier injection efficiency of deep ultraviolet LEDs also needs improvement[10]. On one hand, the presence of numerous dislocations in AlGaN thin films acts as compensation and scattering centers, reducing the carrier mobility and activation efficiency. On the other hand, as the aluminum content of the AlGaN material increases, the donor/acceptor levels continue to deepen, resulting in an increase in activation energy and a gradual decrease in carrier concentration. Therefore, in deep ultraviolet LEDs, the electron mobility and activation concentration of n-type electrons must be much higher than those of p-type holes[11]. The asymmetric mobility and activation concentration of electrons and holes make electrons prone to passing through the quantum well region and recombining with holes in the p-doped region, resulting in parasitic peaks. This problem is particularly severe for LEDs with shorter emission wavelengths. Currently, the general method to prevent electron leakage to the p-side is to introduce a high-aluminum-content p-type AlGaN doping layer, known as the electron-blocking layer (EBL), between the quantum well active region and the p-side. By optimizing the aluminum content and doping of the EBL, a certain degree of electron leakage can be prevented. However, while the EBL blocks electron leakage, it also blocks some hole injection, thereby reducing hole injection efficiency.

3. The Way to Improve AlGaN UV LEDs Efficiency

Researchers around the world have conducted extensive studies aimed at improving the efficiency of deep ultraviolet LEDs. In the field of epitaxial material growth and structural design for UV LEDs, Mohammad et al. presents a numerical analysis of the photoluminescence (PL) performance in DUV-LEDs, The researchers constructed three AlGaN-based DUV-LED structures with different Al compositions in their QWs[12]. The primary finding is that a uniform distribution of carriers across the QWs significantly enhances radiative recombination, which in turn, drastically improves the PL intensity. Specifically, the structure with the thinnest step-shaped QW exhibited a PL intensity six times greater than the first structure and twice as high as the second. This enhancement is attributed to the more uniform electron and hole distribution within the QWs, facilitated by a careful manipulation of the Al content in the heterostructures. Li et al. explores advancements in improving the light extraction efficiency of AlGaN DUV-LEDs using Al/MgF2-based highly reflective films[13]. The integration of highly reflective Al/MgF2 films on the sidewalls and substrates of the LED structure led to a significant improvement in light extraction efficiency. This was achieved by redirecting sideways-travelling photons towards the substrate for more effective light extraction, which in turn increased the luminous intensity by about 10%. Sun et al. (2023) focuses on improving the external quantum efficiency (EQE) of DUV LEDs by developing an ultra-thin AlGaN last quantum barrier (LQB)[14]. The study successfully developed a 1.0 nm thick AlGaN LQB with a high aluminum content. This was achieved by controlling surface desorption kinetics during the growth of MQWs, which significantly suppressed the formation of group-III vacancies. The implementation of the ultra-thin LQB resulted in a 44% increase in light output power to 7.16 mW at 40 mA and a peak EQE of 4.04%. These improvements demonstrate the effectiveness of the ultra-thin barrier in enhancing hole injection and reducing carrier overflow. Qi et al. propose the use of multiple thin p-AlGaN insertion layers within an Al-composition-decreasing (ACD) p-AlGaN HSL[15]. This design aims to enhance the hole injection efficiency. The proposed structure led to significant improvements in both EQE and LOP across various current injection densities. These enhancements are attributed to increased hole-accelerating electric fields and strong intraband tunneling processes facilitated by the ultrathin insertion layers. The improvements are due to polarization-induced sheet charges at the interfaces of the insertion layers, which augment the electric field and improve hole transport via tunneling. The structure also promotes the accumulation of holes at the upper interfaces of the insertion layers, contributing further to enhanced hole injection. Liu et al. explores methods to enhance carrier injection in AlGaN-based DUV LEDs. The study focuses on improving carrier injection efficiency in AlGaN-based DUV LEDs, which is crucial for achieving high device performance and efficiency. By optimizing the design and thickness of the p-GaN layer, the researchers aim to enhance hole injection into the active region of the LED, thereby improving overall device efficiency. Improving carrier injection efficiency has significant implications for the performance of AlGaN-based DUV LEDs, including enhanced light output, reduced forward voltage, and improved device reliability.

In the field of chip design and manufacturing for UV LEDs. Zhang et al. investigates the enhancement of reliability in AlGaN-based DUV LEDs by altering the reflective n-type electrode design[16]. The study compares two types of Cr/Al/Ti/Au electrode structures with different thicknesses of the aluminum layer in a flip-chip configuration. The paper highlights how defects in the electrode, especially from the thicker Al layer, lead to significant device degradation through mechanisms such as leakage currents and luminous attenuation. These defects arise due to the migration of Sn atoms from solder paste into the electrode during the burn-in test, acting as leakage current channels. It was found that the DUV LEDs with the 60 nm thick Al layer retained about 86.8% of their normalized optical power after a 1000-hour burn-in test, showcasing improved reliability and performance compared to the one with the 120 nm thick layer.Yao et al. explores the size-dependent characteristics of Aluminum AlGaN-based DUV μLEDs with emission wavelengths ranging from 277 nm to 304 nm[17]. The μLEDs with smaller mesa dimensions (down to 20 μm) show improved light extraction efficiency and reduced forward voltage compared to larger counterparts. The smallest devices demonstrated some of the highest EQEs reported for DUV μLEDs, with peak EQEs of 2.5% and 4.0% for 277 nm and 304 nm emissions, respectively. Smaller devices benefit from better current spreading, which reduces series resistance and forward voltage. This improvement in current spreading is critical in smaller μLEDs, as it helps mitigate the effects of poor current distribution seen in larger devices. Yan et al. discusses the fabrication of a vertical AlGaN DUV LED[18]. The study focuses on the fabrication of a vertical DUV LED structure, which offers advantages such as improved light extraction efficiency (LEE) and reduced series resistance compared to conventional lateral designs. By implementing a vertical structure, the LED design aims to enhance LEE, which is crucial for achieving high external quantum efficiency (EQE) in DUV LEDs. The vertical design allows for better photon extraction from the active region, leading to improved overall device performance. Liu et al. discusses the development of high-power UVC LED wafers and their integration into low-cost 4-inch substrates[4]. The paper discusses the design considerations and fabrication processes involved in developing UVC-LED chips suitable for integration into 4-inch substrates. This includes optimizing epitaxial growth, chip layout, and packaging techniques. Efficient heat dissipation is crucial for maintaining the performance and reliability of high-power UVC-LEDs. The study explores various thermal management techniques to ensure proper heat dissipation and prolong LED lifespan. By integrating high-power UVC-LEDs into low-cost 4-inch substrates, the researchers aim to improve cost-effectiveness and scalability, making UVC-LED technology more accessible for widespread deployment in various industries. Zhang et al. introduce a novel honeycomb hole-shaped electrode (HHSE) structure for 280 nm AlGaN-based flip-chip DUV-LEDs, aiming to enhance current spreading, optical emission, and thermal dispersion[19]. The finite element method was employed to analyze the current density distribution at an injection current density of 67.5 A/cm^2, revealing that the effective current spreading area was increased by approximately 35.3% in LEDs utilizing the HHSE structure compared to those with conventional finger-shaped electrodes. The study demonstrated that the wall-plug efficiency (WPE) of LEDs with the novel HHSE was improved by 28% at an injection current of 280 mA due to more uniform current spreading, which also facilitated symmetrical optical emission and temperature distribution across the device. This research provides a substantial breakthrough in addressing the current crowding effect attributed to the poor conductivity of high Al content n-AlGaN layers, significantly enhancing the efficiency of AlGaN-based DUV-LEDs.

4. Conclusions

This article explores the current development, challenges, and recent scientific advances in UV LEDs. AlGaN UV LEDs, recognized as an efficient ultraviolet light source, have demonstrated their unique application value across various fields, particularly in environmental disinfection, medical diagnostics, and scientific research, showcasing vast potential for broader applications. Despite this, the commercialization and mass deployment of AlGaN UV LEDs are still constrained by technical and cost-effectiveness issues. This paper discusses in detail the core technical challenges such as enhancing light output efficiency, optimizing thermal management, and extending device longevity. Ongoing material research, innovation in device structures, and improvements in manufacturing processes have led to significant advancements in these areas. However, overcoming more technical hurdles is necessary to achieve wider commercial use. Future research should continue to focus on enhancing the overall performance and economic viability of AlGaN UV LEDs, especially in improving photonic conversion efficiency and reducing production costs. Moreover, interdisciplinary collaboration will be crucial in advancing AlGaN UV LED technology. Close cooperation among fields such as materials science, optoelectronics, physics, and chemistry will offer more opportunities to address existing challenges and explore new applications.

References

1. J. Lang, F. Xu, J. Wang, L. Zhang, X. Fang, Z. Zhang, X. Guo, C. Ji, C. Ji, F. Tan, Y. Wu, X. Yang, X. Kang, Z. Qin, N. Tang, X. Wang, W. Ge, and B. Shen, Adv Elect Materials 2300840 (2024)
2. S. Liu, W. Luo, D. Li, Y. Yuan, W. Tong, J. Kang, Y. Wang, D. Li, X. Rong, T. Wang, Z. Chen, Y. Li, H. Wang, W. Wang, J. Hoo, L. Yan, S. Guo, B. Shen, Z. Cong, and X. Wang, Adv. Funct. Mater. 2008452 (2020)
3. S. Zhou, Z. Liao, K. Sun, Z. Zhang, Y. Qian, P. Liu, P. Du, J. Jiang, Z. Lv, and S. Qi, Laser & Photonics Reviews **18**, 2300464 (2024)
4. S. Liu, Y. Yuan, L. Huang, J. Zhang, T. Wang, T. Li, J. Kang, W. Luo, Z. Chen, X. Sun, and X. Wang, Adv Funct Materials **32**, 2112111 (2022)
5. B. C. Letson, J. W. Conklin, P. Wass, S. Barke, G. Mueller, M. A. J. Rasel, A. Haque, S. J. Pearton, and F. Ren, ECS J. Solid State Sci. Technol. **12**, 066002 (2023)
6. S. Liang and W. Sun, Advanced Materials Technologies **7**, 2101502 (2022)

https://doi.org/10.1051/e3sconf/202456001006

7. H. Amano, R. Collazo, C. D. Santi, S. Einfeldt, M. Funato, J. Glaab, S. Hagedorn, A. Hirano, H. Hirayama, R. Ishii, Y. Kashima, Y. Kawakami, R. Kirste, M. Kneissl, R. Martin, F. Mehnke, M. Meneghini, A. Ougazzaden, P. J. Parbrook, S. Rajan, P. Reddy, F. Römer, J. Ruschel, B. Sarkar, F. Scholz, L. J. Schowalter, P. Shields, Z. Sitar, L. Sulmoni, T. Wang, T. Wernicke, M. Weyers, B. Witzigmann, Y.-R. Wu, T. Wunderer, and Y. Zhang, J. Phys. D: Appl. Phys. **53**, 503001 (2020)
8. M. N. Sharif, M. I. Niass, J. J. Liou, F. Wang, and Y. Liu, Superlattices and Microstructures **158**, 107022 (2021)
9. Y. Qian, Z. Liao, Z. Lv, S. Qi, and S. Zhou, Opt. Lett., OL **48**, 1072 (2023)
10. S. F. Chichibu, K. Nagata, M. Oya, T. Kasuya, K. Okuno, H. Ishiguro, Y. Saito, T. Takeuchi, and K. Shima, Appl. Phys. Lett. **122**, (2023)
11. X. Qiu, Y. Chen, E. Han, Z. Lv, Z. Song, and H. Jiang, Mater. Adv. **1**, 77 (2020)
12. M. A. Hairol Aman, F. A. Ahmad Fajri, A. F. Ahmad Noorden, S. Daud, M. Bahadoran, and M. Z. Abdul Kadir, Phys. Scr. **98**, 015503 (2023)
13. P. Li, X. Huang, Y. Yang, Z. Xia, Q. Wang, H. Chen, and X. Wang, Semicond. Sci. Technol. **39**, 035010 (2024)
14. X. Sun, J. Liu, Y. Huang, C. Liu, X. Zhan, H. Gao, Q. Sun, Q. Xu, and H. Yang, Appl. Phys. Express **16**, 111004 (2023)
15. Y. Qi, W. Tian, M. Liu, S. Li, and C. Liu, IEEE Trans. Electron Devices **71**, 3077 (2024)
16. H. Zhang, W. Zhang, S. Zhang, M. Shan, Z. Zheng, A. Wang, L. Xu, F. Wu, J. Dai, and C. Chen, IEEE Electron Device Lett. **42**, 978 (2021)
17. Y. Yao, H. Li, P. Li, C. J. Zollner, M. Wang, M. Iza, J. S. Speck, S. P. DenBaars, and S. Nakamura, Appl. Phys. Express **15**, 064003 (2022)
18. J. Yan, J. Yuan, Y. Jiang, H. Zhu, H. W. Choi, and Y. Wang, Appl. Phys. Express **15**, 032003 (2022)
19. S. Zhang, F. Wu, S. Wang, H. Zhang, Y. Zhang, L. Xu, J. Dai, and C. Chen, IEEE Transactions on Electron Devices **66**, 2997 (2019)

Construction of Co/CeO_2 catalyst and study on catalytic performance

Long Li , Lifang Zhao,Yunyan Li,Weiwen Wang*

(School of Chemical Engineering,Qiandao University of Science and Technology,Shandong Qingdao 266042, China)

Abstract. High-temperature ammonia decomposition is a promising method for hydrogen production, but high temperatures and low conversion rates limit its wide application. In order to solve this problem, the development of nano-materials is essential to achieve high performance ammonia oxidation decomposition. In this paper, ordered mesoporous Co/CeO_2 catalyst was synthesized by codeposition method and a new process of ammonia hydrogen production was developed and designed without the need for external heating. The catalyst was characterized by XRD, BET and TEM. The results show that the catalyst activity is the highest when Co content is 10wt%. When the molar ratio of NH_3/O_2 is 4:1, the bed temperature increases from 165°C to 405°C, the conversion rate of NH_3 and O_2 is 72% and 81%, respectively, and the productivity of H_2 is 45%. The catalyst also showed good stability and reusability.

1. Introduction

Hydrogen energy is recognized as a clean energy with great potential. Hydrogen energy has the advantages of high calorific value, high density, wide source, abundant yield, high energy utilization rate, and pollution-free products, so it is considered as one of the most promising clean energy.The energy density of hydrogen (140 MJ/kg) is three times that of petroleum and 4.5 times that of coal. Applications of hydrogen energy are leading the way in the field of transport, including cars and other means of transport and as an energy carrier for storing excess electricity generated off-peak. However, hydrogen energy is still facing the problems of poor safety performance, storage and transportation difficulties, and high storage material requirements, which make it difficult to apply on a large scale. Ammonia is a high-quality hydrogen carrier, easier to liquefy and transport. It has high combustion value, green environmental protection, mature synthesis technology and safe and convenient storage and transportation mode. In the context of the increasingly severe environmental protection situation, especially in the context of China actively advocating the realization of "carbon peak" and "carbon neutrality", it is of great significance to explore the use of ammonia fuel as a new energy source to reduce carbon emissions. The development of the process of ammonia decomposition to hydrogen is not synchronized. The production capacity of hydrogen and ammonia is not a significant barrier to meeting the high demand, and current ammonia decomposition systems are not capable of producing enough pure hydrogen in large quantities at low temperatures. Therefore, ammonia decomposition becomes the bottleneck of hydrogen chain, and the current research focus on the process of NH_3 decomposition. Ammonia decomposition for hydrogen production as a green and environmentally friendly technology has received widespread attention[1-3]. Currently, there are many types of catalysts used for ammonia decomposition hydrogen production, including Ru-based, Fe-based, Ni-based, Co-based, and copper-based single metal catalysts, as well as combinations of multiple metals[4-6]. In addition to these metal combinations, materials such as metal nitrides and carbides are also used as catalysts. Although traditional ammonia decomposition hydrogen production catalysts show high efficiency in laboratory conditions, they face a series of challenges in large-scale production. The equilibrium calculation shows that the NH_3 conversion rate can reach 99%. However, it takes a lot of time and energy to reach reaction condition. Therefore, developing catalysts that can quickly initiate and have high H_2 production rates has become a hot research topic. This catalyst will help overcome the application difficulties of traditional catalysts in large-scale production[7-8].

In this paper, we discussed a new method of NH_3 oxidative decomposition for H_2 production. We studied the simultaneous occurrence of ammonia oxidation decomposition reactions, and compared the performance of different catalysts to reveal the reaction mechanism of the catalysts. By studying a new type of catalyst, we have made a contribution to the field. These results provide a direction for the preparation and improvement of efficient catalysts.

* Corresponding author: wwwang@qust.edu.cn

2.Triggering Tests

The catalyst fixed bed evaluation system was used to evaluate the performance of the catalyst (Fig. 1). N_2 gas leakage test was carried out before the experiment to ensure no gas leakage during the whole experiment. Before the experiment, the catalyst was pretreated at 500°C H_2/Ar. Stop heating and wrap the reactor in glass wool to reduce heat loss. The experiment was carried out under the condition of insulation. When the bed temperature drops to 165 °C, gas mixture is introduced into the reactor in a certain proportion. The reduction carrier and the supported metal oxide release heat, thus activating the ammonia oxidation decomposition reaction. The reaction temperature rises rapidly and reaches stability after a period of time. After a period of reaction, O_2 stops and the reaction terminates. After the reaction stops, in the absence of oxygen, when the temperature rises above 165 °C, O_2 is reintroduced to restart the reaction. In subsequent cycles, the ammoxidation decomposition reaction is repeatedly triggered several times. A routine sample was taken from the outlet gas and the gas components were detected. The gas composition was analyzed using gas chromatography. The following formula is used to calculate conversion (Eq. 8-9) and the hydrogen production rate via Eq. 10.

$$NH_3 \rightarrow 1.5H_2 + 0.5N_2 \quad \Delta H = +45.9\ kJ/mol \tag{1}$$

$$NH_3 + 0.25O_2 \rightarrow H_2 + 0.5N_2 + 0.5H_2O \quad \Delta H = -75\ kJ/mol \tag{2}$$

$$NH_3 + 0.75O_2 \rightarrow 0.5N_2 + 1.5H_2O \quad \Delta H = -317\ kJ/mol \tag{3}$$

$$NH_3 + 0.37O_2 \rightarrow 0.76H_2 + 0.5N_2 + 0.74H_2O \quad \Delta H = -135kJ/mol \tag{4}$$

$$NH_3 + 0.095O_2 \rightarrow 1.32H_2 + 0.5N_2 + 0.18H_2O \quad \Delta H = 0kJ/mol \tag{5}$$

$$NH_3 + 0.15O_2 \rightarrow 1.2H_2 + 0.5N_2 + 0.3H_2O \quad \Delta H = -26.7kJ/mol \tag{6}$$

$$NH_3 + 0.2O_2 \rightarrow 1.1H_2 + 0.5N_2 + 0.4H_2O \quad \Delta H = -50.9kJ/mol \tag{7}$$

$$X_{NH_3} = \frac{(M_{NH_3,in} - M_{NH_3,out})}{M_{NH_3,in}} \times 100\% \tag{8}$$

$$X_{O_2} = \frac{(M_{O_2,in} - M_{O_2,out})}{M_{O_2,in}} \times 100\% \tag{9}$$

$$Y_{H_2} = \frac{M_{H_2,out} \times 2}{M_{NH_3,in} \times 3} \times 100\% \tag{10}$$

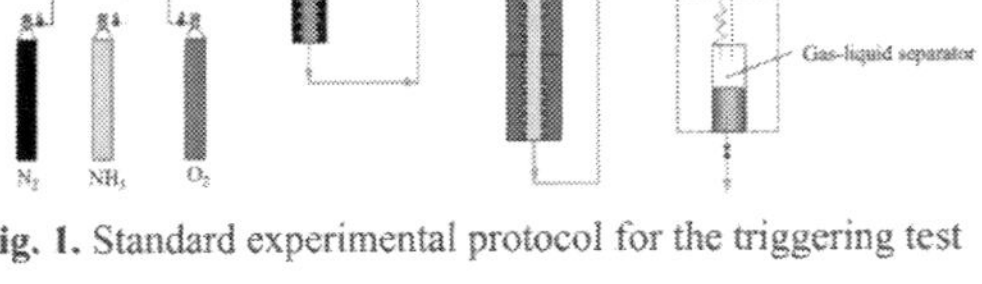

Fig. 1. Standard experimental protocol for the triggering test

3. Results

3.1 Activity Tests

First, the catalyst activity was tested by placing 0.25g of catalyst in a reactor. Before the reaction, N_2 (30mL • min^{-1}) was introduced for 30 min to flush out the air in the device. Then the catalyst was heated to 500°C in an H_2/Ar to remove water and carbon dioxide from the catalyst surface. Turn off the heat, the bed temperature dropped to 165°C. At this point, the gas was switched to supply an ammonia-oxygen mixture (NH_3:O_2=100:25 mL·min^{-1}) as shown in Fig. 2. Sampling bags were used to collect gas samples every 20 s to detect the gas composition .This indicates that ammonia first undergoes oxidation to produce N_2, and then undergoes ammonia oxidation decomposition to produce H_2. Furthermore, the experimental results show that the NH_3 oxidation decomposition is fast, and NO_x and N_2O are not detected . Until 150 s, the formation rate of the gas is almost stable, indicating that the reaction has reached equilibrium. Throughout the reaction process, the heat required for ammonia decomposition comes from the oxidation of NH_3, and the generation speed of hydrogen is faster due to the ammonia's internal heat.

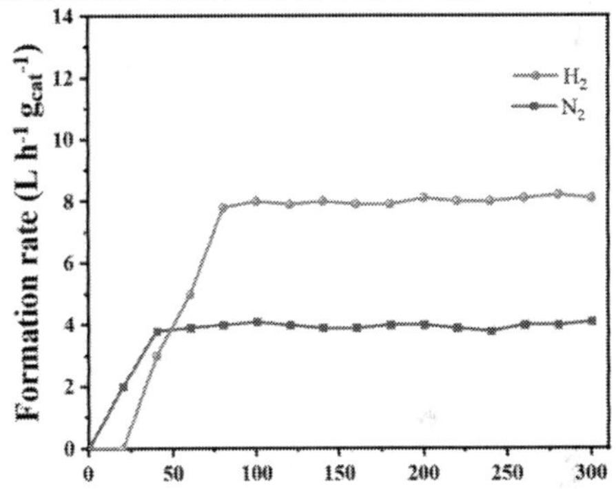

Fig. 2 The formation rate of H_2 and N_2

As shown in Fig. 3, when the reaction gas was fed to the 10wt% Co/CeO_2 catalyst reactor, the catalytic bed temperature remained basically unchanged, and the bed temperature decreased with the passage of time, indicating that no reaction occurred. However, under the same experimental operation described above, when the temperature is reduced to 165 °C, the bed temperature rises rapidly, indicating that the reaction is occur at 165 °C. The spontaneous combustion temperature of 10wt% Co/CeO_2 was determined at 165°C. With the increase of Co load from 1wt % to 10wt%, the spontaneous combustion temperature of catalyst decreased gradually. When Co content exceeds 10%, the spontaneous combustion temperature does not change.

https://doi.org/10.1051/e3sconf/202456001007

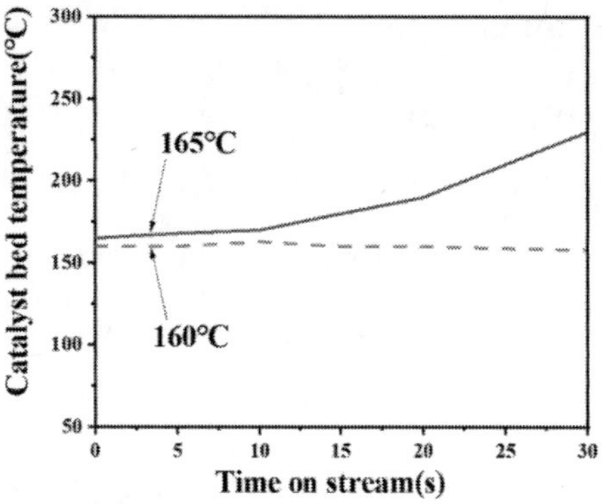

Fig. 3. The bed temperature of 10wt% Co/CeO₂

3.2 Catalysts Characterization

The Co/CeO_2 catalyst was studied by SEM and HRTEM, as shown in Fig. 4. The results show the lattice spacing is between d=0.236 ~ 0.238nm, and the structure is orderly porous. The particles are evenly dispersed, the lattice spacing is obvious, and there is no agglomeration. The active metal can be found successfully attached to the catalyst surface, indicating that we have prepared the Co/CeO_2 catalyst, and the active metal is uniformly distributed on the catalyst surface.

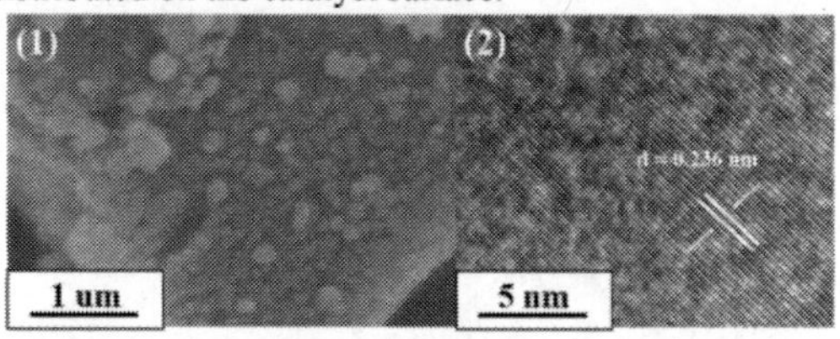

Fig. 4. (1)TEM, (2) HRTEM images of 10wt%Co/CeO_2

The XRD pattern of Co/CeO_2 catalyst particles was analyzed, as shown in Fig 5a. 2θ = 28.6°、32.9°、47.5°、56.3° 、 59.1° 、 69.4° 、 76.7° 、 79.1° respectively corresponding to the peak (111), (200), (220), (311), (222), (400), (331) and (420) plane, This indicates that fluorite structure exists in all catalysts. After loading Co, the characteristic peak of CeO_2 remained unchanged and the structure was intact, indicating that the structure of CeO_2 remained intact. However, no diffraction peaks were observed for Co, which may be due to their small grain size or high dispersion. In addition, the XRD patterns of the catalysts were similar before and after 100 h of reaction, indicating that the catalyst did not undergo obvious sintering before and after reaction, and the structure remained stable. At the same time, the content of Co remained basically unchanged during the reaction, which indicates that the catalyst has excellent stability.

The N_2 physical adsorption isotherm is exhibited in Fig.5b. BET was 42.5 m^2/g and a hysteresis loop of H3, indicating that Co/CeO_2 catalyst has a larger specific surface area and a wider channel pore structure. P/P_0 = 0.4~1, indicating high porosity of the catalyst. It can be seen providing a larger reaction space and exposing more active sites。

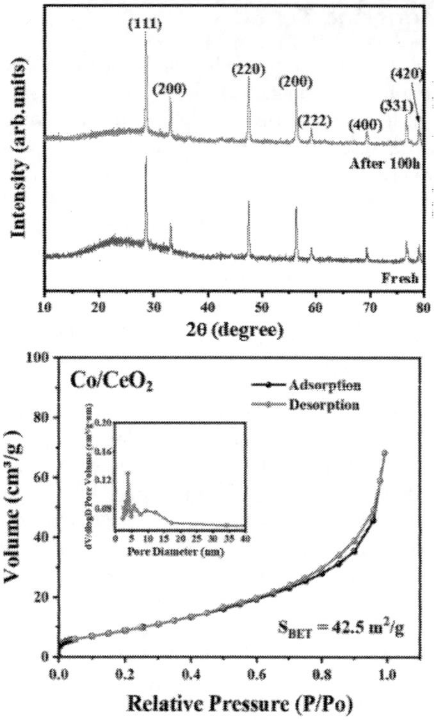

Fig. 5. (a) XRD before and after reaction. (b) BET

3.3 Cyclic Experiment and Stability

We conducted an ammonia oxidation decomposition cycle experiment at 165°C to test the stability of the catalyst (Fig. 6a). The specific process of the experiment was as follows: When the temperature dropped to 165°C, the feed was switched to an ammonia-oxygen mixture, and the reaction started. The reaction was stopped after 120 min of oxygen supply, and the reaction was completed. At this time, the bed temperature began to drop rapidly to 165°C, O_2 was supplied again, the reaction was restarted. This process was repeated five times (Fig. 6b). We took five samples at each sampling point and took the average value as the experimental result. The ammonia conversion rate was 70.1% (± 2.1%), the oxygen conversion rate was 80% (±1.6%), the hydrogen yield was 43.5% (± 1.8%), and the conversion rate and hydrogen yield were not reduced during the experiment. The experimental results show that the Co/CeO_2 catalyst exhibits significant stability and repeatability . We conducted 100 h stability test (Fig. 6c). This indicates that the catalyst has good stability, which is beneficial for practical production and reducing costs (Table 1). The XRD analysis compared the catalyst before and after the reaction, confirming the catalyst's stability. It is very important in commercial application.

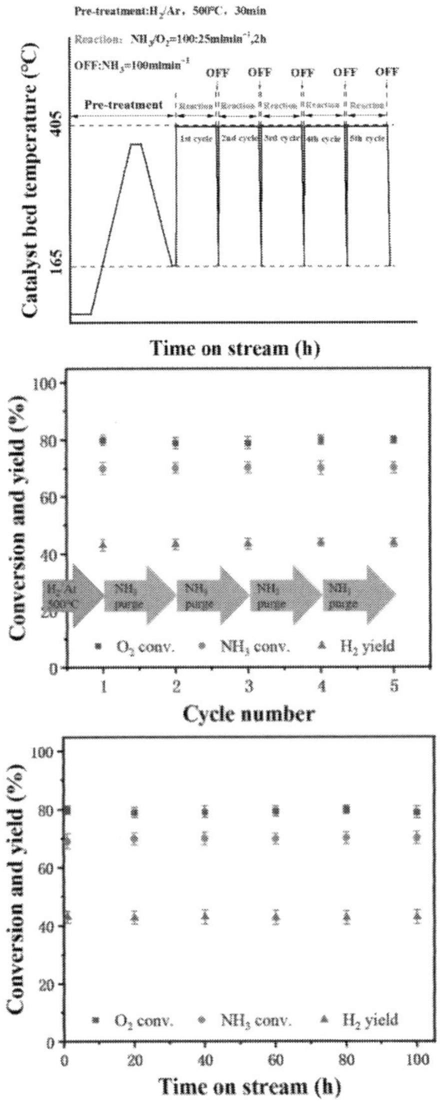

Fig. 6. (a) Cycle tests, (b) Catalytic activity (c) Catalyst stability

Table 1. Amount of Co in Co/CeO_2 before and after 5 cycles and stability tests

Treatment	Mass % of Co
Fresh	10
After 5 cycles	9.88
After 100 h	9.82

4.Conclusion

This study discovered a new process for oxidizing ammonia into hydrogen without the need for external heating, which can be initiated immediately upon the addition of ammonia and oxygen without complex processes. The Co/CeO_2 catalyst showed excellent catalytic performance, with NH_3 and O_2 conversion rates of 72% and 81%, respectively, and a hydrogen yield of 45% when the NH_3/O_2 ratio was 4:1. The catalyst also showed good stability and reusability, and the method for preparing the catalyst was simple and cost-effective, significantly reducing production costs. The hydrogen production process has the advantages of low start-up, fast heating, high energy utilization rate and high hydrogen production rate. We expect that the process can be widely used in commercial applications.

References

1. Awad O I, Zhou B, Harrath K, et al.Characteristics of NH_3/H_2 blend as carbon-free fuels: A review [J]. International Journal of Hydrogen Energy, 2023, 48(96): 38077-100.
2. Wan Z, Tao Y, Shao J, et al. Ammonia as an effective hydrogen carrier and a clean fuel for solid oxide fuel cells [J]. Energy Conversion and Management, 2021, 228: 113729.
3. G. Chehade, I. Dincer, Progress in green ammonia production as potential carbonfree fuel, Fuel 299 (2021) 120845.
4. Selvakumaran D, Pan A, Liang S, et al. A review on recent developments and challenges of cathode materials for rechargeable aqueous Zn-ion batteries [J]. Journal of Materials Chemistry A, 2019, 7(31): 18209-36.
5. York R, Bell S E. Energy transitions or additions? Why a transition from fossil fuels requires more than the growth of renewable energy [J]. Energy Research & Social Science, 2019, 51: 40-3.
6. Braff W A, Mueller J M, Trancik J E. Value of storage technologies for wind and solar energy [J]. Nature Climate Change, 2016, 6(10): 964.
7. Markard J. The next phase of the energy transition and its implications for research and policy [J]. Nature Energy, 2018, 3(8): 628-33.
8. Bockris J O M. The hydrogen economy: Its history [J]. International Journal of Hydrogen Energy, 2013, 38(6): 2579-88.

Effect of Pyrolysis Temperature and Time on Liquid Smoke Yield in Coconut (*Cocos nucifera L.*) Shells Pyrolysis Process

Rosdanelli Hasibuan*, Yunus Widodo Limbong, Risma Fazillah, *and* Viqry Pramananda

Chemical Engineering Department, Faculty of Engineering, University of Sumatera Utara, Almamater Kampus USU Street, Padang Bulan, Medan 20115, Indonesia

Abstract. Biomass pyrolysis is an emerging method to manage solid biomass from agricultural activity, which transforms solid organic materials thermally to create solid, liquid, and non-condensable gases as its product. This research was conducted to understand the effect of pyrolysis temperature and time on the yield of liquid smoke from coconut shell biomass. In this work, the pyrolysis was performed with temperatures of 350°C, 450°C, and 550°C, with a pyrolysis time of 2, 3, and 4 hours. Based on this work, it is known that both pyrolysis temperature and time affected the yield of liquid smoke. The liquid smoke obtained at temperature pyrolysis of 350°C for 4 hours yields 46,5%, with density of 1.025 g/ml, viscosity of 1,555 cP, and pH of 2. GC-MS characterization revealed the 20 components in obtained liquid smoke, including organic acids, carbonyl compounds, phenolic compounds, and another compound. Acetic acid is the dominant component, up to 51.40%.

1 Introduction

One new method for producing liquid smoke is biomass pyrolysis, a rapidly expanding field of study in bioresource technology. By using heat to break down solid organic materials, biomass pyrolysis produces liquid, solid, and non-condensable gases [1]. To create these compounds, pyrolysis requires fast heating in the absence of oxygen. The amount of biomass utilized, the kind of pyrolysis, and process factors including pre-treatment settings, pyrolysis temperature, and heating rate all affect how much of each of these products is produced [2].

Water evaporation and the degradation of cellulose, hemicellulose, and lignin are the four steps in the pyrolysis process. Between 180°C and 350°C, hemicellulose and cellulose decompose to form carboxylic acids and carbonyl compounds, whereas lignin decomposes to form phenols between 300°C and 500°C. In addition to phenols, acids, and carbonyls, biomass pyrolysis frequently yields unwanted substances including polycyclic aromatic hydrocarbons (PAHs), which are created at temperatures ranging from 500°C to 900°C and some of which are naturally occurring. Others, on the other hand, are the outcome of incomplete combustion. Unwanted PAHs may be eliminated during the purification process, and the liquid smoke's flavor and color intensity can be altered [3].

Condensing the pyrolysis vapour product yields liquid smoke, which is composed of oxidized organic components such phenol, aldehyde, ketone, and carboxylic acid. Liquid smoke gets its unique flavor and smell from these ingredients, which also have antibacterial and antioxidant properties [2]. At present, liquid smoke has been employed for a variety of applications, such as natural preservatives for food products [4][5], fertilizer for plant [6], pesticide [7], natural rubber coagulant [8][9], and etc.

Coconut shells are an example of easily found biomass in Indonesia. According to the Indonesian Ministry of Agriculture [10], there is 3,355,535 Ha of coconut plantations in Indonesia in 2021, with 2,877,504 tons of coconut production generated. Ahmad et al. [11] states that a fully ripe coconut has 35% fiber, 12% hard shell, 28% white flesh, and 25% coconut water. Biomass from coconut shells is produced during the processing of coconuts to make various items. Valuing the shells into value-added goods is an alternative approach to managing solid waste from Indonesia's coconut industry. This is particularly effective when using the shells as sources of multipurpose liquid smoke.

The purpose of this work is to explore how temperature and time during pyrolysis affect the amount of liquid smoke produced from coconut shells. In order to determine the composition of the liquid smoke that was collected, GC-MS characterization had been carried out along with quality analyses of density, viscosity, and pH.

2 Methodology

Coconut shell was the material employed in this study. A digital scale, pycnometer, pH meter, viscosimeter Ostwald, filter paper, measuring cup, and gas

* Corresponding author: rosdanelli@usu.ac.id

 https://doi.org/10.1051/e3sconf/202456001008

chromatography-mass spectrometry (GC-MS) were among the tools utilized. Figure 1 displays the pyrolysis reactor design. The coconut shells underwent thorough separation and cleaning to remove any contaminants. Next, the coconut shells were weighted up to 2 kg and subjected at various pyrolysis time (2, 3, and 4 hours) and different temperatures (350, 450, and 550 degrees Celsius). After pyrolysis, the liquid smoke was collected and its yield, density, viscosity, pH, and chemical content were examined.

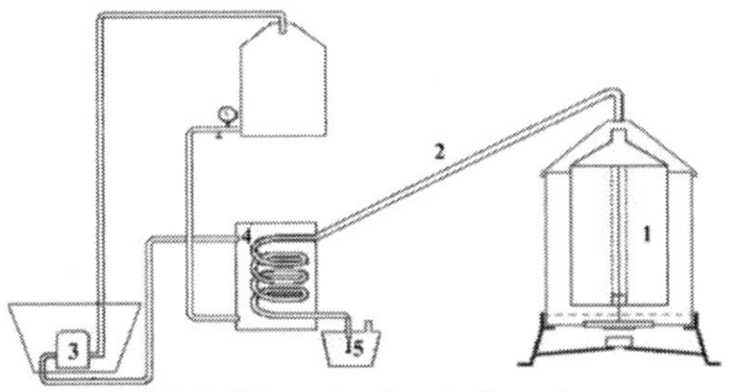

Fig. 1. Schematic of pyrolysis reactor

Note: (1) pyrolysis reactor; (2) condenser; (3) cooling water pump; (4) cooling water tank; (5) liquid smoke tank

3 Results and Discussion

3.1 Effect of Pyrolysis Temperature and Time on the Liquid Smoke Yield

Yield analysis was done to see the percentage of resulting liquid smoke from the coconut shells after pyrolysis. Fig 2 shows the effect of pyrolysis temperature and time on the yield of liquid smoke. In general, increasing pyrolysis temperature and time impacts liquid smoke yield. The yield of liquid smoke produced at 350°C ranges from 36.30-46.50%. There was variation in the liquid smoke yield at 450°C, namely 43.70%, 47.70%, and 43.50%. Ultimately, the liquid smoke production climbed to 46.50–51.00% when the pyrolysis temperature was raised to 550°C.

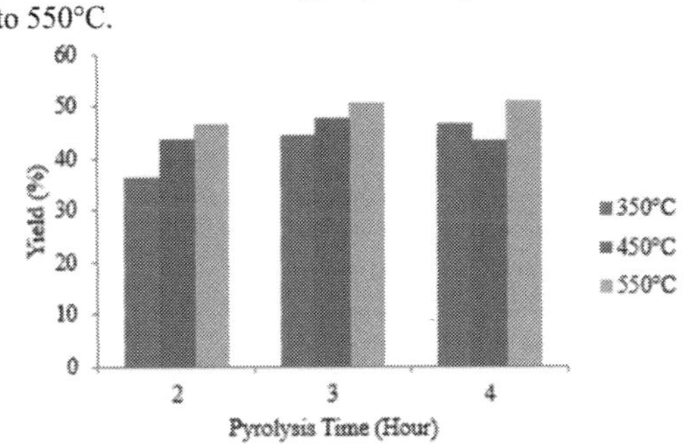

Fig. 2. Effect of pyrolysis temperature and time on the yield of liquid smoke

Elevated temperatures and extended pyrolysis times result in a more thorough breakdown of the starting components, producing more liquid smoke. [12]. Temperature is one of the critical variables affecting pyrolysis yield. Liquid smoke yield consistently increases at the beginning of the process, increasing temperature to a certain maximum temperature, after which the yield drops when the temperature rises further. At temperatures higher than optimum values, the liquid smoke yield is reduced, and the production of non-condensable gases is increased. On the other hand, liquid smoke yield has been observed to increase slightly with an increase in retention time [13].

3.2 Effect of Pyrolysis Temperature and Time on the Liquid Smoke Density

The quantity of components present in the liquid smoke is indicated by its density or specific gravity, which has an impact on the liquid smoke's weight [14]. For combustion applications, liquid smoke density is crucial since it represents the quality of atomization, vaporization, ignition, and combustion properties [15]. The effect of pyrolysis temperature and time on the liquid smoke density is depicted in Fig 3. The liquid smoke density decreases when the temperature is raised but increases when the pyrolysis time is extended. Specifically, the density of liquid smoke ranges from 1.024-1.027 g/ml at a pyrolysis temperature of 350°C, 0.933-1.021 g/ml at a pyrolysis temperature of 450°C, and 0.917-0.931 g/ml at pyrolysis temperature of 550°C.

According to the Japanese Quality Liquid Smoke Standard, the density requirement for liquid smoke is > 1.005. On the other hand, Indonesian National Standard (SNI) [16] set the density requirements for grade I-liquid smoke at 1.005-1.050 g/ml. Based on this research, only five samples of liquid smoke met this density requirement: all liquid smoke obtained at a temperature of 350°C and liquid smoke obtained at a temperature of 450°C for 3 and 4 hours.

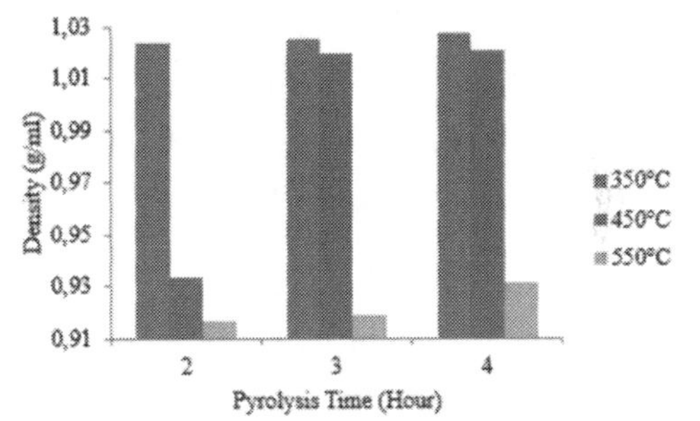

Fig. 3. The effect of pyrolysis temperature and time on the density of liquid smoke

3.3 Effect of Pyrolysis Temperature and Time on The Viscosity of Liquid Smoke

Fig 4 presents the viscosity of liquid smoke at various temperature and time of pyrolysis. Based on Fig 4, the viscosity of liquid smoke is ranging from 1.395-1.555 cP at 350°C, 1.265-1.398 at 450°C, 1.036-1.115 cp at pyrolysis temperature of 550°C. It's clearly seen that increasing temperature will produce the liquid smoke with lower viscosity. Higher temperatures during pyrolysis promote a more severe breakdown of biomass or pyrolysis vapour, leading to the synthesis of smaller molecules or lower molecular weight substances [15].

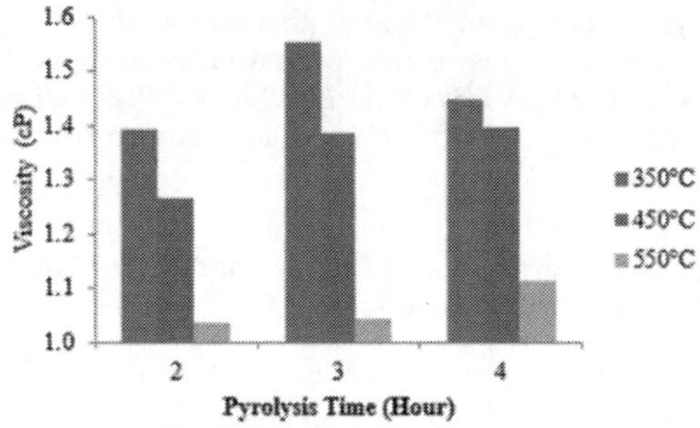

Fig. 4. Viscosity of liquid smoke

3.4 pH of Liquid Smoke

One of the quality criteria for the liquid smoke is the pH level. The measurement of liquid smoke pH aims to establish the degree of pyrolysis decomposition and to generate the smoke's natural acid. If the liquid smoke has a low pH, it shows good quality and can be used as a food preservative. The durability, shelf life, and organoleptic qualities of the liquid smoke are all impacted by the low pH value overall [14].

In this research, the pH measurement results show that the produced liquid smoke has an acidic pH of about 2-3. This pH value had met the Japanese Quality Liquid Smoke Standards, ranging from 1.5 to 3.7. More specifically, liquid smoke produced at 350°C has a pH of 2 and meets the SNI No. 8985:2021 [16]. The liquid smoke product is acidic from this work, based on the resultant pH. At temperatures higher than 200°C, pyrolysis is an exothermic reaction that generates heat. The decomposition rate is increasing at this point, and the biomass components that break down throughout the pyrolysis process significantly impact the liquid smoke product's acidity [17].

3.5 Composition of Liquid Smoke

In this work, the liquid smoke collected after 4 hours of pyrolysis at 350°C was submitted to a composition analysis. Twenty components have been determined through GC-MS analysis, as Fig. 5 and Table 1 demonstrate. Temperature may have an impact on how the chemicals in raw materials break down during pyrolysis. As the temperature rises, the average rate at which chemical bonds break increases [18]. From an application standpoint, the components were categorized into three functional groups: carbonyls, phenols, and carboxylic acids. This helped to depict the chemical composition of liquid smoke. These functional classes are not the only products; there are also hydrocarbons, lactones, and alcohols [19].

The obtained liquid smoke in this work contains aldehyde, organic acids, carbonyl, phenolic, aromatics, and another compound, with the highest component being acetic acid, 51.40%. In another investigation, the main ingredient in liquid smoke derived from coconut shells that were pyrolyzed for eight hours was acetic acid, which was found to be 51.6% at 200°C, 50.88% at 250°C, and 39.98% at 300°C [3]. Carbonyl compounds and acetic acid are the results of cellulosic pyrolysis. Furfural, furan, and its derivatives are the products of hemicellulose pyrolysis. Finally, phenol will be produced as lignin breaks down [12]. Acetic acid can act as an antimicrobial. Several factors, including concentration, pH, ambient factors, contact time, contact temperature, and species of bacteria, might affect the antibacterial activity of acetic acid. On the other hand, the development of the odour, texture, and colour of liquid smoke products is significantly influenced by other substances, such as phenols, that are present in liquid smoke [18]. Furthermore, phenolic compounds have antibacterial and antifungal effects of eliminating bacteria that cause rot and break down proteins into amino acids, stopping the offensive odour. Because they are capable of making cell membranes more permeable, phenols act as bactericides by inactivating vital enzymes and damaging or inactivating genetic material. [7]. After being refined using adsorption or distillation techniques, the liquid smoke from this research exhibits significant potential for use as a food preservation agent.

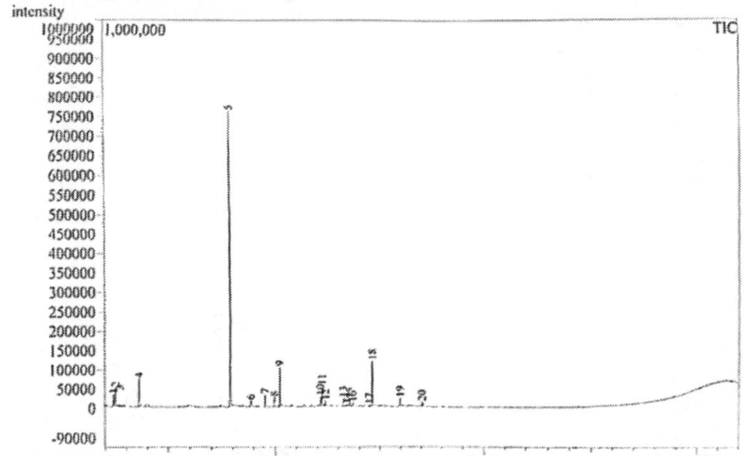

Fig. 5. Chromatogram of liquid smoke

Table 1. Composition of obtained liquid smoke

Component	Area (%)
2,2-dimethoxybutane	1.98
1-Hydroxy-2-Butanone	3.27
Propanal (CAS) Propionaldehyde	0.66
Acetic acid, Hidroxy	6.77
Acetic acid	51.40
2-Cyclopenten-1-one,2 hydroxy-3-methyl-(CAS) Coryclon	1.26
Phenol, 2-methyl-(CAS) o-Cresol	2.28
Phenol, 3-methyl- (CAS) m-Cresol	2.25
Phenol, 2-methoxy- (CAS) Guaiacol	6.60
2-Methoxy-4-methylphenol	1.87
1,2 Benzenediol (CAS) Pyrocathecol	4.31
III-Imidazole, 1-(1-oxooctadecyl)-	1.01
Benzene, (methylsulfinyl)- (CAS) Methyl phenyl sulfoxide	2.35
1,3-Benzenediol (CAS) Resorcin	1.09
Phenol, 4-ethyl-2-methoxy- (CAS) p-Ethylguaiacol	1.37
4 Methyl catechol	0.72
2-Methoxy-5-methyl pyrazine	0.67
Phenol, 2,6-dimethoxy- (CAS) 2,6-Dimethoxyphenol	7.92
Benzene, 1,2,3-trimethoxy- (CAS) 1,2,3-Trimethoxybenzene (CAS) Methylsy	1.36
Ethanone,1-(2,6-dihydroxy-4-methoxyphenyl)-(CAS) 2,6-Dihydroxy-4-methyl	0.85

4 Conclusion

The outcomes of this investigation indicate that the yield and characteristics of liquid smoke derived from coconut shells, particularly its density, viscosity, and pH, were impacted by both temperature and yield. After 4 hours of pyrolysis at 350°C, 46.5% of the liquid smoke was produced. It had a density of 1.025 g/ml, a viscosity of 1,555 cP, and a pH of 2. GC-MS analysis identified 20 components in the liquid smoke, with acetic acid contributing 51.40% of the total.

References

1. X. Xin, K. Dell, I. A. Udugama, B. R. Young, and S. Baroutian, *J. Clean. Prod.*, **294**, 1–12, 2021.
2. S. Maulina, R. Amalia, and E. R Kamny, in *E3S Web of Conferences*, 2020, 1–3.
3. W. A. Rizal *et al.*, in *IOP Conference Series: Earth and Environmental Science*, IOP Publishing, 2020, 1–7.
4. M. Faisal, A. Gani, and F. Mulana, *F1000Research*, **8**, 240, 1–19, 2020.
5. M. Faisal and A. Gani, *Int. J. GEOMATE*, **15**, 47, 145–150, 2018.
6. Sriharti, A. Indriati, and S. Dyah, *Asian J. Appl. Sci.*, **08**, 01, 1–11, 2020.
7. M. Faisal, T. Chamzurni, and H. Daimon, *Int. J. GEOMATE*, **14**, 43, 36–41, 2018.
8. D. A. Triawan *et al.*, in *IOP Conference Series: Earth and Environmental Science*, IOP Publishing, 2022, 1–6.
9. Evahelda, R. F. Astuti, S. N. Aini, and Nurhadini, in *IOP Conference Series: Earth and Environmental Science*, IOP Publishing, 2021, 1–6.
10. Kementerian Pertanian, "Statistik Perkebunan Unggulan Nasional 2021-2023," 2023.
11. R. K. Ahmad, S. A. Sulaiman, S. Yusup, S. S. Dol, M. Inayat, and H. A. Umar, *Ain Shams Eng. J.*, **13**, 1–13, 2022.
12. S. Maulina and F. Silia, in *IOP Conference Series: Materials Science and Engineering*, IOP Publishing, 2018, 1–5.
13. S. Mutsengerere, C. H. Chihobo, D. Musademba, and I. Nhapi, *Renew. Sustain. Energy Rev.*, **104**, 328–336, 2019.
14. W. E. Triastuti, P. A. Budhi, E. Agustiani, R. A. Hidayat, R. Retnoningsih, and A. A. Nisa', in *IPTEK Journal of Proceedings Series*, 2019, 114–117.
15. C. Paenpong and A. Pattiya, *J. Anal. Appl. Pyrolysis*, 1–28, 2016.
16. Badan Standardisasi Nasional, "Crude Asap Cair Lignoselulosa sebagai Bahan Baku." Badan Standardisasi Nasional, 1–15, 2021.
17. N. Izza *et al.*, in *Atlantis Highlights in Engineering*, Atlantis Press International BV, 2023, 113–127.
18. M. Faisal, A. Gani, F. Mulana, H. Desvita, and S. Kamaruzzaman, *Rasayan J. Chem.*, **13**, 1, 514–520, 2020.
19. R. Hadanu and D. A. N. Apituley, *Makara J. Sci.*, **20**, 3, 95–100, 2016.

Comparative Heat Transfer Performance of TPMS Structures: Spotlight on Fisch Koch S versus Gyroid, Diamond and SplitP Lattices

Abdallah Barakat, Yi Pan, BeiBei Sun*

School of Mechanical Engineering, Southeast University, Nanjing, China

Abstract. This study assesses the thermal behavior of different TPMS configurations, specifically examining the Fischer Koch S lattice in comparison to Gyroid, Diamond, and SplitP lattices. Utilizing numerical simulations alongside experimental validation, the research delves into the efficacy of different TPMS designs in enhancing convective heat transfer and fluid dynamics. Results indicate that the Fischer Koch S lattice outperforms other models, exhibiting enhanced convective heat transfer rates and increased Nusselt numbers due to greater areas of contact between the fluid and the solid surfaces and optimized flow patterns. Nevertheless, it also demonstrates significant pressure drops at high Reynolds numbers. These findings underscore the necessity for further research to refine performance based on these insights.

1 Introduction

As industries continue to grapple with rising power consumption, the demand for enhanced thermal management solutions grows. Conventional cooling techniques face constraints due to increased power densities and smaller design spaces [1], [2]. Consequently, the interest in triply periodic minimal surface (TPMS) structures is growing, fueled by their exceptional thermal and mechanical characteristics, including notable surface area-to-volume ratios and impressive thermal conductance. [3], [4], TPMS structures are proving crucial in enhancing thermal management across electronics, aerospace, automotive, and other industries.

In 2003, Slaughter [5] introduced TPMS structures for applications in heat and mass transfer, highlighting their reduced weight and excellent thermal efficiency, sparking interest in their adoption for heat exchangers. Advancing this research, Thomas et al. [6] explored the thermal transfer properties across various TPMS configurations, such as Diamond, Gyroid, Schwarz-P, F-RD, C(I-Y**), and Neovius, focusing on their Nusselt numbers and frictional characteristics. They observed that the Diamond structure was highly efficient in heat transfer but experienced significant pressure drops, whereas the Schwarz-P was less effective. Rathore et al. [7] analyzed these structures within porous environments, examining Diamond, Gyroid, Iwp, and Primitive models to assess differences in permeability, tortuosity, and pressure dynamics, where the Primitive Type 2 showed minimal tortuosity and drag. Subsequent studies by Alketan et al. [8] on solid and sheet configurations of Diamond and Gyroid TPMS structures underscored the Gyroid-sheet's superior thermal efficiency, attributed to the increased surface area and intricate design. This finding aligns with Li et al.'s [9] observation of a 15–100% improvement in thermal efficiency of Gyroid and Diamond sheet heat exchangers in turbulent flows, also noting a significant increase in friction factors which enhanced fluid mixing. Khalil et al. [10] compared Gyroid and Diamond in both solid and sheet formats, finding that solid configurations had lower pressure drops, while sheet-networks showed lower thermal resistance but higher-pressure loss. Diamond solid-networks achieved optimal thermal efficiency. Liang et al. [11] studied TPMS cross-flow heat exchangers, finding Gyroid and Diamond designs outperformed Primitive despite similar pressure drops. More recently, Tang et al. [12] tested TPMS heatsinks examining the Gyroid, Diamond, and IWP models, observing significant convective enhancements for Diamond and Gyroid compared to standard heatsinks. Collectively, these researchs corroborates the effectiveness of TPMS configurations in improving thermal transfer, though detailed analysis on specific design parameters and their impacts remains scant.

Prior research has extensively documented the efficacy of various TPMS structures in improving heat transfer. However, while these studies offer broad insights into the capabilities of TPMS designs, detailed analyses specific to Fischer Koch S (FK-S) lattice structure are scant. This structure, despite its potential to contribute significantly to the field, has not been as thoroughly investigated as others. This paper focuses on comparing Fischer Koch S with three other TPMS structures, seeking to highlight its unique properties and possible advantages in thermal management applications.

* Corresponding author: bbsun@seu.edu.cn

2 Numerical simulation

The geometric modelling of FK-S, Gyroid, Diamond, and SplitP TPMS structures was conducted using nTopology software, as shown in Fig. 1. The study featured four 20mm cubic units configured into a $20mm \times 80mm$ rectangle, balancing cost and computational efficiency against precision. An 80mm extension ensures proper fluid flow at both inlet and outlet, shown in Fig. 2, with specific geometrical details listed in Table 1. Simulation operation conditions maintained the heating surface at 373.15 K, air inflow speed between 0.2 and 1.4 m/s at 300 K, and free outflow at the exit, with no-slip adiabatic conditions on other surfaces.

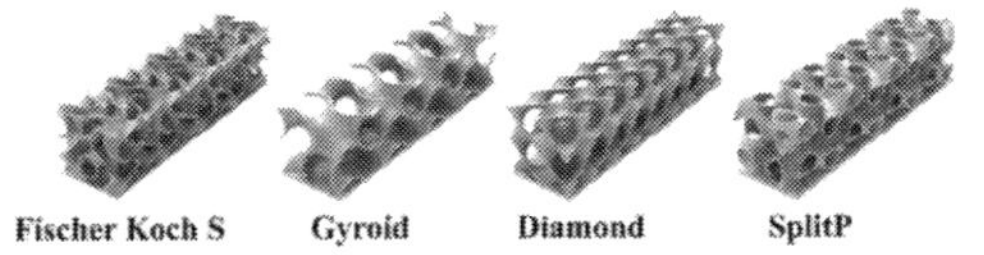

Fig. 1 Geometric models of the TPMS structures

The analysis of hydraulic and thermal behaviors within TPMS was performed using ANSYS Fluent, employing the *SST k-ω* realizable model for its efficacy in predicting complex flow behaviors like separation and recirculation, crucial for assessing TPMS configurations. The simulation resolved steady- state, three-dimensional equations of continuity, energy, and momentum, using a Coupled scheme for Pressure-Velocity coupling and advanced discretization methods for accurate flow and heat distribution.

The simulation's computational domain was crafted from an STL geometry file via nTopology, refined in SpaceClaim, and meshed using Fluent's Fault-tolerant meshing.

Table 1. Geometrical characteristics of the FK-S, Gyroid, Diamond and SplitP models.

Parameters	FK-S	Gyroid	Diamond	SplitP
Surface area	1.84×10^4	1.01×10^4	1.26×10^4	1.68×10^4
Volume	3808	1547	1853	2094
Thickness	0.5	0.5	0.5	0.5
D_h	6.12	12.02	9.61	7.11
Porosity	0.881	0.952	0.942	0.935

Following the computational setup, the hydraulic characteristics of the TPMS structures were analyzed by calculating the Reynolds number, friction factor, and pressure gradient. The Reynolds number was calculated using the following equation:

$$Re = \frac{\rho u D_h}{\mu \, \varepsilon} \tag{1}$$

where ρ is the fluid density, D_h the hydraulic diameter, u the fluid velocity, ε the porosity and μ the viscosity. The hydraulic diameter, D_h, was computed from:

$$D_h = \frac{4V_{void}}{A_s} \tag{2}$$

where V_{void} represents the void volume, and A_s is the wetted surface area.

The TPMS structures' heat transfer capabilities were evaluated through the convective heat transfer coefficient (h) and Nusselt number (Nu), using:

$$h = \frac{\rho u A' C \, (T_{out} - T_{in})}{A'' \, \Delta T_{LMTD}} \tag{3}$$

where, A' and A'' are the sectional inlet and heating surface areas respectively, C is the specific heat capacity, and ΔT_{LMTD} is the logarithmic mean temperature difference, calculated by:

$$\Delta T_{LMTD} = \frac{T_{out} - T_{in}}{ln\left(\frac{T_h - T_{in}}{T_h - T_{out}}\right)} \tag{4}$$

The Nusselt number was derived as:

$$Nu = \frac{h \, D_h}{\lambda} \tag{5}$$

where λ is the thermal conductivity. The Darcy friction factor, f, another vital dimensionless measure, was defined by:

$$f = \frac{2\Delta p \varepsilon^2 D_h}{L \rho u^2} \tag{6}$$

where, Δp and L are the pressure drop and TPMS length, respectively. Finally, overall thermal performance was assessed using the ratio (j/f), where j is the Chilton-Colburn factor, a measure that relates mass and heat transfer coefficients, calculated as:

$$j/f = \frac{Nu}{Re \; Pr^{1/3}} \tag{7}$$

where Pr is the Prandtl number.

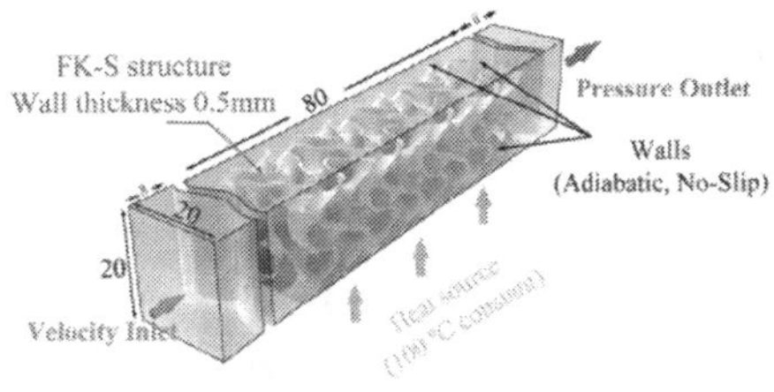

Fig. 2. Simulation Domain and Boundary Conditions.

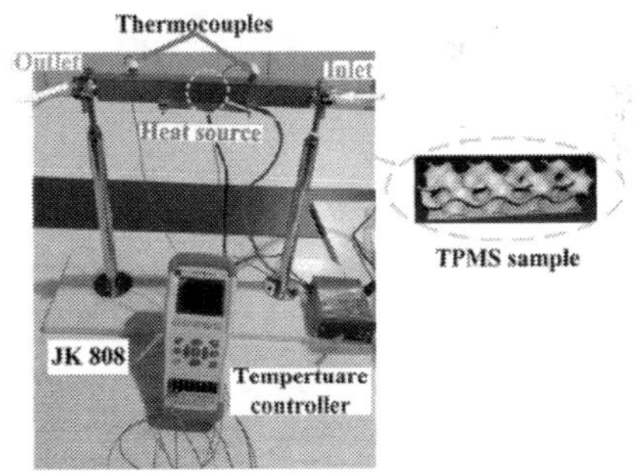

Fig. 3 Experiment setup and TPMS structure sample

3 Experiment

The experimental setup (see Fig. 3) included a 120W blower, a straightened flow channel, a temperature-controlled heating plate, a multi-channel handheld temperature tester (JK808), a pair of GG-K-30 thermocouples, a flow meter, and various valves. Air served as the coolant. The setup included a TPMS structure for testing, a locking plate, and a heating plate. These components were 3D-printed using selective laser melting (SLM) technology for the Gyroid model and locking plates, and Multi Jet Fusion (MJF) powder sintering technology for the airflow channel.

Aluminum (AlSi10Mg) alloy and high-performance Nylon were used as materials, respectively. The surrounding temperature was kept around 298 K, with airflow rates controlled by adjusting the valves.

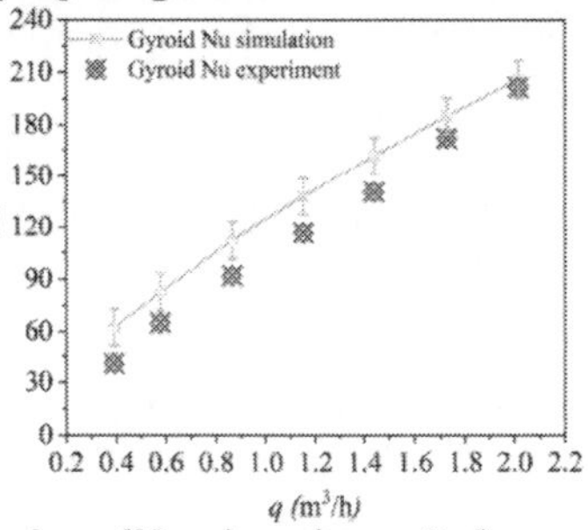

Fig. 4 Comparison of Nusselt number results from experimental and numerical simulations

The experimental Nusselt number were compared with corresponding numerical simulations. Results (see Fig. 4) showed consistent trends across varying flow rates, with deviations of up to 3% in Nusselt number at higher flow rates. Larger deviations at lower flow rates were attributed to increased sensitivity to minor errors or variations, manufacturing tolerances, and surface roughness introduced by 3D printing. Numerical simulations generally tended to be higher than experimental results, possibly due to random errors and contact thermal resistance. Despite differences, Consistency between experimental and numerical results validates the accuracy of the simulations.

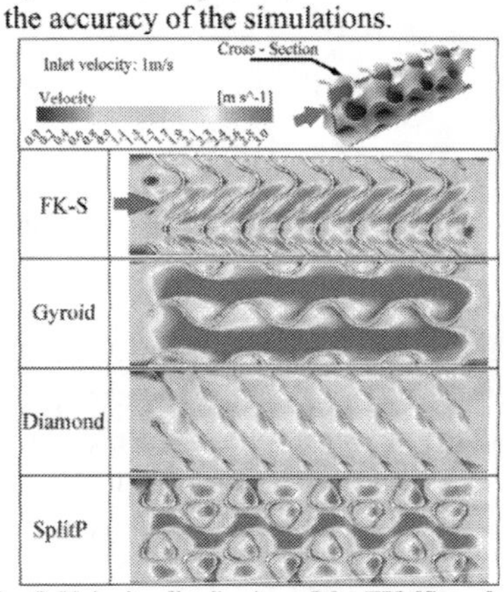

Fig. 5. Velocity distribution of the TPMS models.

4 Result and discussion

4.1 Fluid Flow Behavior in TPMS structures

Fig. 5 depicts the velocity distribution around various TPMS models at a cross-section. The complex curved surfaces of these models significantly influence the flow patterns. Notably, different TPMS structures exhibit unique flow characteristics due to the vast differences in surface configurations. An analysis of velocity contours reveals that the Gyroid structure maintains the highest velocity across both sections, followed by the FK-S, SplitP, and Diamond structures. Both the Gyroid and FK-S structures demonstrate a streamlined flow, encountering minimal resistance, which suggests a lower blockage ratio and promotes efficient fluid transport.

Two primary flow modes are noted: the first features high flow rates concentrated along main and major streamlines in structures like Gyroid and SplitP, optimizing flow but possibly impacting convective heat transfer efficiency. The second mode, seen in FK-S and Diamond structures, shows a more even velocity distribution due to fluid being channeled through intricate surface-created non-flow channels, minimizing fluid stagnation in cooler main flow regions and improving contact with the surface [12].

4.2 Pressure drop and friction factor

Fig. 6 (a) and (b) compare pressure drop (*ΔP/L*) and friction factor (*f*) across TPMS structures at varying Reynolds numbers. Pressure drop increases with Reynolds number due to enhanced inertial forces, particularly noticeable in Fig. 6 (a). FK-S structure exhibits the highest pressure drop, attributed to its low porosity and hydraulic diameter, with SplitP structure following closely. Diamond and Gyroid structures demonstrate the lowest pressure drops, around 440 Pa/m at flow rate 2 m³/h. However, this trend is not mirrored in the friction factor behavior (Fig. 6 (b)). Gyroid models report the highest friction factors, peaking at 6.33 at Reynolds number of 300, likely due to higher hydraulic diameter. Conversely, FK-S structure records the lowest friction factor at 3.55 at the same Reynolds number, correlating with findings by Ho, et al. [13].

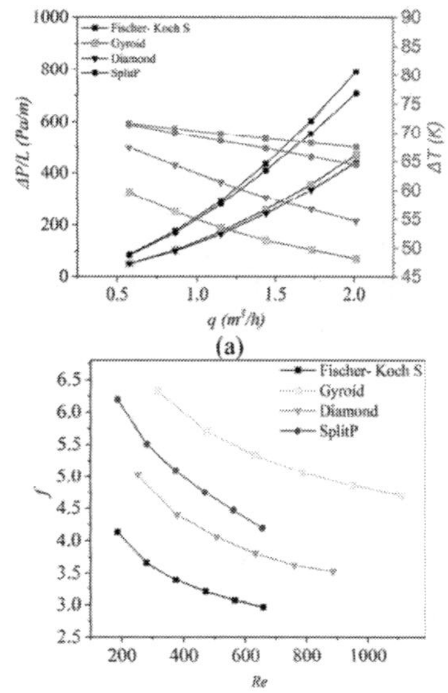

(a)

(b)

Fig. 6. Comparison of (a) Pressure drop and (b) Friction factor.

4.3 Evaluation of heat transfer performance

Fig. 7 displays the temperature distribution across various TPMS models, highlighting significant temperature gradients that enhance heat transfer. The intricate design of these models facilitates periodic mixing and efficient fluid transmission. Among these, the FK-S model, characterized by a higher surface area as shown in Table 1, displays the highest outlet temperatures (see Fig. 6 (a)), marked by a prominent red hue at its outlet, followed by the SplitP, Diamond, and Gyroid models.

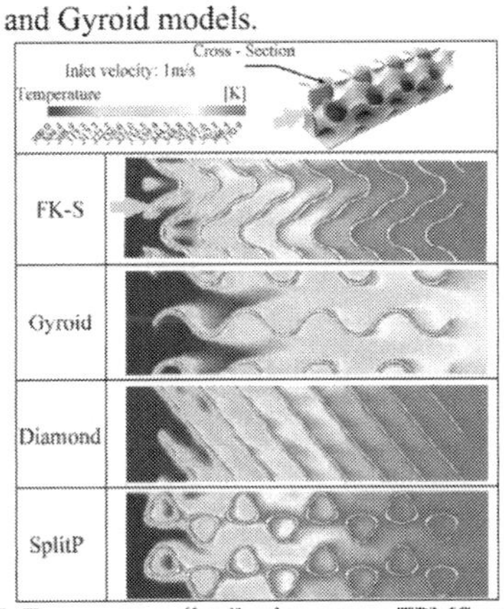

Fig. 7. Temperature distribution across TPMS models.

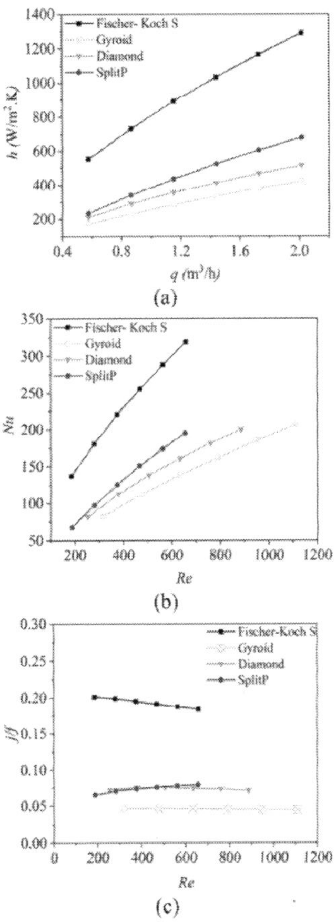

Fig. 8. Comparison of (a) Convective heat transfer coefficient, (b) Nusselt number, and (c) Effectiveness (*j/f*).

Fig. 8 (a) and (b) present an analysis of the convective heat transfer coefficient and the Nusselt number for TPMS models at different flow rates, corresponding to *Re* between 60 to 1200. In Fig. 8 (a), the results reveal a rapid increase in the convective heat transfer coefficient with flow velocity, with the FK-S model exhibiting the highest coefficient. This is attributed to its larger surface area, as indicated in Table 1, surpassing the Gyroid, Diamond, and SplitP models by 82%, 46%, and 9%, respectively. At flow rate 2 m^3/h, the FK-S model's convective heat transfer coefficient is 90%, 150%, and 204% higher than that of the SplitP, Diamond, and Gyroid models, respectively. Meanwhile, Fig. 8 (b) illustrates that the Nusselt number increases with rising *Re* for all models, with the FK-S model demonstrating the highest numbers, surpassing the SplitP, Diamond, and Gyroid models by 61.5%, 92%, and 120%, respectively. Fig. 8 (c) shows the ratio *j/f* to evaluate heat transfer performance of the TPMS structure. The FK-S structure exhibited the highest comprehensive heat transfer performance due to its high heat transfer and low specific pressure drop, followed by SpliP, Diamond and Gyroid.

5 Conclusion

This study examined various TPMS structures, focusing on the FK-S lattice. It showed that the FK-S lattice has superior thermal management capabilities, achieving high convective heat transfer coefficients and Nusselt numbers. However, challenges such as notable pressure drops at higher Reynolds numbers were observed. Analysis of flow patterns revealed unique characteristics of each TPMS structure, with the Gyroid and FK-S structures showing streamlined flows. Friction factor behavior varied among TPMS structures, with the Gyroid model reporting the highest values and the FK-S lattice recording the lowest. In summary, this research highlights the potential of the FK-S lattice in advanced thermal management, but further refinement is needed to address challenges and optimize its performance.

References

1. Y. Shabany, Heat transfer: Thermal management of electronics.(2009).
2. A. L. Moore and L. Shi, **vol. 17**, (2014).
3. S. Catchpole-Smith, R. R. J. Sélo, A. W. Davis, I. A. Ashcroft, C. J. Tuck, and A. Clare, Addit Manuf, **vol. 30**, (2019).
4. Z. Cheng, R. Xu, and P. X. Jiang, Int J Heat Mass Transf, **vol. 170**, (2021).
5. V.B. Slaughter, patent, U.S. 7,866, 377 (2011).
6. J. Iyer, T. Moore, D. Nguyen, P. Roy, and J. Stolaroff, Appl Therm Eng, **vol. 209**, (2022).
7. S. S. Rathore, B. Mehta, P. Kumar, and M. Asfer, (2022).
8. O. Al-Ketan, M. Ali, M. Khalil, R. Rowshan, K. A. Khan, and R. K. Abu Al-Rub, J Therm Sci Eng Appl, **vol. 13**, (2021).
9. W. Li, G. Yu, and Z. Yu, Appl Therm Eng, **vol. 179**, (2020).
10. M. Khalil, M. I. Hassan Ali, K. A. Khan, and R. Abu Al-Rub, Case Stud. Therm. Eng, **vol. 38**, (2022).
11. D. Liang, C. Shi, W. Li, W. Chen, and M. K. Chyu, Int J Heat Mass Transf, **vol. 201**, (2023).
12. W. Tang et al., Int J Heat Mass Transf, **vol. 201**, (2023).
13. J. Y. Ho, K. C. Leong, and T. N. Wong, Int. J. Therm. Sci, **vol. 137**, (2019)

Research on unsteady characteristics of valveless piezoelectric micropump with planar conical tube

Shengchuan Cai[1,*]

[1]School of Energy and Building Engineering, Shandong Huayu University of Technology, Dezhou 253000, Shandong, China

Abstract. The model of valveless piezoelectric micropump with planar conical tube is established in this paper, and dynamic mesh technology is used to simulate the unsteady characteristics of the micropump at different pressure amplitudes and driving frequencies for different diffusion angles. The elements affecting the output performance of the micropump are analyzed. The research indicates that the longer the conical tube, the smaller the flux that flows into the cavity of the micropump during inhalation. While the longer the conical tube, the greater the flux that flows out of the cavity of the micropump during expulsion. The net output flux of the valveless piezoelectric micropump increases with the length of the conical tube. The rectification efficiencies of micropump with tube length of 1mm, 3mm, and 5mm are 8.77%, 14.58%, and 18%, respectively.

1 Introduction

As an important component of MEMS[1], microfluidic systems[2] can realize the functions of supplying trace amounts of liquid and controlling the direction of fluid transport.With the rapid development of micromachining technology, microfluidic systems with miniaturization, automation, integration and intelligence have become a hot topic of attention and research in the academic and business communities of various countries. As the core driving and executing components of microfluidic systems, the development level of micropump directly affects the overall performance of microfluidic systems. Valveless piezoelectric micropump uses the inverse piezoelectric effect of the piezoelectric materials to drive reciprocating motion of piezoelectric oscillator, causing a changes in the volume of cavity of the micropump, and resulting in increasing or decreasing of the pressure in the cavity of the micropump, thus achieving the directional transportation of fluid. It has advantages of good controllability, simple structure, fast response, easy integration, and high energy exchange efficiency. Valveless piezoelectric micropump have been widely used in electronic device cooling, biochips, aerospace, medical devices, microchemical analysis, environmental detection and other fields[3-5]with its unique working mode and good performance.

He et al.[6] conducted simulation of the micropump using electro-liquid-solid coupling method.Both the back pressure and flux of the micropump show a tendency increasing firstly and then decreasing as the frequency increases, as the voltage was 200V and the frequency was 50Hz, the flux was up to 4.023mL/min. Bo et al.[7] studied the effect of phase difference of voltage on the performance of the dual cavity series micropump. The results showed that when the diameters of the cavity of the micropump were equal, the optimal phase difference increased stepwise with the increase of excitation frequency. However, it increased linearly with increasing excitation frequency when the diameters of the cavity of the micropump were not equal. Singh et al.[8] conducted comprehensivly studies of the piezoelectric micropump with divergent/convergent tube. The net flow had a tendency to increase firstly and then decrease with the increase of the parameters, and the net flux continues to rise as the diameter of cavity of the micropump becomes larger.

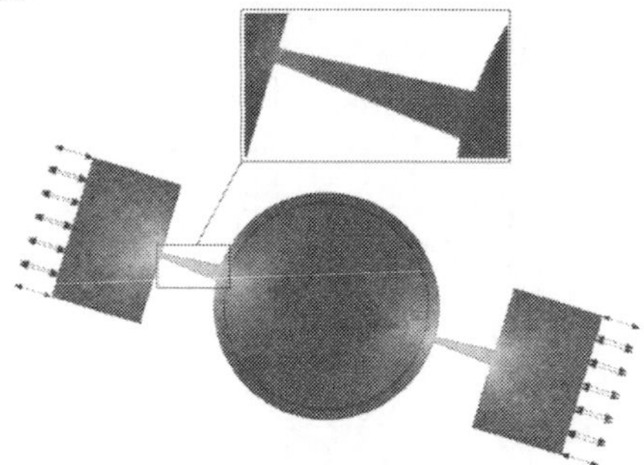

Fig. 1. Grid diagram of the computational region of valveless piezoelectric micropump with planar conical tube

2 Numerical calculation

The grid diagram of the computational region of valveless piezoelectric micropump with planar conical tube is shown in figure 1. The piezoelectric oscillator is defined as the moving grid region, while the other computational regions are set as the stationary grid region. The instantaneous displacement variation condition[9] is used

* Corresponding author: shengchuancai@163.com

to set dynamic grids for the computational region of piezoelectric oscillator. The import and export of the valveless piezoelectric pump are set as equal-pressure boundary conditions, with the relative pressure set to zero, and no-slip boundary conditions set on the rest of the walls.

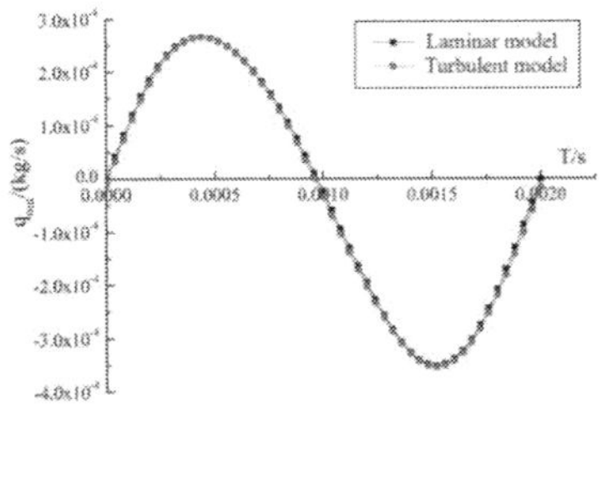

(a) θ=10°

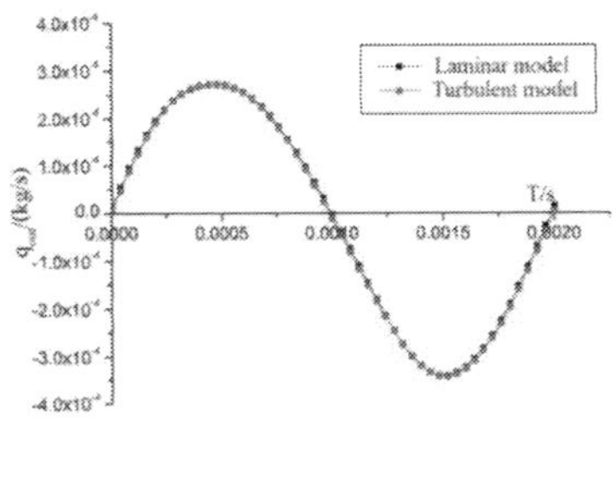

(b) θ=20°

Fig. 2. The variation of instantaneous flux with time at export of the micropump at f=500Hz

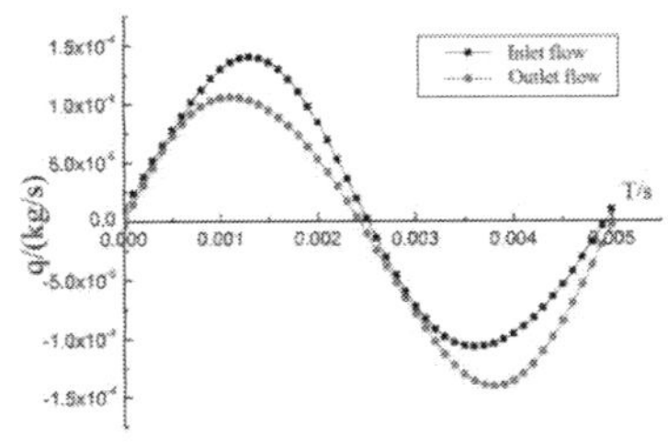

Fig. 3. The variation of instantaneous flux with time at the import and export of the micropump

Turbulent model and laminar model are adopted to calculated flow model of the micropump numerically. Figure 2 gives the variation of instantaneous flux with time at the export of the micropump with various angles of diffusion at driving frequency of 500Hz. Relative deviation of the maximum instantaneous flux of the micropump with an diffuser angle of 10°is 0.23%, and relative deviation of the total output flow is 9.46%. Relative deviation of the maximum instantaneous flux of the micropump with an diffuser angle of 20° is 0.15%, and relative deviation of the total output flow is 6.07%. The laminar and turbulent models have little effect on the simulation results of the micropump. The micropump with different parameters is simulated under different conditions, and the maximum Reynolds number of flow is obtained in the range of 460~2360. Based on the Reynolds number range and referencing literature[10], the turbulent model is used to simulate the flow in the micropump.

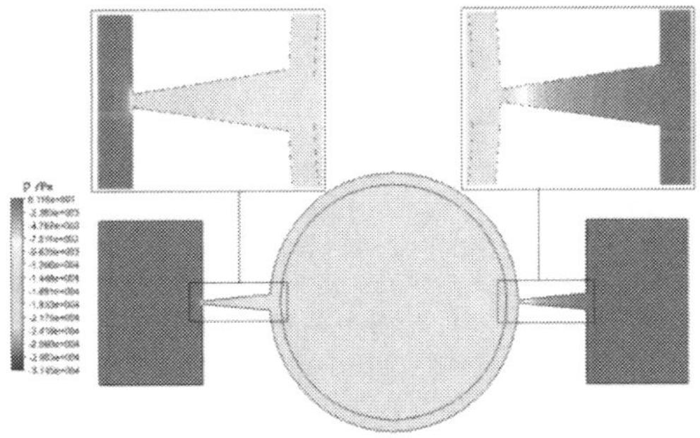

(a) Pressure

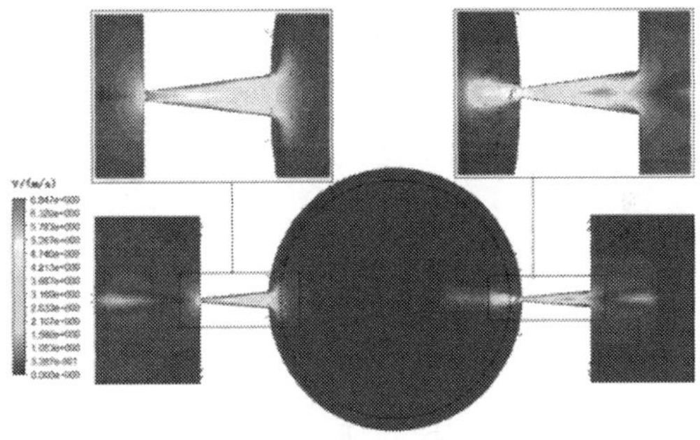

(b) Velocity

Fig. 4. Pressure and velocity distribution of intermediate section of the valveless piezoelectric micropump during inhalation at 1/4T

3 Analysis and discussion of results

3.1 import and export flow analysis

The variation of instantaneous flux with time at the export and import of micropump in one cycle is shown in figure 3. At the initial time, the piezoelectric oscillator moves upward from negative maximum displacement, the volume of the cavity of the micropump increases, resulting in the pressure decreases gradually. Fluid flows into the cavity of the micropump, and the flux that flows into the cavity of the micropump from the export conical tube smaller than that from the import conical tube. At 1/4T, the piezoelectric oscillator reaches the equilibrium position, and the flow velocity in the conical tubes reaches its maximum value. Figure 4 gives the pressure and velocity distribution of intermediate section of the valveless piezoelectric micropump at this time. It can be seen that the pressure gradient within the export conical tube is larger than that in import conical tube, and flow is stable. Vortices are generated due to sudden contraction as fluid flows into the cavity of micropump from the export conical tube. At 1/2T, the piezoelectric oscillator reaches the maximum forward displacement, the volume of cavity of the micropump has reached its maximum value. The flux from tubes is approximately zero, and the suction process is completed. The volume of the cavity of the micropump decreases as the piezoelectric oscillator moves downward, which result in increasing of the pressure gradually. The flux that flows out of the cavity of the micropump from the export conical tube is larger than that from the import conical tube. At 3/4T, the piezoelectric oscillator reaches the equilibrium position, and the flow velocity in the import and export conical tubes reache its maximum value. Figure 5 gives the pressure and velocity distribution of intermediate section of the valveless piezoelectric micropump at this time. It can be seen that the pressure gradient within the import conical tube is greater than that in the export tube, and there are vortices generated in the import tube, which increased the pressure loss. At t=T. Piezoelectric oscillator returns to the negative maximum displacement, and the volume of cavity of the micropump reaches its minimum value. The instantaneous flux from the conical tubes is approximately zero, the micropump completes a working cycle and realizes the directional transportation of fluid.

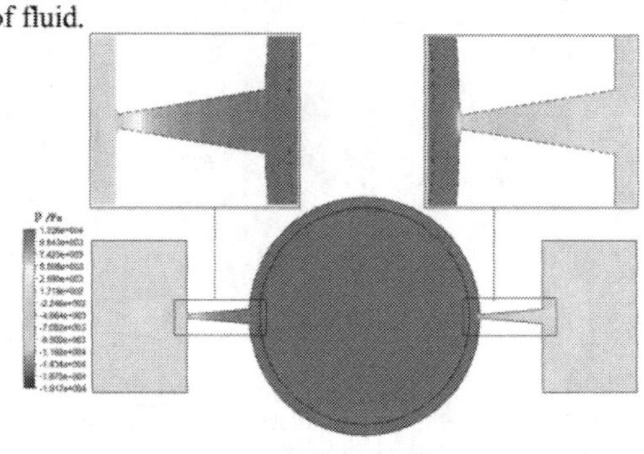

(a) Pressure

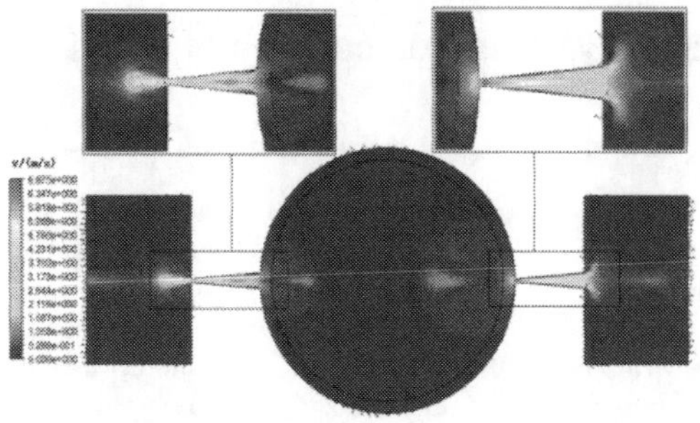

(b) Velocity

Fig. 5. Pressure and velocity distribution of intermediate section of the valveless piezoelectric micropump during expulsion at 3/4T

3.2 The effect of the tube length

The characteristics of the piezoelectric micropump are studied by changing the conical tube length, which are taken as 1mm, 3mm and 5mm respectively, with other parameters held constant. The instantaneous flux variation at the export of micropump with different tube lengths is showed in figure 6. The instantaneous flux that flows into the cavity of the micropump decreases with increasing the conical tube length during inhalation. The maximum instantaneous fluxs of conical tube with the length of 1mm, 3mm, and 5mm are 0.1097g/s,0.1065g/s, and 0.106g/s, respectively, and the total fluxs are 0.0353g/s, 0.0333g/s, and 0.032g/s, respectively. The instantaneous flux that flows out of the cavity of the micropump increases with increasing the conical tube length during expulsion. The maximum instantaneous fluxs of the tube with length of 1mm, 3mm, and 5mm are 0.1349g/s, 0.1402g/s, and 0.1404g/s, respectively, and the total fluxs are 0.0423g/s, 0.0447g/s, and 0.046g/s, respectively. The above analysis reveals that the longer the conical tube, the smaller the flux that flows into the cavity of the micropump during inhalation. While the longer the conical tube, the greater the flux that flows out of the cavity of the micropump during expulsion. The rectification efficiencies of micropump with conical tube length of 1mm, 3mm, and 5mm are 8.77%, 14.58%, and 18%, respectively.

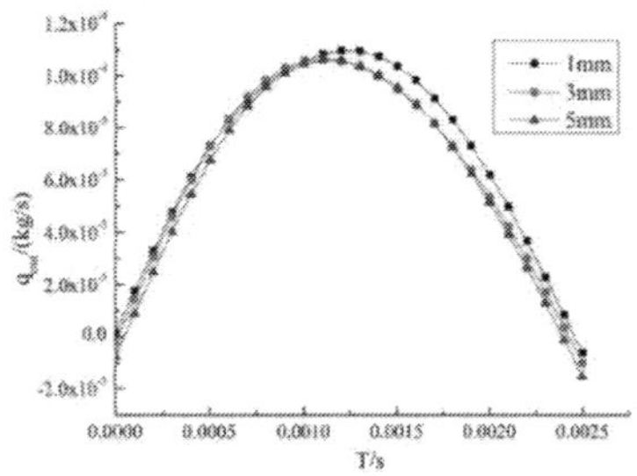

(a) During inhalation

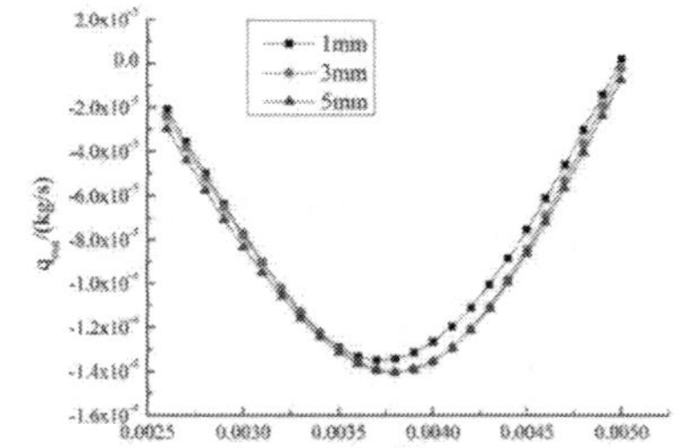

(b)During expulsion

Fig. 6. Instantaneous flux variation curves at the export of the valveless piezoelectric micropump with different tube lengths

Figure 7 gives distribution of the velocity in the intermediate section of the export conical tube at 1/4T during inhalation. The pressure gradients in the 3mm and 5mm conical tubes are equivalent and greater than that in the 1mm conical tube, and the conical tube with 1mm has a high velocity and smooth flow without backflow. There are obvious vortices generated as the conical tube length increases, and the influence range and duration of the vortices increase with increasing the conical tube length. The vortices motion not only causes significant pressure loss, but also reduces the effective flow area of the fluid from the tube. The loss of flow resistance in the contraction direction also increases with increasing the conical tube length. Therefore, flux that flows into the cavity of the micropump from the export conical tube decreases with the increase of conical tube length. The fluid flows out of the cavity of the micropump along the diffusion direction of conical tube during expulsion. The shorter the tube length, the more intense the change of pressure gradient in the tube, and the larger the scale of the vortices formed in the export buffer cavity, so flux that flows out of the cavity of the micropump from export conical tube increases with the increase of tube length. Because of the difference of flow resistance, the net output flux of the micropump increases with the increase of tube length within one working cycle.

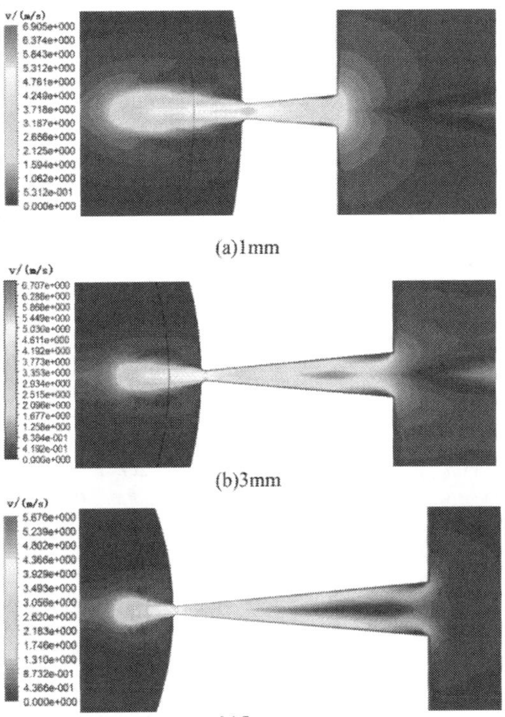

Fig. 7. Velocity distribution of intermediate section of the planar conical tube in the suction process at 1/4T

4 Conclusion

The unsteady characteristics of the micropump with planar conical tube is researched by simulation, and the factors affecting the output performance are analyzed. The following conclusions have been obtained:

(1) The longer the conical tube, the smaller the flux that flows into the cavity of the micropump during inhalation. While the longer the conical tube, the greater the flux that flows out of the cavity of the micropump during expulsion. The rectification efficiencies of micropump with conical tube length of 1mm, 3mm, and 5mm are 8.77%, 14.58%, and 18%, respectively.

(2) The flux that flows into the cavity of micropump from export conical tube decreases with increasing the tube length during inhalation, while the flux that flows out of the cavity of micropump from export conical tube increases with increasing the conical tube length during expulsion. The output flux of the micropump increases with the increase of conical tube length within one working cycle.

Acknowledgments

This work was supported by the Green and low-carbon smart heating and cooling technology characteristic laboratory of the Shandong Huayu University of Technology.

References

1. L.D.Wang, C.Liu, The development trend of MEMS science and technology.*Journal of Dalian University of Technology*. **40**(5):505-508. (2000)
2. S.Shoji, M.Esashi, Microflow devices and systems. *Journal of micromechanics and microengineering*. **4**:157-171. (1994)
3. X.W.Jiao, W.Q.Huang, C.S.Zhao, The piezoelectric pump technology and its application. *Micromotors*. **38**(5):66-69. (2005)
4. J.W.Kan, Z.G.Yang, G.M.Cheng, Research on piezoelectric pump and its development.*Optics and Precision Engineering*.**10**(6):619-625.(2002)
5. J.H.Zhang, J.Z.Lu, Q.X.Xia, Working principle and flux characteristics of valveless piezoelectric pump with Y-shape tubes for transportingcells and macromolecule. *Chinese Journal of Me-chanical Engineering*. **44**(9): 92-99.(2008)
6. X.H.He, H.Yang, S.Yang, Multi-field coupling simulation of valveless piezoelectric micropump with wall attached jet. *Journal of drainage and irrigation machinery engineering*. **39**(05):439-444. (2021)
7. B.H.Kim, I.C.Kim, Y.J. Kang, Effect of phase shift on optimal operation of serial-connected valveless micropumps. *Sensors and Actuators A Physical*. **209**:133-139.(2014)
8. A.K.Sen, S.Singh, N.Kumar, D.George, Analytical modeling, simulations and experimental studies of a PZT actuated planar valveless PDMS micropump. *Sensors and Actuators A Physical*. **225**: 81-94.(2015)
9. M.Jain, B.Puranik, A numerical investigation of effects of cavity and orifice parameters on the characteristics of a synthetic jet flow. *Sensors and Actuators A Physical*. **165**(2): 351-366.(2011)
10. Z.X.Li, D.X.Du, Experimental study on flow characteristics of liquid in circular micro-tubes. *Microscale Thermophysical Engineering*. **7**:253–265. (2003)

E3S Web of Conferences **560**, 01011 (2024) https://doi.org/10.1051/e3sconf/202456001011
ICEMEE 2024

Study on Pore Structure Characteristics of tight Sandstone Reservoirs --Taking Chang 8 in XX block of Ordos Basin as an example

Binhu Han, Wenbo Qiu, Zhifei Sun, Dingjun Wang, Yicang Liu*

The 11th Oil Production Plant of CNPC Changqing Oilfield Company, Qingyang City, Gansu Province, 745000, China

Abstract: The XX block of Ordos Basin has recently achieved a major exploration breakthrough, however the reservoir is more heterogeneous and the single well productivity is more different. Based on drilling, core, analysis and testing data, the rock characteristics, physical characteristics and microscopic pore structure characteristics of the area are comprehensively analyzed. A set of reservoir classification criteria with strong applicability is established. The results show that the reservoir lithology in the study area is complex, and the types and geometric forms of reservoir space are diverse. The porosity is 0.15%~13.65%, and the permeability is) 0.001~83.5 ($\times 10^{-3} \mu m^2$. The distribution range of physical properties is wide, and the overall performance is low porosity and ultra-low permeability. The heterogeneity is strong, and a few large pores have an important influence on reservoir permeability. A set of reservoir classification and evaluation criteria is established. The study area mainly develops type II reservoirs. The classification and evaluation criteria can provide reference for the later exploration and development of the study area.

1 Introduction

The reservoir characteristics have an important influence on the enrichment of oil and gas resource [1, 2]. Therefore, it is of great significance to study the reservoir characteristics for the screening of favorable areas for production and construction. The classification of tight sandstone reservoirs in the Ordos Basin has been systematically developed since the 1980s. The evaluation criteria used at present are mainly derived from the Triassic and Permian reservoirs in the early exploration results of the basin, which have poor adaptability to the reservoirs in the study area. Based on the data of drilling, logging, core analyzing and testing, this paper synthesizes the macroscopic characteristics and microscopic pore structure of reservoirs and establishes a set of reservoir classification criteria to provide a reliable geological basis for the efficient exploration and exploitation of reservoirs [3].

2 Geological background

The Ordos Basin is located in the western part of the North China Platform, with a long rectangular shape [4]. The XX oil field is located in Gansu Province. XX oilfield is a sedimentary basin with stable ascending and descending type (Fig. 1). Under the relatively stable sedimentary environment of lake facies and delta front facies, the sedimentary rotation is obvious, the marker layer is clear, and the lithologic thickness of the sedimentary body has certain rules to follow. The work area ranges from Xiaoguan in the south to Jiyuan in the north, from Shancheng in the west to Qiaochuan in the east, and the main mining horizon is Chang 8_1^2. The top structure of Yanchang Formation is a west-dipping monocline, and the structure is uplifted. There is an east-west-trending nose uplift, which runs through the whole area [5]. The width of the nose is about 2 ~ 5km, and the uplift amplitude is about 10 ~ 20m.

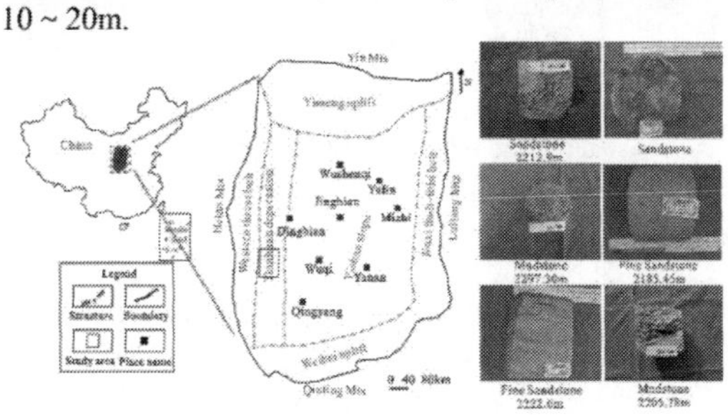

Fig. 1. XX Block Location Chart

3 Method

The petrological characteristics of the reservoir mainly include the mineral composition of rock debris particles, the types and characteristics of interstitial materials, as well as the characteristics of debris rounding, sorting and arrangement. It not only determines the characteristics of reservoir rock skeleton, but also affects the pore structure and combination characteristics of reservoir. It is the

* Corresponding author: lyc_cq@petrochina.com.cn

premise of sedimentary diagenesis research and the decisive factor of reservoir oiliness research [6].

3.1 Rock types and detrital components

According to the statistics data (Fig.2) : The Chang 8 is mainly composed of gray and dark gray fine-grained lithic feldspar sandstone, followed by feldspar lithic sandstone, and the main particle size distribution is 0.07 ~ 0.25 mm. The sandstone has good sorting, and the debris particles are mostly sub-round to sub-prism [7, 8]. The cementation types are mainly pore, enlargement-pore and film-pore. In the sandstone composition content, the quartz content is the highest, the content is more than 30%; followed by feldspar debris, the content is concentrated in 25 % ~ 30 %; the content of rock debris is about 20 % ~ 25 %. The content of interstitial material accounts for about 10% .

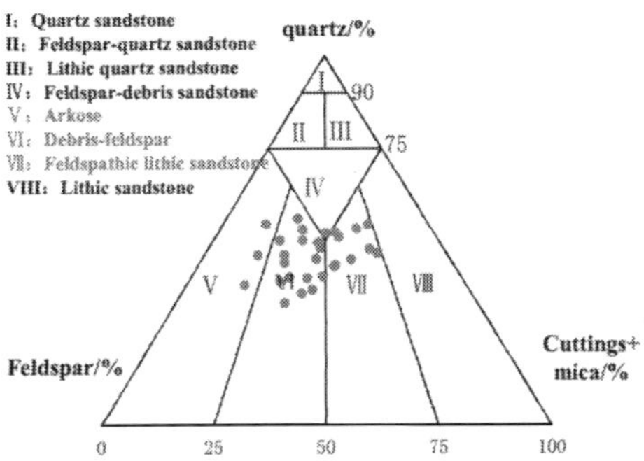

Fig. 2. Lithology diagram of the study area

On the whole, the sandstone detrital composition of the Chang 8 reservoir in the study area accounts for about 87 % of the total rock content on average, of which the average quartz content is 33.1 %, the average feldspar content is 26.3 %, the average debris content is 24.8 %, and the average mica content is 4.2 %. The average interstitial content was 11.6 %. In general, sandstone has the characteristics of high quartz, high feldspar and high debris, and the content of interstitial material is relatively high.

3.2 Characteristics of interstitial material

The content of Chang 8 sandstone interstitial material in the study area is 11.9 %. The interstitial material is mainly composed of clay minerals (mainly chlorite and hydromica) (6.6%), carbonate cements (4.3%) and siliceous cements (1.0%). The authigenic clay minerals are mainly chlorite (4.4%) and hydromica (2.2%). The carbonate cement is dominated by iron-containing cement iron calcite, with a content of 4.3 % (Fig.3).

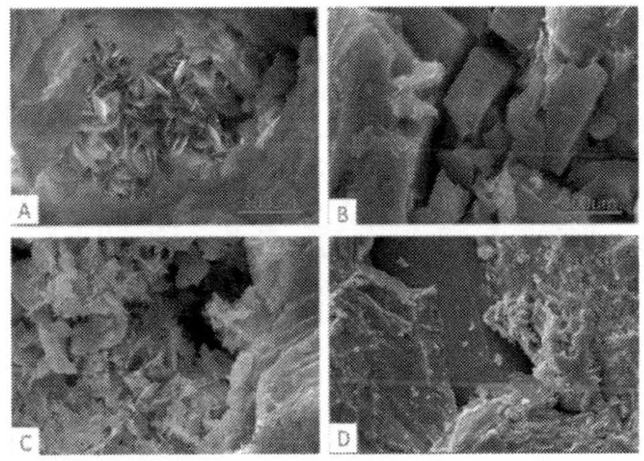

Fig. 3. Typical interstitial materials in the study area (A: L301, C8, 2673.1m, chlorite; B: L340, C8, 2575.3m, calcite; C: L340, C8,2575.3m, hydromica; D: L340, C8,2575.3m, siliceous)

4 Discussion

By studying the microscopic pore structure, the occurrence state of reservoir pores and throats can be gotten. It is beneficial to the next comprehensive reservoir evaluation and favorable area prediction.

4.1 Reservoir pore characteristics

4.1.1 Reservoir pore structure characteristics

The pore types of sandstone reservoirs in the study area mainly include intergranular pores, feldspar dissolved pores, lithic dissolved pores and matrix dissolved pores. The connectivity is relatively good in this area. Feldspar dissolved pores are well developed in the reservoir space of this area, and the shape is reticular or irregular intragranular dissolved pores. The contribution of lithic dissolution pores to the pore properties of reservoir sandstone is small, but the contribution to the permeability of reservoir sandstone cannot be ignored

The average face porosity of Chang 8 reservoir in the study area is 4.9 %, of which the intergranular pore content is 3.13 %, the feldspar dissolved pore content is 1.13 %, the cuttings dissolved pore content is 0.15 %, and the matrix micropore accounts for 0.49 %. (Fig.4)

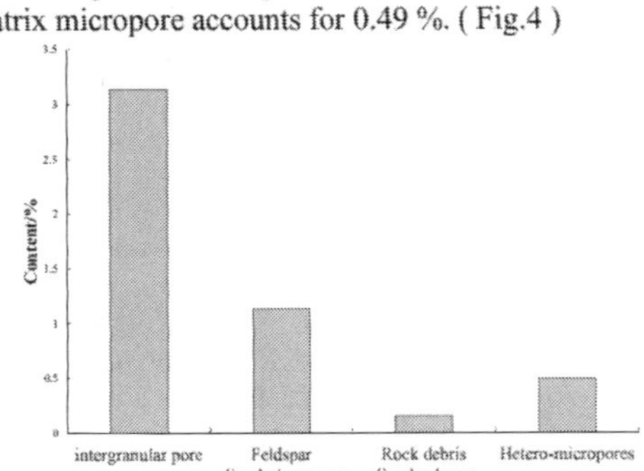

Fig. 4. Statistics of pore types in the study area

4.1.2 Reservoir pore size and distribution

It can be got that the Chang 8 is dominated by small pores; the second is fine pores and macropores, with the contents of 27.75% and 15.16%, respectively.

4.1.3 Reservoir pore combination type

The pore types of Upper Paleozoic reservoirs in the study area are mainly micropores, intercrystalline pores-dissolved pores, dissolved pores and other types. However, it occurs not only in a single pore type in the storage space, but in most cases in a combination of pore types [1]. The combination of different types of pore types in the study area is analyzed, and the pore combination in the study area is divided into micropores, intercrystalline pores-micropores, dissolved pores-intercrystalline pores, dissolved pores-intergranular pores, dissolved pores, etc. Several combination types (Fig. 5).

The main characteristics of the solution pore-intergranular pore combination are as follows: the size distribution of reservoir mineral particles is uniform, the particle shape is mostly subcircular-subedge, the contact mode is mainly point and point-line, and the cementation type is contact-pore cementation and contact cementation [6]. This combination type is the best one of the six combination types.

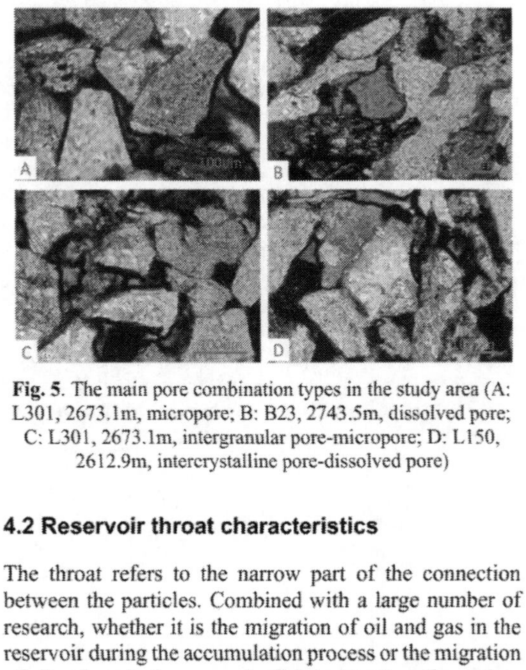

Fig. 5. The main pore combination types in the study area (A: L301, 2673.1m, micropore; B: B23, 2743.5m, dissolved pore; C: L301, 2673.1m, intergranular pore-micropore; D: L150, 2612.9m, intercrystalline pore-dissolved pore)

4.2 Reservoir throat characteristics

The throat refers to the narrow part of the connection between the particles. Combined with a large number of research, whether it is the migration of oil and gas in the reservoir during the accumulation process or the migration of oil and gas in the reservoir after the later transformation, the movement ability of the fluid will be affected by the size and distribution of the throat. The development of the throat is mainly controlled by the shape and size of the rock particles, as well as the contact relationship and cementation type of the particles. In general, the more uniform the shape of rock particles, the larger the throat space; the larger the rock particles, the larger the throat space; the more sparse the contact between the particles, the greater the throat space; the simpler the cementation type, the smaller the reverse.

There are four main types of throat types in previous research results: necking type, point type, flake or curved flake, and tube bundle throat (Fig.6).

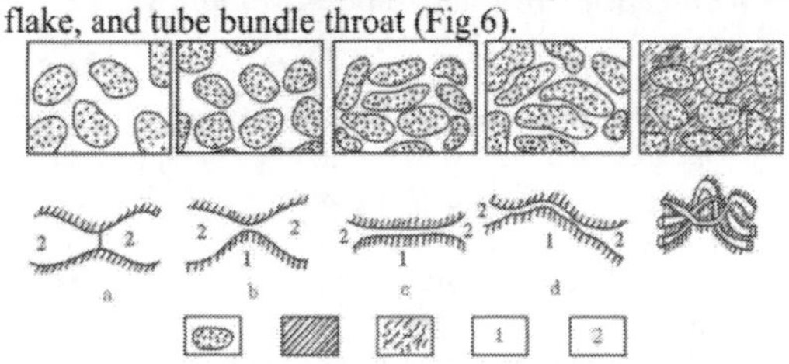

Fig.6 Pore throat type of clastic rock (according to Luo, 1986) (a: necking throat ; b:point throat ; c:flaky throat ; d:curved flaky throat ; e:tubular throat)

The types of throats are mainly flakes and curved flakes, and there are bundle-like and point-like throats (Fig.7).

Fig.7. Typical pore throat types in the study area (A: CT scanning analysis of L339; B: CT scanning analysis of L301)

4.3 Reservoir pore throat combination characteristics

There are both pores and throats in the tight sandstone, and the throats connect the pores, which constitute the reservoir space of the reservoir.

4.3.1 Pore throat structure parameters

At present, the mercury injection experimental analysis method is the most commonly used and effective method to characterize the pore structure. The parameters involved are displacement pressure, median pressure, sorting coefficient, coefficient of variation, mean coefficient, skewness, maximum pore throat radius, median radius, maximum mercury saturation and mercury removal efficiency, etc., and reservoir classification can be performed according to curve characteristics and parameters (Table 1).

Table 1. Classification and Evaluation Criteria for Formation Reservoir

Type	I	II	III	IV	Average
Porosity/%	5.38	11.17	1.82	6.04	6.10
Permeability/mD	0.10	0.42	0.05	0.26	0.21
Average throat radius/ μm	0.15	0.60	0.32	0.12	0.30
Median throat radius/ μm	0.03	0.35	0.06	0.01	0.11
Maximum throat radius/ μm	0.49	2.86	1.86	0.77	1.50
Median pressure/Mpa	26.22	2.12	106.50	12.82	36.91
Median radius/μm	0.03	0.35	0.01	0.06	0.11
Displacement pressure/Mpa	1.84	0.33	1.19	0.60	0.99
Maximum S_{Hg}/%	64.27	81.14	51.07	74.70	67.79
Mercury withdrawal efficiency/%	16.68	25.81	24.77	22.11	22.34

4.3.2 Classification of pore-throat combination

The type II has the highest degree of development. The sandstone with such capillary pressure curve has a higher maximum mercury saturation, which can reach about 70 ~ 80 %, and the average mercury removal efficiency is 55 % (Fig.8).

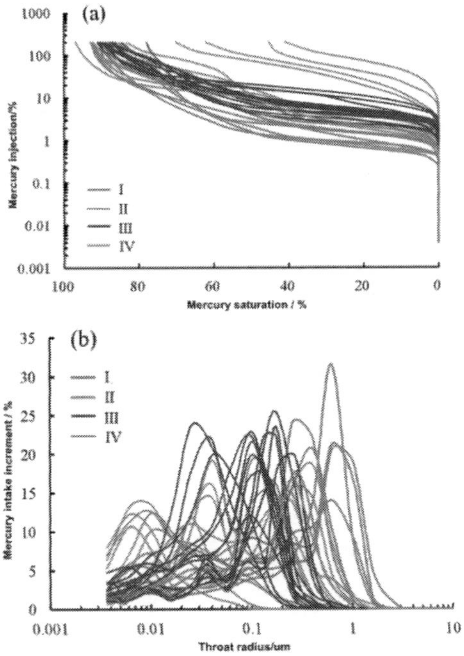

Fig. 8. Mercury injection data of well L150 (A: Capillary pressure curve; B: Distribution of pore throat size)

4.3.3 Nuclear magnetic resonance movable fluid

Nuclear magnetic resonance experiment is an effective non-destructive technology. There is no process of non-wetting phase displacing wetting phase, but the spin signal of hydrogen nucleus in rock pores is measured by rock samples under low-frequency magnetic field, which can fully reflect the distribution characteristics of fluid in pore throat space.

Through the nuclear magnetic resonance experiment of 31 sandstone samples, the curve distribution of saturated water T2 spectrum and T2 spectrum after centrifugation of each sample was obtained (Fig. 9). According to the classification of pore throat structure by high pressure mercury intrusion, the NMR saturated water T spectrum curves of the corresponding samples are also divided into four categories for analysis. It is found that the transverse relaxation time T2 of each sample is mainly distributed between 0.3 and 1000 ms. The shape of T2 spectrum curve is mainly single peak, left high and right low double peak and left low and right high double peak, and there is a big difference between the shape of saturated water T2 spectrum curve of type I and type I pore throat structure and the size distribution curve of high pressure mercury intrusion pore throat of the corresponding sample.

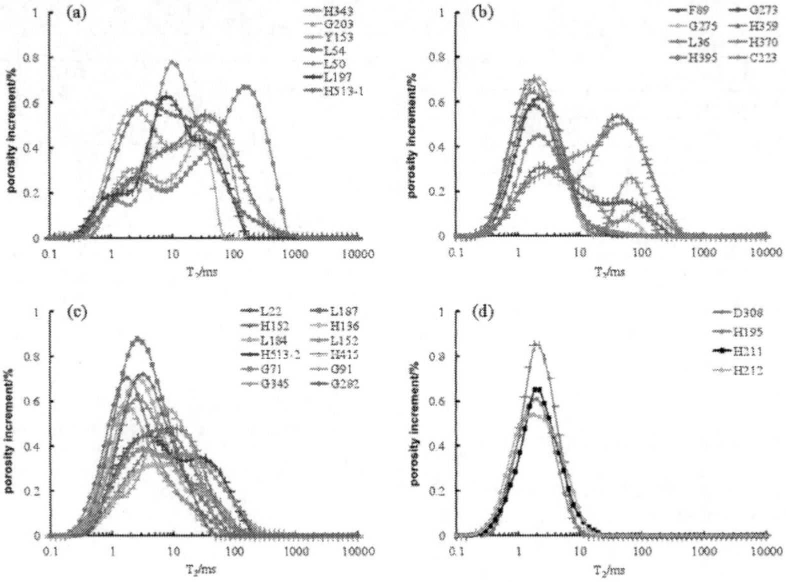

Fig. 9. Mercury injection data of wells (A: I; B: II; C: III; D: IV)

5 Conclusions

(1) The lithology of Chang 8 reservoir in the study area is mainly gray fine-grained lithic feldspar sandstone.

(2) Among the detrital components, quartz has the highest content, with an average of more than 30%, followed by feldspar, debris, interstitial material and mica.

(3) Pores are mainly intergranular pores (3.50%), accounting for 74.78% of the total surface porosity; followed by feldspar dissolution pores (0.70%), accounting for 14.95 % of the total face rate.

Acknowledgements

This work is financially supported by the scientific research project of Shaanxi Provincial Department of Science and Technology (No.2024JC-YBQN-0563); the Postdoctoral Research Project in Shaanxi Province, China (No.2023BSHYDZZ131).

References

1. Song F, Kong Q, Su N, et al. Formation mechanism of the upper Paleozoic tight sandstone gas reservoir in the Daniudi gas field, Ordos Basin, China [J]. Frontiers in Earth Science, 2024, 12.
2. He J, Niu X, Meng Q, et al. Analysis of seepage mechanism and productivity model of multi-stage fractured horizontal well in tight sandstone gas reservoir [J]. Petroleum Science and Technology, 2023, 41(21): 2055-2070.
3. Xiao Z, Chen S, Yin X, et al. Geochemical characteristics and source of Chang 8 crude oil in the east part of the Shanbei Slope, Ordos Basin, China [J]. Interpretation-a Journal of Subsurface Characterization, 2021, 9(2): T349-T56.
4. He J, Liu Z, Zhang H, et al. Study on development methods of different types of gas wells in tight sandstone gas reservoirs [J]. Scientific Reports, 2023, 13(1).
5. Zang J, Liu J, He J, et al. Characterization of the pore structure in Chinese anthracite coal using FIB-SEM tomography and deep learning-based segmentation [J]. Energy, 2023, 282.
6. Kessel B. Lower Paleozoic sequence stratigraphy, deposystems and paleogeography of northwestern Ordos basin, North China [M]. 2006.
7. Van Alstine J M. An integrated tectono-sedimentary analysis of the Ordos basin and its sub-basin components through the Triassic and Jurassic [M]. 2010.
8. Zhang L, Liu C, He X. Discovery of Kilometer-Scale Uplift and Exhumation Related to the Late Indosinian Movement in the Northern Ordos Basin, North China [J]. Acta Geologica Sinica-English Edition, 2019, 93(1): 231-2.

Numerical simulation study on combustion characteristics of square hollow oil pool fire in ship cabins

He Li [1*], Kun Li [2], Qing Huang [1], Enwei Wang [1]

[1] Marine Design & Research Institute of China, Shanghai 200011, China
[2] China University of Geosciences, Wuhan, 430070, China

Abstract: To study the combustion phenomenon of square hollow fire sources in ship cabins, PyroSim software was used to establish numerical models of different inner and outer edge lengths of the same fire source area. FDS software was used to numerically simulate fire temperature and smoke spread, observe and record the flame and smoke spread characteristics in cabins under different working conditions, and analyze the temperature distribution of flame plumes and the temperature field characteristics inside the cabin. The results indicate that: During the combustion process of a square hollow fire source in a ship's cabin, a typical flame convergence phenomenon occurs, and the flame temperature of the fire source combustion first increases significantly and then gradually stabilizes. The smaller the inner edge length of a square hollow, the higher the combustion temperature of the fire source, the faster the heating rate, the more obvious the phenomenon of fire convergence, and the more intense the flame combustion. As the inner edge length increases, the influence of flame combustion at the center of the square decreases, and the peak temperature of the flame decreases with a smaller amplitude and tends to concentrate.

1 Introduction

With the continuous development of industrialization and urbanization, the degree of harm caused by fires in the complex environment of ships has been increasing year by year. Among them, there are countless cases of ship fire accidents caused by combustible liquid leakage. In general, after a liquid fuel leak, it will flow freely on the ground and form a fire upon encountering a fire source. However, in actual ship fires, the flow of combustible liquids is often limited by the ship's boundaries and various obstacles, making them distributed in a square shape around the obstacles. After being ignited, more complex square hollow fire sources are formed. The combustion duration of a square hollow fire source is long and the thermal radiation is strong, which can easily lead to the recurrence of disasters and accidents, and may cause serious harm to personnel, equipment, and the environment on board ships.

Some domestic and foreign scholars have conducted relevant research on the basic combustion model of fire sources with boundary constraints. Li Cong et al. [1] found through a fixed outer diameter annular n-hexane combustion test that the flame area, perimeter, and other factors of the annular flame source gradually decrease with the increase of the inner to outer diameter ratio. Sun Zhiyou et al. [2] studied the combustion characteristics of square kerosene pool fires with different side lengths under no wind conditions. The study found that the presence of wall height resulted in the combustion rate of the fire source being lower than the theoretical value. The experimental value of the combustion rate showed a monotonically decreasing relationship with the characteristic size of the oil pool. Liu et al. [3-4] investigated the combustion matrices of 3x3, 7x7, and 15x15 square heptane, and found that different distances between ignition sources and matrix sizes can affect the thermal feedback of the flame itself and the thermal conversion of surrounding ignition sources, thereby affecting the combustion of multiple ignition sources. WANG C J et al. [5] conducted experimental research on the combustion rate of a square hollow converging fire source and found that typical flame fusion phenomena occur during the combustion process, and this phenomenon leads to a significant increase in the combustion rate and flame height of the fire source. Sun Hao et al. [6] conducted a series of combustion tests on annular fire sources with different inner and outer diameters of equivalent areas, and studied the effects of intermediate air flow entrainment, fuel thickness, and intermediate obstacle size on the combustion characteristics of annular fire sources. The experiments found that different sizes of hollow areas would affect the combustion rate, flame height, and flame fusion time of annular fire sources, further confirming the phenomenon of fire convergence. DENG L et al. [7] studied the probability of flame merging and pulsation frequency of annular flame sources with different inner and outer diameters, and found that as the flame burns, the initial "annular flame" will eventually merge. Based on size analysis, they proposed a function fitting correlation between different heat release rates under seamless

*e-mail:1536921537@qq.com

conditions and the size of the annular oil pool.

It can be seen that many researchers have conducted extensive research on the basic combustion models of boundary limited ignition sources, including combustion parameters such as combustion rate, flame height, and thermal radiation. However, the study of the combustion behavior of square hollow ignition sources is not yet sufficient. This article will use FDS numerical simulation method to study the combustion characteristics of square hollow ignition sources in ship cabins, focusing on numerical simulation research and analysis of the combustion characteristics of square hollow ignition sources from different inner and outer edge lengths.

2 Model establishment and grid partitioning

2.1 Model establishment and working condition setting

In order to explore the combustion characteristics of square fire sources, based on previous experimental research on circular fire sources [8], five square hollow fire sources with the same fire source area but different inner and outer side lengths were designed. The specific operating conditions are shown in Table 1。

Table 1 Working condition settings

working condition	Inner edge length a/m	Outer edge length A/m	Fire source area /m^2
1	1	2	3
2	1.5	2.3	3
3	2	2.65	3
4	2.5	3.04	3
5	3	3.5	3

The selected simulation area has a length, width, and height of 8.0, 8.0, and 5.0 meters respectively. The boundary conditions for the windows on both sides and the front door are set to "Open". As shown in Figure 1, the red square hollow area represents the combustion surface of the fire source, with kerosene as the fuel. The heat release rate per unit area is 2270kW/m2. The combustion reaction of kerosene adopts the default vortex dissipation combustion model of FDS, and calculates the radiation heat conduction by solving the radiation heat transfer equation of the gray body. The ambient temperature is 20 °C and the atmospheric pressure is 101.325kPa. A temperature slice was set up through the center of the square fire source to determine the phenomenon of flame combustion, and a total of 30 thermocouples were set up to measure the temperature changes of flame combustion.

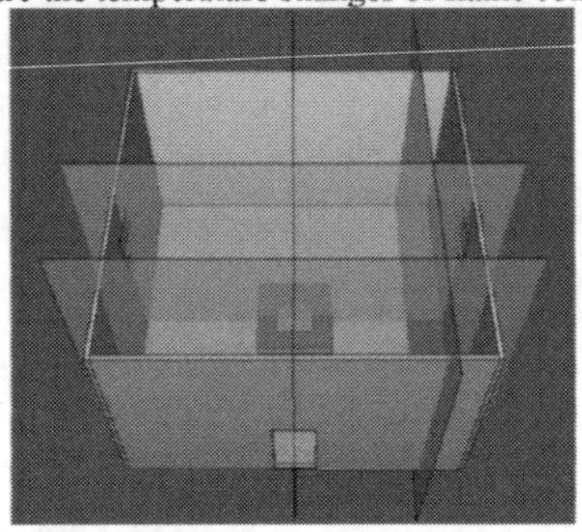

Figure 1 Numerical simulation model diagram

2.2 Numerical calculation model for cabin combustion

Burning releases heat, causing the temperature of the gas near the flame to increase and the density to decrease, forming a flow under buoyancy. This flow phenomenon is compressible and has a velocity much lower than the speed of sound, therefore it belongs to thermally driven low Mach number compressible flow, which can be described by the following control equation:

Continuity equation:

$$\frac{\partial \rho}{\partial t}+\frac{\partial \rho u_j}{\partial x_j}=0 \tag{1}$$

Momentum conservation equation:

$$\frac{\partial \rho u_i}{\partial t}+\frac{\partial \rho u_i u_j}{\partial x_j}=\rho f_i-\frac{\partial \rho}{\partial x_i}+\frac{\partial \tau_{ij}}{\partial x_j} \tag{2}$$

Energy conservation equation:

$$\frac{\partial \rho h}{\partial t}+\frac{\partial \rho h u_j}{\partial x_j}=\frac{\partial p}{\partial t}+u_j\frac{\partial p}{\partial x_j}+\varepsilon-\frac{\partial \dot{q}_j''}{\partial x_j}+\dot{q}''' \tag{3}$$

Gas state equation:

$$p=\rho RT \tag{4}$$

Composition conservation equation:

$$\frac{\partial \rho Y_\alpha}{\partial t}+\frac{\partial \rho Y_\alpha u_j}{\partial x_j}=\frac{\partial}{\partial x_j}(\rho D_\alpha\frac{\partial Y_\alpha}{\partial x_j})+\dot{m}_\alpha''' \tag{5}$$

In the formula: ρ is the density (kg/m3); u is Speed (m/s); f_i is Volume force (N/m3); P is Pressure (Pa); τ_{ij} is Shear stress (Pa); h is enthalpy (J/kg); ε is Dissipation function (W/m3); $\dot{q}''$ is Heat flux caused by conduction, radiation, and diffusion (W/m^2); $\dot{q}'''$ is Heat release rate per unit volume of combustion (W/m^3); T is Temperature (K); R is Gas constant (J/(kg · K)); D_α is Diffusion coefficient of component α (m^2/s); Y_α is Mass fraction of component alpha (kg/kg); $\dot{m}_\alpha'''$ is Unit volume consumption rate or generation rate of component α (kg/(m3 · s))。

2.3 Grid independence verification

In FDS, D */δ x is generally used for grid quality evaluation, and grid independence directly affects the results of numerical simulation. Among them, D * is the characteristic diameter of the fire source, δ x is the grid size, and when D */δ x takes a value of 4-16, the calculation accuracy and speed are better, The calculation formula for D * is:

$$D^* = \left(\frac{Q}{\rho_\infty C_p T_\infty \sqrt{g}} \right)^{\frac{2}{5}} \quad (6)$$

In the formula: Q is the heat release rate of the fire source, KW; ρ∞ is the air density, generally taken as 1.2kg/m^3; Cp is the specific heat of air, taken as 1 kJ/(kg · K); T ∞ is the ambient air temperature, taken as 293K; G is the gravitational acceleration, taken as 9.81m/s^2.

For a fire scenario with a heat release rate of 6810kW, after calculation, if the characteristic diameter D * of the fire source is 2.07m, then 1/16D * -1/4D * is 0.13m-0.52m. Perform grid independence analysis for typical operating conditions within the above range, taking into account both numerical accuracy and simulation time. Select six different grid sizes of 0.15m, 0.20m, 0.25m, 0.30m, 0.35m, and 0.40m, and compare the temperature curves of the same measuring point over time within 900 seconds, as shown in Figure 2. It can be seen that when the grid size is 0.20m, 0.25m, and 0.30m, there is not much difference in the temperature curve. Therefore, the grid size is finally set to 0.25m x 0.25m x 0.25m. Due to the same combustion area of the fire source in each operating condition, the heat release rate remains unchanged, and this grid independence verification is applicable to all operating conditions.

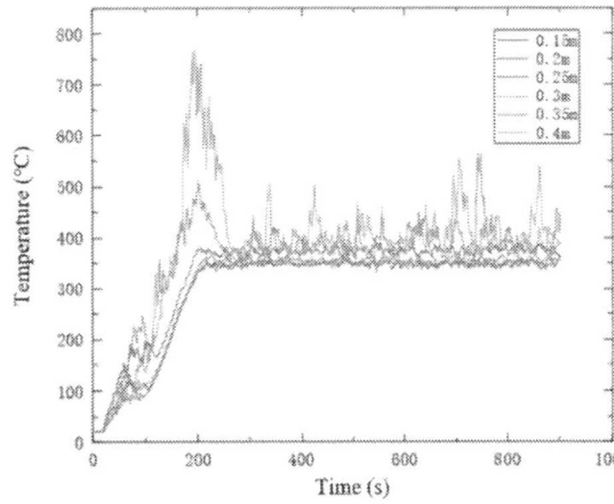

Figure 2 Temperature comparison of different grid sizes

3 Results and analysis

3.1 Analysis of flame and smoke propagation characteristics inside the cabin

3.1.1 Analysis of Flame Combustion Characteristics in Cabins

Figure 3 shows the flame combustion situation inside the cabin under different operating conditions. It can be seen that at 35 seconds, the flame began to burn, and around 50 seconds, the square hollow combustible material continued to burn and the flame gradually increased. At 100 seconds, the flame continued to burn, and when the inner edge length was small, it showed a phenomenon of flames converging towards the middle obstacle. At 195 seconds, the fire increased, and the flames burned to the windows on both sides of the cabin. At 300 seconds, the flame tended to stabilize, showing convergence and burning to the front door of the cabin.

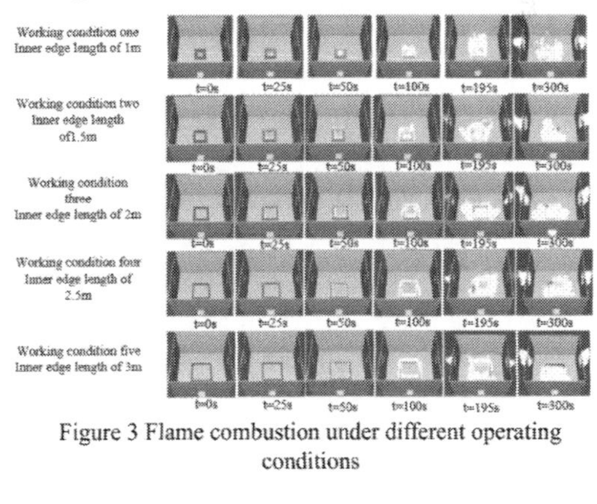

Figure 3 Flame combustion under different operating conditions

3.1.2 Analysis of Smoke Spread Characteristics in Cabins

Figure 4 shows the smoke spread in the cabin under different operating conditions. It can be seen that at 25 seconds, the flame did not start to burn but smoke appeared at the square hollow area. Around 50 seconds, the combustible material in the square hollow area began to burn and the smoke gradually spread to all sides. At 85 seconds, the flame increased and the smoke spread to the entire cabin. At 120 seconds, the smoke concentration increased and the visibility inside the cabin significantly decreased, but flame combustion could still be seen. At 300 seconds, the flame combustion stabilized, and the smoke completely covered the entire cabin, with visibility almost zero.

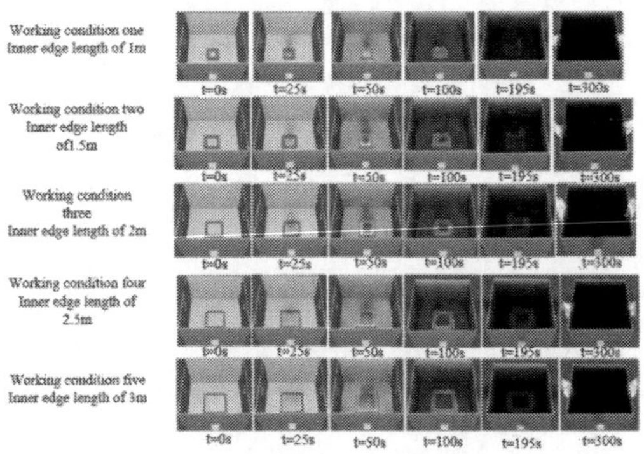

Figure 4 Smoke spread under different operating conditions

3.2 Temperature distribution of flame plume

Set up 30 measuring points above the fire source in the cabin to record the temperature at different positions during flame combustion. Among them, THCP1-13 is a measuring point that increases by 0.4m from a height of 0.2m to the center of the fire source, THCP14-27 is a measuring point uniformly arranged along the X-axis at a height of 2.6m from the center of the fire source, THCP28-30 is a measuring point located near the cabin doors and windows.

Figure 5 shows the temperature variation curves of any 8 flame measurement points under various operating conditions. It can be seen that the flame temperature of the square hollow flame source combustion first increases significantly under all operating conditions, reaches its peak at 200s, and gradually stabilizes. When the inner edge length of the square hollow does not exceed 2m, the flame temperature can reach a maximum of 600 ℃, and when the inner edge length is greater than 2m, the flame temperature basically stabilizes at 350 ℃. After analysis, it was found that the smaller the inner edge length of a square hollow, the more obvious the phenomenon of flame convergence, and the more intense the flame combustion, resulting in a greater temperature change at the bottom measuring point. As the inner edge length increases, the influence of flame combustion on the center of the square decreases, and the temperature difference at each measuring point is small and the change tends to be concentrated.

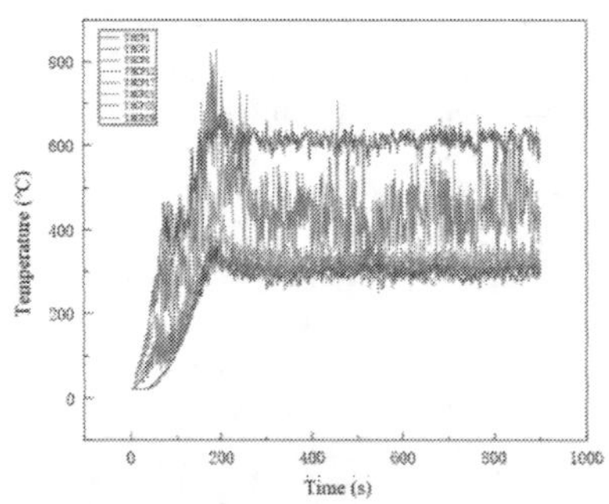

(a) Inner edge length of 1m

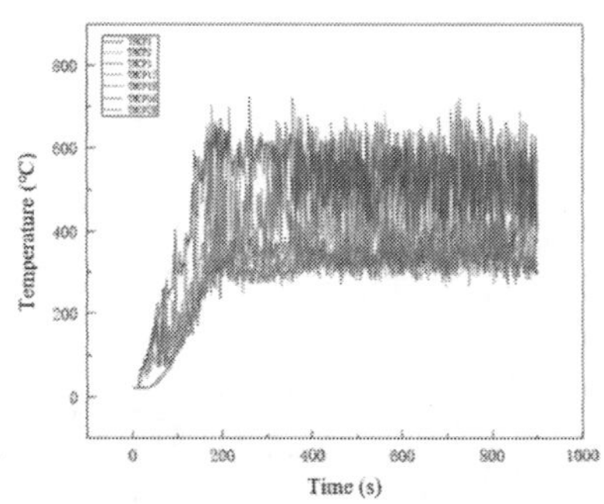

(b) Inner edge length of 1.5m

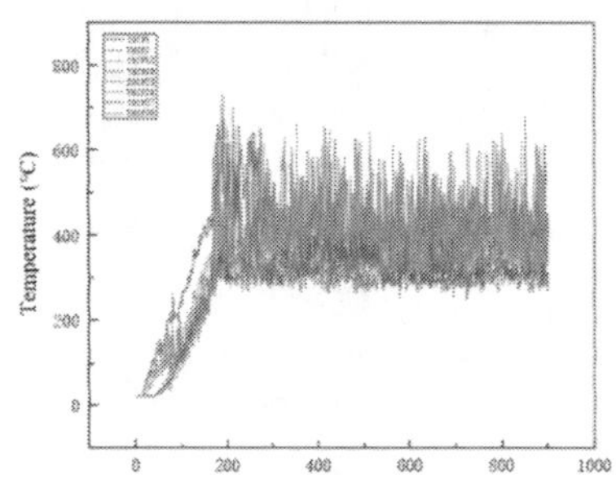

(c) Inner edge length of 2m

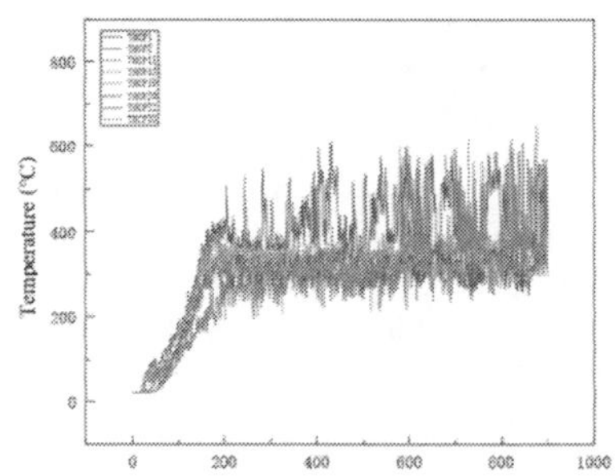

(d) Inner edge length of 2.5m

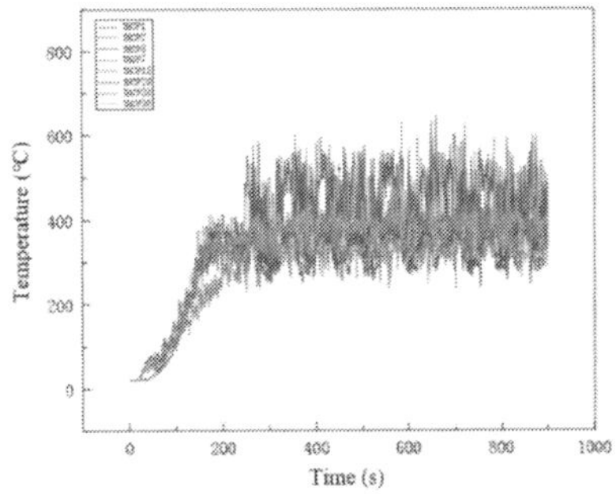

(e) Inner edge length of 3m

Figure 5 Temperature variation curves at different measurement points under different working conditions

Figure 6 shows the temperature variation curves of the same measuring point under different operating conditions during the combustion of the fire source. Measurement points THCP01 and THCP07 were selected at heights of 0.2m and 2.6m from the center of the fire source, THCP28 at a distance of 1m from the cabin window, and THCP30 at a distance of 1.5m from the cabin door for plotting and analysis. It can be seen that under different operating conditions, the temperature of all measurement points showed a significant increase before stabilizing during the combustion of a square hollow fire source. When the distance from the center of the fire source is closer, the smaller the inner edge length of the square fire source, the higher the temperature rise and speed of flame combustion in a short period of time. As the inner edge length increases, the peak temperature of the measuring point decreases and the subsequent sharp drop time shortens. At a distance from the fire source, the temperature change at the same measuring point is less affected by the length of the inner side of the square hollow, and remains stable at around 350 °C.

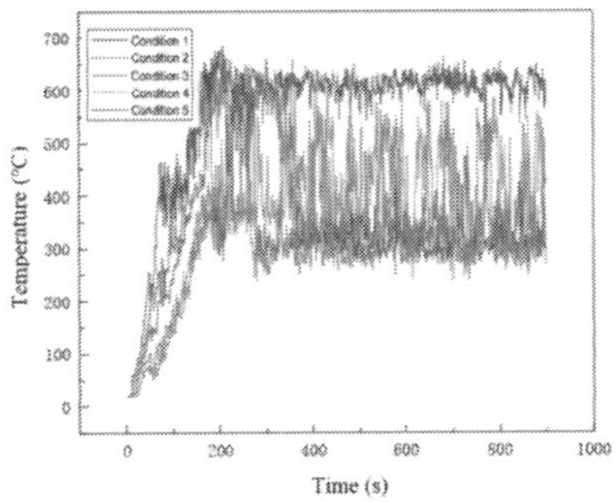

(a) THCP 01

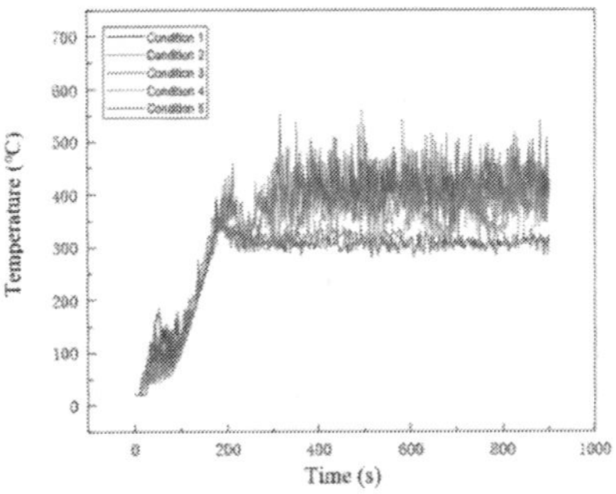

(b) THCP 07

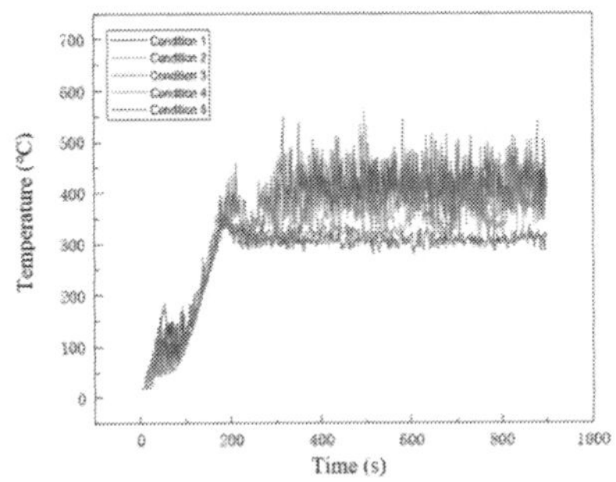

(c) THCP 28

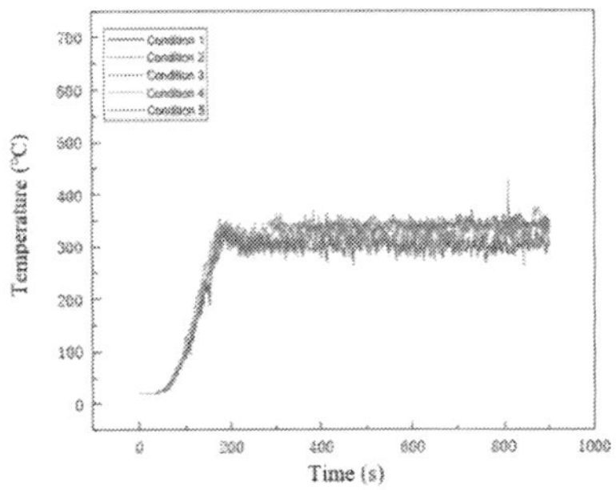

(d) THCP 30

Figure 6 Temperature variation curve of the same measuring point under different working conditions

3.3 Temperature field characteristics inside the cabin

Figures 7 and 8 respectively show X under different operating conditions Temperature slicing in the Y direction. From the graph, it can be seen that at 25 seconds, the temperature at the square hollow before flame combustion reaches 55 °C; At 50 seconds, the temperature of the square hollow further increases, and the smaller the inner edge length, the higher the temperature. The flame center temperature in operating condition one reaches 250 °C; At 100s, the temperature inside the cabin gradually increased and reached 110 °C. From the X-direction slice, it can be seen that as the inner edge length increases, the phenomenon of fire convergence weakens and the highest temperature of the fire source decreases.

From the Y-direction slice, it can be seen that due to the right side opening door, the fire source tilts to the left; At 195s, the flame continued to burn, and the temperature inside the cabin further increased to 340 ℃. The temperature of the fire source reached its maximum value on the opposite side of the door. The temperature of the fire source in working conditions one, two, and three reached a maximum of 570 °C, and the temperature of the fire source in working conditions four and five reached a maximum of 520 °C; At 300s, the flame converges and the combustion tends to stabilize.

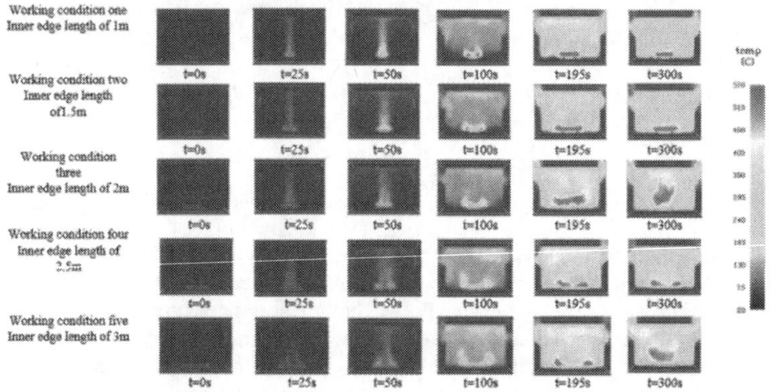

Figure 7 Temperature slicing in the X-direction under different operating conditions

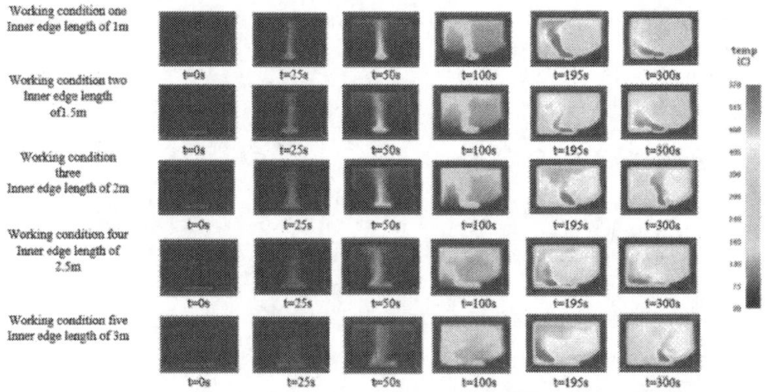

Figure 8 Temperature slicing in the Y direction under different operating conditions

4 Conclusion

(1) During the combustion process of a square hollow fire source in a ship's cabin, a typical phenomenon of flame convergence occurs, and this phenomenon leads to a significant increase in the combustion temperature and rate of the fire source. (2) The flame temperature of the square hollow flame source combustion first increases significantly and gradually stabilizes after reaching its peak. The smaller the inner edge length of a square hollow, the higher the combustion temperature of the fire source, the faster the heating rate, and the more obvious the phenomenon of flame convergence towards the middle obstacle, resulting in more intense flame combustion(3) As the inner edge length increases, the influence of flame combustion at the center of the square decreases, and the peak temperature of the flame decreases with a smaller amplitude and tends to concentrate.

References

1. Li Cong, Ma Xun, Yang Rui, et al. Characteristics of n-heptane annular oil pool fire based on shape and texture features [J]. Journal of Tsinghua University (Natural Science Edition), 2021,61 (6): 502-508
2. Sun Zhiyou, Zhuang Lei, Lu Shouxiang. Study on the combustion characteristics of square kerosene pool fires [J]. Fire Science, 2008, (02): 93-98
3. Liu N, Liu Q, Lozano J S, Shu L F, Zhang L H, Zhu J P, et al. Global burning rate of square fire arrays: Experimental correlation and interpretation[J]. Proceedings of the Combustion Institute, 2009, 32(2):2519-2526.
4. Liu N, Liu Q, Lozano J S, Zhang L H, Deng Z H, Yao B, et al. Multiple fire interactions: A further investigation by burning rate data of square fire arrays[J]. Proceedings of the Combustion Institute, 2013, 34(2):2555-2564.
5. WANG C J, GUO J, DING Y M, et al. Burning rate of merged pool fire on the hollow square tray[J].Journal of Hazardous Materials,2015, 290:78-86
6. Sun Hao. Experimental Study on Combustion Characteristics of Circular Oil Pool Fire [D]. Hefei University of Technology, 2018

7. DENG L, TANG F, MA X. Experimental study on flame merging probability and pulsation frequency of annular hydrocarbon pool fires with various inner and outer diameters[J].Process Safety and Environ - mental Protection,2021,146:473-480.

8. Fang Wei, Zhang Shaogang, Liu Jiahao, et al. Simulation study on temperature and height of flame plume during annular oil pool combustion [J]. Fire Science and Technology, 2023,42 (06): 762-767

Optimization of Emergency Resource Dispatch for Overseas Long-distance Natural Gas Pipeline Leaks

Xinyi Li[1,2,a], Jinqiu Hu[1,2,b*], Shaohua Dong[1,2,c], Xingtao Li[3,d], Lumeng Jiang[1,3,e], Shengyang Yu[3,f]

[1]College of Safety and Ocean Engineering, China University of Petroleum (Beijing), Beijing, China
[2]Laboratory of Oil and Gas Safety and Emergency Technology , Ministry of Emergency Management, Beijing, China
[3]China National Oil and Gas Exploration and Development Co., Beijing, China

Abstract: Overseas long-distance natural gas pipelines usually span across multiple regions or even countries, facing more severe environmental conditions and more prone to leakage and other accidents, and the study of the optimization method of emergency resource scheduling for overseas long-distance natural gas pipeline leakage accidents can help to reduce the casualties and property losses. Therefore, this paper establishes an emergency resource scheduling model in line with the emergency disposal mechanism of overseas long-distance natural gas pipeline leakage accidents, takes into account the variability of the operating ability of maintenance personnel, and improves the accuracy of the scheduling program to a certain extent, and also considers the actual environment of overseas long-distance natural gas pipelines, takes into account the impact of landslides on the emergency scheduling time, and calculates the probability of landslides occurring in the landslide point by using the interpolated value of rainfall The rainfall interpolation is used to calculate the probability of landslide at the landslide point, and the emergency dispatching time is corrected to make the dispatching program more in line with the actual demand. The analysis of the results shows that the emergency resource scheduling time is dominant in the case of small amount of maintenance and repair work and short total maintenance and repair time.

1 Introduction

Overseas long-distance natural gas pipelines usually span several regions or even countries[1], face harsher environmental conditions, and are more prone to accidents such as leakage[2]. The construction and operation of transnational pipeline projects involves a variety of factors such as politics, economics, law, engineering, environmental protection, etc., and in the event of an accident, it is a major issue concerning national energy security. Therefore, studying the optimization method of emergency resource dispatching for overseas long-distance natural gas pipeline leakage accidents can help reduce casualties and property losses.

At present, there are many experts and scholars studying the optimization of emergency resource scheduling, including the optimization of emergency resource scheduling for sudden geologic disasters[3,4], the optimization of emergency resources for sudden traffic accidents[5,6], and the optimization of emergency resource scheduling for sudden industrial accidents[7,8] and so on.

By considering the supportive role of multi-resource synergy on distribution network resilience, a unified decision-making method for post-disaster mobile emergency resource scheduling and distribution network restoration is proposed with the goal of maximizing the expectation of load restoration effect[3]; by using spatio-temporal network constraints for scheduling vehicles, and with the goals of maximizing the actual demand rate of the various affected points, minimizing the cost of emergency material transportation and minimizing the risk of vehicle transportation, a multi-objective emergency material scheduling planning model is established[4]; in order to make the known highway damage points be quickly disposed of, and at the same time to ensure that the remaining emergency resources at the emergency point can be quickly disposed of on the subsequent damage time and location of the random highway damage points, the establishment of a multi-objective emergency resource assignment model[5]; in the context of emergency resource dispatching of the municipal road as the background of the hierarchical analysis of the various factors affecting the use of the shortest path of the Dijkstra algorithm for modeling, solving the optimal path of emergency resource dispatching for municipal transportation road network[6]; In order to discharge the water in the coal mine in the shortest time, a multi-objective emergency dispatching model with the shortest emergency response time and the smallest equipment delay ratio was established based on the consideration of the existence of the differences in the operation rate of the pumping machine[7]; Considering the

[a]lxy_anquan@163.com, *Corresponding author: [b]hjq_anquan@163.com
[c]dsh_anquan@163.com, [d]lixingtao@cnpcint.com, [e]jianglumeng@cnpcint.com,
[f]yushengyang@cnpcint.com

timeliness, adequacy and economy of emergency resource dispatching for emergencies and other characteristics, the emergency resource dispatching strategy was studied[8].

Therefore, this paper establishes an emergency resource scheduling model in line with the emergency response mechanism of overseas long-distance natural gas pipeline leakage accidents, and optimizes the emergency resource scheduling with the optimization objective of the shortest total emergency response time. The model takes into account the differences in the operating ability of maintenance personnel, and improves the accuracy of the scheduling scheme to a certain extent by calculating the impact of the maintenance time of different maintenance personnel on the emergency response time. Meanwhile, the model also takes into account the actual environment of overseas long-distance natural gas pipelines, and considers the impact of landslides on the emergency response time, and calculates the probability of landslides at landslides by interpolating the rainfall to make emergency response scheduling time correction, so that the scheduling program is more in line with the actual needs. In order to facilitate the solution of the problem, three assumptions are made. The same, resource point maintenance and repair rate is consistent, and the emergency resource dispatch time is related only to the distance from the emergency point to the incident point and landslide delays, with the total number of transportation vehicles remaining constant.

2 Emergency Resource Dispatch Model

The accident emergency process includes the following parts, first of all, the accident point personnel issued an emergency call, the emergency point personnel to receive the emergency call, and then issued a vehicle scheduling order, the rescue personnel to receive the scheduling order, the vehicle will be driven from the current location to the emergency location to achieve the emergency resource scheduling, and finally arrive at the accident point to carry out the emergency repair.

The emergency resource scheduling process for overseas long-distance natural gas pipeline leakage accidents includes three steps: receiving scheduling orders, emergency resource scheduling, and long-distance natural gas pipeline leakage repair, and the total emergency response time is the total time spent from the receipt of the scheduling order to the completion of the long-distance natural gas pipeline leakage repair and disposal.

After receiving the dispatching order, the maintenance and repair personnel of each emergency point will start at the same time with the repair equipment and arrive at the accident point along the respective dispatching routes. Let the total amount of maintenance tasks for M, maintenance personnel collection $s=\{s_1,s_2\dots s_n\}$, maintenance rate for $v=\{v_1,v_2\dots v_n\}$, due to the maintenance personnel arrival time is not the same, so there is a difference in maintenance time, maintenance time $t=\{t_1,t_2\dots t_n\}$, where $t_1 \geqslant t_2 \geqslant \dots \geqslant t_n$, the emergency repair time T_1 is t_1. the amount of repair tasks for M is shown in the following formula.

$$M=\sum_{i=1}^{n} v_i \times t_i \tag{1}$$

Where v_i is the repair speed of the ith maintenance personnel, t_i is the repair time of the ith maintenance personnel, and M is the total maintenance task volume.

Overseas long-distance natural gas pipeline through the mountainous areas of complex geological structure, steep terrain, gullies and ravines complex geological conditions[9], once a landslide occurs, the traffic will be interrupted, in order to mitigate the impact of landslides on the scheduling of work, so the emergency resource scheduling routes to carry out risk assessment. Based on the statistics of landslide accidents in a certain place, the results are shown in the Table 1.

Table 1 Statistics on rainfall and landslide incidents

Rainfall level	24h rainfall r/mm	Probability of landslide P/%
Light rain	0~10	1.6
Moderate rain	10~25	9.5
Heavy rain	25~50	31.3
Rainstorm	50~100	52.3
Heavy rainstorm	100~250	77.2

Take the middle value of each group of precipitation, approximate the value of the probability of the occurrence of landslides in the group, and use the interpolation method to obtain the equation of precipitation and landslide probability.

$$p=-6\times10^{-4}r^3+4.61\times10^{-2}r^2-1.574\times10^{-1}r^1+1.3117 \tag{2}$$

Where r is the 24h precipitation and p is the probability of landslide occurrence.

Let a certain dispatch route passes through k highway landslide points, the landslide probability of the landslide point is p_i, and the total probability of route landslide P is calculated as follows.

$$P=1-(1-p_i)^k \tag{3}$$

Where p_i is the landslide probability of the landslide point and P is the total probability of route landslide.

If the total probability of landslide is greater, the greater the probability of emergency resource delay and the longer the total duration of emergency resource dispatch. The total duration T_2 of emergency resource dispatching is calculated as follows.

$$T_2=(1+P)\times\frac{l}{V} \tag{4}$$

Where P is the total probability of route landslides, l is the total length of the route, and V is the vehicle traveling speed.

Based on the total emergency time is the total time spent from the start of receiving the dispatching order to the completion of the repair and disposal of the long-distance natural gas pipeline leakage, the formula for calculating the total emergency time T is shown below.

$$T=T_1+T_2 \tag{5}$$

Where T is the total emergency time, T_1 is the emergency time for repair and maintenance, and T_2 is the total time for emergency resource dispatching.

3 Problem Description and Assumptions

In order to prevent overseas long-distance natural gas pipeline leakage accidents from further evolving into out poisonings, explosions, fires and other accidents, which seriously endanger personal safety[10], we take the shortest total emergency response time as the optimization objective for the optimization of emergency resource scheduling.

In order to facilitate problem solving, the following assumptions are made:

(1) The maintenance personnel at the same emergency resource point have the same skill level and the same maintenance rate. The maintenance equipment of the same emergency resource point is identical. The maintenance rate is consistent and there is no delay in maintenance such as damage during the maintenance time.

(2) There are and only a limited number of landslide points that have the probability of landslides in the entire route of emergency resource dispatch, and in the rest of the locations there is no risk of landslides and all vehicles have the same speed except for the location of the landslide point, and the time of emergency resource dispatch is only related to the distance of the emergency point from the point of the accident and the delays caused by the landslide point.

(3) One vehicle can only transport one maintenance worker and the equipment he/she uses, and the sum of transportation vehicles at all emergency points remains constant.

4 Calculation example analysis

In an overseas long-distance natural gas pipeline leakage accident occurred in a place, the type of accident is two small holes leakage, the accident point will be recorded as point A, a total of three emergency resource points have leakage accident maintenance and repair personnel and maintenance and repair equipment, the three emergency response locations were recorded as B, C, D, respectively, three, its distance l_B, l_C, l_D with point A were 10km, 15km, 35km, the speed of the vehicle is 20km/h, the total number of vehicles 4, 24h rainfall is 30*r*/mm, the road map is shown in the following Fig. 1, where the black dots are respectively indicated the accident point The number of maintenance personnel and work efficiency at points B, C and D are shown in the following Table 2

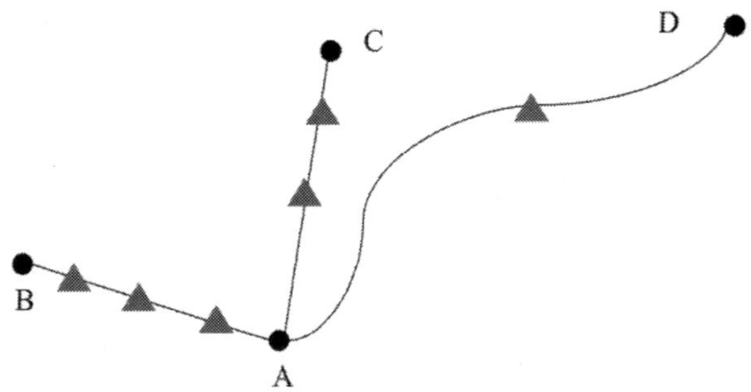

Figure 1 Roadmap of incident and emergency locations

Table 2 Number and efficiency of maintenance personnel at the emergency site

Location	Number of maintenance staff	Efficiency of maintenance staff
B	2	Pinhole 0.5h
C	3	Pinhole 0.4h
D	3	Pinhole 0.1h

At t_0 moment B, C, D three emergency point personnel to receive emergency calls, respectively, issued vehicle scheduling commands, rescue personnel to receive scheduling commands, a total of four emergency rescue vehicles from B, C, D were traveling to the emergency location to achieve the emergency resource scheduling, arriving at the point of the accident immediately after the emergency dimension repair.

Using the formula precipitation and landslide probability formula to calculate the probability of landslide point is 21.8797%, According to the formula for the total probability of a route landslide P respectively, it can be derived that the total probability of landslide on route B is 52.3249%, the total probability of landslide on route C is 38.9722%%, and the total probability of landslide on route D is 21.8797%, and further calculations can be made to obtain the total time on route B is 0.7616h, the total time on route C is 1.0423h, and the total time on route D is 2.1329h.

The exact representation of M in this calculation example can be calculated from the formula for the total time T as shown in the following equation.

$$M = \frac{(T-0.7616)}{0.5} \times s_b + \frac{(T-1.0423)}{0.4} \times s_c + \frac{(T-2.1329)}{0.1} \times s_d \tag{6}$$

Where M is the total amount of maintenance tasks, take 2; T is the total time, s_b, s_c, s_d are the maintenance personnel in B, C, D three emergency locations, through the calculation of the shortest total maintenance time is 1.1937h, s_b, s_c, s_d are 1, 3, 0.

The analysis of the results shows that the emergency resource scheduling time is dominant in the case of small amount of maintenance and repair work and short total maintenance and repair time, the landslide has less impact on the emergency resource scheduling time when the 24h rainfall is 30*r*/mm.

5 Conclusion

(1) This paper establishes an emergency resource scheduling model that conforms to the emergency response mechanism of overseas long-distance natural gas pipeline leakage accidents. The model takes into account the differences in the operational capabilities of maintenance and repair personnel, and considers the maintenance and repair time of maintenance and repair personnel in the total emergency response time, which improves the accuracy of the scheduling program to a certain extent.

(2) The model in this paper also considers the actual environment of overseas long-distance natural gas pipeline, overseas long-distance natural gas pipeline through the mountainous areas, and the complex geological structure, steep terrain, gullies and ravines complex geological conditions, taking into account the impact of landslides on the emergency scheduling time, the use of rainfall interpolation to calculate the probability of landslides occurring in landslides, emergency scheduling time correction, so that scheduling program is more in line with the actual needs.

(3) The analysis of the results shows that in the case of small amount of maintenance and repair work and short total maintenance and repair time, the emergency resource scheduling time is dominant, the landslide has less impact on the emergency resource scheduling time when the 24h rainfall is 30*r*/mm, and the distance occupies a dominant position in the impact on the emergency resource scheduling time, therefore, it is necessary to consider the actual situation to decide the emergency resource scheduling plan for the emergency point.

Acknowledgments

This work was supported by the Natural Science Foundation of China (52074323), China National Petroleum Corporation (CNPC)-China University of Petroleum (Beijing) Strategic Cooperation in Science and Technology Special Project: Research and Application of Key Technology for Integrity of Overseas Long-distance Pipelines of "Belt and Road" (ZLZX2020-05) .

Reference

1. A.J. Brito, A.T. de Almeida. Multi-attribute risk assessment for risk ranking of natural gas pipelines[J]. Reliability Engineering & System Safety, 94(2), 2009: 187-198.
2. Dinesh Herath, Faisal Khan, Ming Yang. Risk-based winterization to prevent hydrate formation in northern harsh environment[J]. Ocean Engineering, 119, 2016: 208-216.
3. H.Y. Chen, Y. Xu, Y. Wang, et al. Unified Decision-making for Post-disaster Mobile Emergency Resource Scheduling and Distribution System Restoration [J]. SHANDONG ELECTRIC POWER, 2024, 51(03): 1-7.
4. X. Zhang, H.C. Wang, Y.D. Li, et al. Research on emergency resources scheduling considering demand under sudden flood disaster [J]. Journal of Safety Science and Technology, 2023, 19(12): 187-192.
5. C.Y. Ma, B. Wang. Multi-objective Optimization of Highway Earthquake Disaster Emergency Resource Scheduling[J]. Highway Engineering, 2023, 48(05): 181-186+198.
6. J.X. Yan, M. Liu, N. Zhang, et al. Route Analysis of Emergency Resource Dispatch for Urban Road Traffic Emergencies[J]. Automobile Applied Technology, 2022, 47(06): 157-161.
7. C.Y. Ma, B. Wang. Multi-objective Optimization of Emergency Resource Scheduling for Coal Mine Flooding Accident[J]. Coal Mine Machinery, 2022, 43(09): 125-127.
8. H.W. Zhao, X.K. Zhao, Y.Y. Wang, et al. Emergency Resource Dispatching Strategy for Sudden Accidents in Chemical Industry Parks[J]. Journal of Shenyang University(Natural Science), 2023, 35(05): 449-458.
9. H.L. Zhu, W.Q. Zhang, P.F. Zhang, et al. Research on the Automatic Welding Construction Technology of Long Distance Pipeline in Mountainous Areas[J]. Electric Welding Machine, 2023, 53(12): 107-112.
10. Y.C. Song. Oil and gas gathering pipeline leakage problems and repair methods[J]. Plant Maintenance Engineering, 2023(18): 79-80.

 https://doi.org/10.1051/e3sconf/202456001014

Selection and Economic Evaluation of Production Process for Titanium Slag Smelting in Yunnan

Shihong Huang [1], Yan Cui [2], Liang Zhang [1] and Xianhao Long [3*]

[1]Office of Science and Technology, Kunming University, 650214, Kunming, Yunnan, China
[2]College of Chemistry and Chemical Engineering, Kunming 650214, University, Kunming China
[3]Graduate Department, Kunming University, 650214, Kunming, Yunnan, China

Abstract. Given the significant titanium resources in Yunnan Province, discuss the economic viability of their development and utilization. This paper presents an overview of the titanium ore resources and the process of titanium slag smelting in Yunnan Province, examines the rationale for selecting a DC arc furnace for titanium slag smelting in Yunnan province. The process offers advantages such as high production capacity, comprehensive utilization of flue gas, energy efficiency, and low electrode consumption. It also assesses input, consumption, and benefits. The project delivers significant social and economic benefits with strong risk mitigation capabilities.

1 Introduction

Titanium dioxide is one of the best materials in the world for producing white pigments. It is widely used in coating, paper making, plastic, chemical fiber, printing ink, rubber, enamel and other industries. The coating industry is the largest user of titanium dioxide. In 2020, the global total demand for titanium dioxide will be about 5.31 million tons. Since 2009, the output of titanium dioxide in China has been ranked first in the world. In 2023, the total output of titanium dioxide in China was 4.16 million tons, accounting for about half of the global output[1]. In 2021, the total demand for titanium materials in high-end fields such as high-end chemical industry (PTA), aerospace, shipbuilding and ocean engineering in China increased by 20542 t, a year-on-year increase of 28.5%. This shows that China's titanium industry is still in an accelerated period of industrial upgrading[2]. The industrial production methods of titanium dioxide include sulfuric acid method and chlorination method. Chlorination method leads the development trend of titanium dioxide production technology because of its technical advantages, such as high rutile content, less impurities and high whiteness, which can be used in coatings with better whiteness. Although China is also increasing the production capacity of chlorination process, in 2021, there were 42 full-process titanium dioxide production enterprises with normal production conditions in China's titanium dioxide industry, with a total production capacity of 4.487 million tons. The production capacity of titanium dioxide by chlorination process accounted for less than 10%, which was about 376,000 tons, an increase from the previous year. The domestic sulfuric acid process still occupied the leading position in the titanium dioxide production market[3].

Most of the high-titanium slag enterprises in Yunnan are small private enterprises with small scales, most of the equipment capacities being below 2200kVA, and they have low levels of automation with high labor intensity. These enterprises also have high energy consumption, unstable product quality, poor comprehensive utilization, and significant environmental pollution. To meet the market's demand for high-quality titanium dioxide, it is necessary to fully utilize Yunnan's abundant titanium ore resources and study the selection of a reasonable technological route to maximize the economic value of the mineral resources[4].

2 Status of titanium resources in yunnan

Yunnan province is rich in ilmenite reserves, with high titanium grade, few impurities and easy mining and dressing. According to the data, there are 30 proven ilmenite deposits, including 15 large-scale deposits and 15 small and medium-sized deposits, with a reserve of about 55 million tons, of which 32.65 million tons are easily mined ilmenite deposits on the surface. The reserve of titanium ore calculated by TiO_2 is 389.3 kilotons. The location of minerals is relatively concentrated. The reserves of ilmenite in Wuding, Lufeng, Luquan and Fumin counties account for more than 70% of the total reserves in Yunnan Province, but the resource advantages of titanium minerals in Yunnan Province have not been well transformed into industrial value[5].

The titanium ores in Wuding, Lufeng, Luquan and Fumin counties in central Yunnan are mainly eluvial titanium placer, and the reserve of B+C+D grade titanium ores is about 30 million tons. The residual diluvial titanium placer in central Yunnan is concentrated in Wuding, Luquan and Fumin counties, with titanium

* Corresponding author: longxh_kmu@qq.com

dioxide grade of 5.1% ~ 12%[6]. See Table 1 for its main economic characteristics.

Table 1. Main Economic Characteristics of Titanium Sand Mines in Central Yunnan

Project	Economic characteristics
Reserve	16.4 million tons
Average (lowest - highest) grade	80kg/m^3 (20-151 kg/m^3)
Average (maximum) thickness of ore body	7-8m (above 10m)

3 Technical route selection

There are many methods to prepare titanium rich materials, which can be divided into the methods to produce titanium slag and artificial rutile according to the final product; According to the production process, it can be divided into pyrometallurgical smelting and Wet smelting. Pyrometallurgical smelting also includes alkaline roasting process, aluminothermic reduction method, alkaline decomposition method, molten salt electroreduction process and magnesiothermic reduction process[7]. The wet process includes partial reduction hydrochloric acid leaching, partial reduction sulfuric acid leaching, total reduction rust leaching, ferric chloride leaching and other chemical separation methods. Common methods are electric furnace smelting method, hydrochloric acid leaching method, sulfuric acid leaching method, reducing rust method, etc[8].

The cost comparison between DC arc furnace and three electrode AC arc furnace is shown in Figure 1. In actual production, the single-electrode DC arc furnace has the following advantages over the 3-electrode AC arc furnace[9]:

(1) stable arc because the current direction is always consistent;

(2) The current fluctuation is 20-30% lower than that of AC furnaces, and the impact on the power grid is relatively small.

The environmental protection advantages of single-electrode DC electric arc furnace smelting process mainly include:

(1) The tail gas of electric furnace can be used as raw material preheating energy, and the titanium concentrate can be preheated to above 800℃ before entering the electric furnace for smelting, which can save energy consumption by 14-21%, reduce production cost and reduce environmental pollution caused by tail gas emission;

(2) The inside of the lifting hollow graphite electrode is hollow, and raw materials directly enter the high-temperature smelting pool in the furnace from the inside of the electrode, thus reducing the flying loss of fine materials, improving the smelting utilization rate and reducing the dust content in the tail gas. DC electric arc furnace can treat fine powder, while AC electric arc furnace needs granulation for raw materials[10].

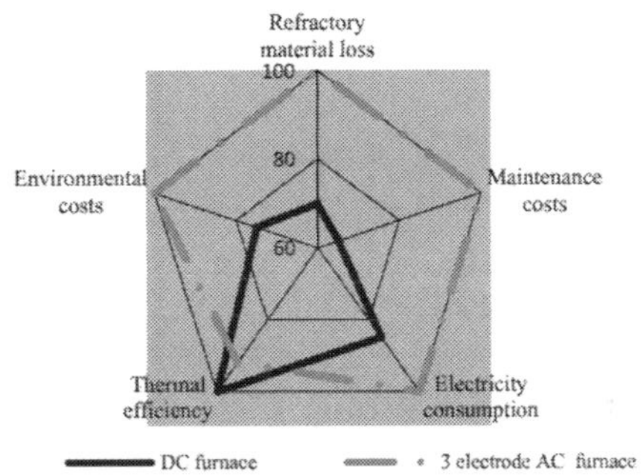

Fig 1. Cost comparison between DC furnace and 3 electrode AC furnace

4 Process flow

4.1. Raw material

The composition and particle size of raw titanium concentrate are shown in Table 2 and Table 3. The raw material is mainly titanium ore from a certain area, with titanium dioxide content higher than 45% and low impurities such as Si, Al, Mn and V, especially low Mg and Ca content.

Table 2. Composition of titanium concentrate (%)

Chemistry component	TiO_2	FeO	Fe_2O_3	Al_2O_3	MgO	SiO_2	MnO	CaO	V_2O_5
Mass fraction	49.00	30.26	17.90	0.49	0.91	0.50	0.70	0.00	0.21

Table 3. Particle size of titanium concentrate

Particle size(μm)	>0.180	0.180-0.076	0.076-0.032	<0.032
Composition(%)	37.123	57.590	4.876	0.144

4.2. Production process

See Figure 2 for the schematic diagram of titanium slag smelting by fully enclosed DC electric arc furnace, and Figure 3 for the complete production process flow chart of titanium slag factory.

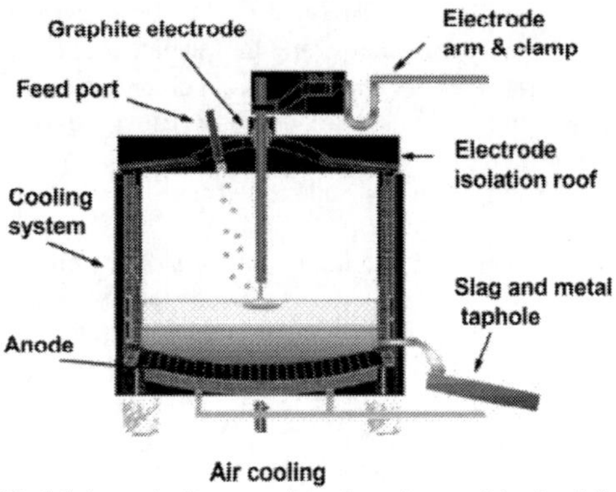

Fig.2 Schematic diagram of titanium slag smelting by fully enclosed DC furnace

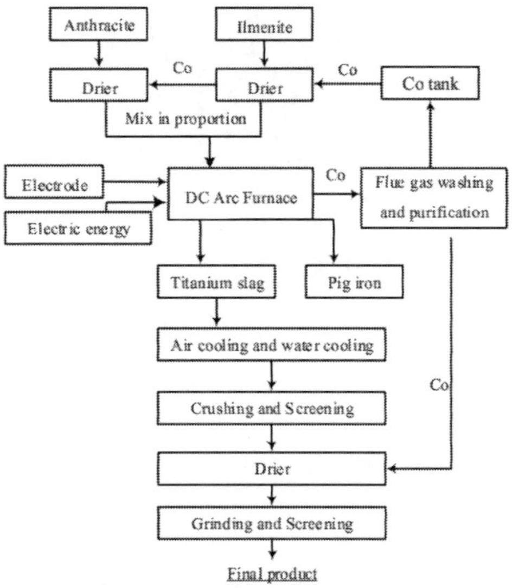

Fig.3 Flow chart of the whole process of titanium slag smelting in large-scale fully enclosed DC furnace

(1) Raw material preparation

In order to reduce the transshipment of raw and auxiliary materials, both titanium concentrate and low ash anthracite are purchased with qualified materials with satisfactory quality and particle size. Titanium concentrate and low ash coal are transported into the factory by car and sent to the raw material shed for storage; Electrodes are transported into the factory by car and directly discharged into the electrode storage room. Low-ash anthracite with qualified particle size is dried in a vibration drying fluidized bed to remove water. After classification, raw materials with particle size of+5 ~ 25mm are transported to the silo for standby, raw materials with particle size of -5mm are sent to the molten iron treatment system, and the tail gas is purified and discharged.

(2) Production process of melting titanium slag in DC furnace

Titanium concentrate and low ash anthracite are transported to the bunker and added to the electric furnace at a uniform speed according to the ratio. The lifting electrode moves up and down in the electric furnace by crane to control the arc length. Using anthracite as reducing agent, titanium concentrate is smelted in electric furnace at high temperature. After TiO_2 is reduced and enriched, rich titanium slag is obtained, and iron oxide in raw materials is gradually reduced to iron. The flue gas containing more than or equal to 80% CO produced by smelting enters the flue gas system and is stored for users after cooling, dust removal and washing.

(3) Treatment process of titanium slag and pig iron

A. Discharge

The molten pig iron and titanium slag liquid obtained by reduction smelting exist in layers at the bottom of the furnace body due to the difference in specific gravity. According to the improbability of specific gravity, it is regularly discharged into the molten iron bag and slag tank from the outlet and the outlet respectively, and then transported into the pig iron treatment workshop and the casting workshop respectively. The slag mouth should be opened with an oxygen gun. The highest temperature of titanium slag discharge port can reach 1750°C, and the discharged titanium slag is transported out by tank car. The highest temperature of taphole can reach 1550°C, and the molten iron is discharged into the preheated ladle on the tank car. After the titanium slag and hot iron are discharged, the slag outlet and iron outlet are sealed with mud gun. The air outlet adopts air collecting hood for air exhaust, and then exhausts through bag dust removal.

B. Pig iron treatment section

In the pig iron workshop, the ladle transported by the tank car is first weighed, and then the temperature of the molten iron is measured to meet the requirements, and then it is transported to the molten iron insulation area. The molten iron is cast into iron ingots, transported to cold water for spraying and cooling, and then concentrated.

C. Titanium slag treatment section

① Cooling

The high-temperature titanium slag products are discharged into the slag tank of the tank car, transported to the lifting area, and then transported to the cooling area by crane. Cooling in the air for 15-17 hours in the cooling zone; Then transport the slag tank to the water cooling area with a crane, pour out the titanium slag in the slag tank, spray water for cooling for 96 hours, and finally transport the titanium slag to the storage area with a loader.

② Crushing, screening and drying

After the ingot forming and cooling, the titanium slag needs to be broken and stored in different particle sizes according to user requirements. The titanium slag is pushed onto the grid screen and crushed (coarse crushed) to 500mm particle size with a hydraulic hammer. Titanium slag particles < 500mm screened by the grid screen are sent to the titanium slag drying section by a belt conveyor. The crushing and screening section adopts a bag-type dust removal system with an air volume of $2450m^3/h$, and the collected titanium slag dust enters the product section.

③ Grinding and sieving

First, the dried titanium slag is firstly screened by a double-layer sieve to remove materials less than 2mm. Titanium slag larger than 2mm is transported to a closed-circuit two-stage cone crusher by a fully enclosed conveyor for grinding and classification. Fine-grained titanium slag < 2mm is screened by a three-layer screening machine, and materials with a particle size of < 0.85mm are screened out; titanium slag with a particle size of > 0.85mm is transported to the grinding and screening section of a grinding machine by a conveyor. All titanium slag with particle size less than 0.85mm enters a two-stage classification and screening system for classification treatment, and titanium slag products with particle sizes of 0.106mm ~ 0.85 mm and < 0.106 mm are obtained. The fraction of 0.106mm ~ 0.85 mm obtained by this classification system accounts for 80% or more of the material, and the fraction less than 0.106 mm accounts for 20% or less.

The titanium slag with different particle sizes after classification is weighed, classified and stored. Classify and package according to customer's granularity requirements. Table 4 shows the main technical indexes.

Table 4. The main technical indexes of titanium slag products smelted by large dc arc furnace

Economic index	Value
Titanium slag content	TiO_2: 86% ~ 92%
Major impurity	MgO: <1.5%
Energy consumption	950 ~ 1200kWh/t ilmenite
Reductant consumption	130-150kg/t ilmenite
Ash	About 4~5% of the material
Electrode consumption	4.4kg/MWh
Melting temperature	1700-1750℃
Pig iron impurity	Ni: 0.1-0.2%, Mo: 0.1-0.12%
Tail gas	CO: 60-65%, H_2: 14-16%, CO_2: 1-2%

5 Economic evaluation

After the project is completed and put into production, it is assumed to produce 80000t/a titanium slag and 47251t/a pig iron as by-product. According to the current market price, titanium slag is 11000 yuan/t (tax inclusive price) and pig iron is 3000 yuan/t (tax inclusive price). The product price in the table is exclusive of VAT. The value-added tax rate of products is 17%. See Table 5 for sales revenue calculation.

Table 5. Calculation of sales revenue

Product name	Unit price (yuan/t)	Output (t)	Income (thousand yuan)
Titanium slag (containing 87% TiO_2)	11000	80000	880000
By-product iron (Fe≥ 97.25%)	3000	47251	141753
Total	—		1021753

6 Conclusion

(1) This project can replace the imported titanium slag products of the chloride process, and promote the development of titanium industry by utilizing Yunnan's rich ilmenite resources and cheap hydropower resources, with good economic benefits in reality.

(2) The planned titanium slag smelting project combines the characteristics of titanium resources in Yunnan Province and the existing production situation of high-titanium slag, and chooses the technology of smelting titanium slag with large-scale closed DC electric arc furnace, which has continuous production and can achieve the purposes of large capacity, comprehensive utilization of flue gas, energy saving, low electrode consumption and reduced environmental pollution. The production process is short, reliable, high in automation level and less in pollution.

(3) The project has good social and economic benefits and strong anti risk ability.

Acknowledgments

The work of this paper is supported by the following projects: High-Level Youth Talent Special Support Plan of Kunming City (C202014002), Yunnan Province's "Xingdian Talent Support Plan" (XDYC-QNRC-2022-0126), Talent introduction project of Kunming University (YJL20013).

References

1. An, Z.S., Chen, Y., Zhao, W., Huai, J. (2024) Report on China Titanium Industry Progress in 2023. Titanium Industry Progress, 41(02): 41-48. https://doi.org/10.13567/j.cnki.issn1009-9964.2024.02.008
2. An, Z.S., Chen, Y., Zhao, W. (2022) Report on China Titanium Industry in 2021. Iron Steel Vanadium Titanium, 43(04): 1-9. https://doi.org/10.7513/j.issn.1004-7638.2022.04.001
3. Chen, G., Qi, X.Z., Zhao, J. (2022) Report on Analysis of China Titanium Dioxide Industry's Economic Performance in 2021 and Future Development. China Coatings, 37(04): 11-17. https://doi.org/10.13531/j.cnki.china.coatings.2022.04.002
4. Lei,T., Zhu, C.J., Zhang Q., Zhang, J.J. (2005) Study On The Present Situation of Titanium Resource in Yunnan and its Exploitation Countermove. China Mining Magazine, (10): 26-29.
5. Liu,Y.Q. Ding, H. Xie, D. (2014) Titanium resource and the environment effects of sponge titanium production in China. Earth Science Frontiers, 21(5): 28-31. https://doi.org/10.13745/j.esf.2014.05.023
6. Xu, Z.C., Wang, D., Feng, Z.X. (2017) Development Status of Titanium Industry in Yunnan Province. Hot Working Technology, 46(18): 30-34+38.

https://doi.org/10.14158/j.cnki.1001-3814.2017.18.007

7. Sampath, A. J., Wickramasinghe, N. D., de Silva, K. N., & de Silva, R. M. (2023) Methods of Extracting TiO_2 and Other Related Compounds from Ilmenite. Minerals, 13(5), 662. https://doi.org/10.3390/min13050662
8. Tahooni Bonab, S., Abdollahi, H., Abbaspour, A. (2024) The Hydrometallurgical Approach in the Production of a High Content of Titanium Dioxide (TiO_2) from Ilmenite, Direct Hydrochloric and Sulfuric Acid Leaching and Pre-Treatment Methods: A Review. Mineral Processing and Extractive Metallurgy Review, 1–17. https://doi.org/10.1080/08827508.2024.2346654
9. Huang, S.H. (2016) Element Distribution and Production Process during Titanium Slag Smelting in Circular Closed Direct Current Arc Furnace. Kunming: Kunming University of Science and Technology, PhD dissertation.
10. Jones, R. T. (2004) Economic and environmentally beneficial treatment of slags in DC arc furnaces. VII International Conference on Molten Slags, Fluxes and Salts, Johannesburg. Pp. 363-376.

Impact of Simulated Natural Reshaping of Abandoned Mining Terrain in Plateau Valleys on Plant Community Structure and Function

Jichuan Chen*

Xichang University, College of Tourism and Urban Planning, Xichang, Sichuan, China

Abstract: In this article, a technique called simulated natural reshaping was described, which is used to restore abandoned mining terrain in plateau valleys. The research results showed that research techniques were effective in improving soil quality, increasing biomass, enhancing species diversity, and strengthening ecosystem stability. Simply put, it accelerated the progress of ecological restoration by simulating the restoration process of natural ecosystems. Through this method, a significant increase in nitrogen, phosphorus, and potassium content in the soil, as well as an increase in biomass, can be observed. The Shannon diversity index increased from 1.34 to 1.9, and plant coverage recovered to 80%. Of course, this research has some limitations, such as not considering the possible impact of other environmental factors on the recovery effect. Anyway, the experimental results showed that the simulated natural reshaping technology effectively promoted the ecological restoration of abandoned mining areas, and enhanced the function and stability of the ecosystem.

1. Introduction

In China, abandoned mining areas often face serious ecological damage, which can have profound impacts on the surrounding environment and biological communities. In the current context, effective ecological restoration technologies are crucial for the sustainable management and ecological restoration of these regions. The research technology accelerates the process of ecological restoration by simulating the restoration process of natural ecosystems, and provides a potential solution. The aim of this study is to evaluate the application effect of research technology in abandoned mining areas of plateau valleys, and to provide scientific basis and guidance methods for similar ecological restoration projects.

The application effect of research technology in the ecological restoration of abandoned mine terrain is comprehensively evaluated. During the experimental phase, four experiments are conducted to evaluate the improvement of the research algorithm on the natural environment. The experimental results show that the total nitrogen, available phosphorus, and exchangeable potassium content in the soil have significantly increased, and plant coverage can also quickly recover after human interference. These achievements not only demonstrate the effectiveness of simulated natural reshaping technology, but also reveal its specific performance on different ecological indicators.

The article first introduces the research background and the theoretical basis of simulated natural reshaping technology. In the methodology stage, soil hierarchy construction, plant planting and configuration, and microbial environment creation are introduced. During the experimental phase, the experimental results are analyzed and discussed; comparative data before and after the application of the technology is presented; in-depth analysis of the data is conducted. The final conclusion summarizes the research findings and provides suggestions for the application in ecological restoration practices, while also pointing out the limitations of the research and future research directions.

2. Related Works

Previous research has mainly focused on different ecological restoration methods, such as vegetation reconstruction, soil improvement, and hydrological regulation. For example, Zhang Zeyu et al. explored the spatial configuration of vegetation communities in abandoned mining areas based on the theory of near natural restoration, starting from the relationship between site types and vegetation community characteristics. This provided a theoretical basis and reference for vegetation restoration in the ecological restoration process of abandoned mining areas [1]. Lei Shaogang proposed that ecological restoration in open-pit mining areas faces problems such as low self-sustaining capacity of the reconstructed ecosystem, high maintenance costs, and obvious mining traces after restoration. To address these issues, he emphasized the necessity of near natural

*215315268@qq.com

ecological restoration and proposed its theoretical connotation and technical framework [2]. Wenying Y U provides reference for the protection of wetland ecological environment and related planning and management policies [3]. Dong L used survey data from a typical fire that occurred on May 6, 1987 and its adjacent unburned areas to study the long-term effects of different fire intensities on the structure, biodiversity, and carbon storage of natural larch forests in the Greater Khingan Mountains. Research has shown that after nearly 30 years of natural restoration, forest diameter structure and species diversity have been significantly restored. However, forest carbon storage has not yet recovered to pre fire levels, and active human intervention is still needed to promote the process of forest stand restoration [4]. Kondratenko L was committed to restoring vegetation around the Borodinsky coal mine area, using two land reclamation areas from 2011 to spring/summer 2021 as an example. The analysis of vegetation in the region showed that the spontaneous plant diversity indicators of plot 1 and plot 2 were considered moderate in both layers [5]. In Daskalova G N's view, land abandonment and rural population reduction were accelerating globally. Given this scale, there is an urgent need to develop a vision on how to achieve a balanced benefit of biodiversity conservation, ecosystem services, and people's livelihoods on abandoned land [6]. However, although these methods have achieved local success, they still face problems such as slow recovery speed, high cost, and unstable effectiveness in special environments such as plateau valleys.

In response to the above issues, methods for simulated natural remodeling have emerged in recent years, which accelerate ecological restoration by simulating natural succession processes. For example, Xian Kanghua et al. used Illumina high-throughput sequencing technology to sequence and analyze the bacterial and fungal sequences in the rhizosphere soil of Huazhong Lou. They studied Huazhong Lou with growth years of 3, 5, 7, and 9 years, respectively, to explore the relationship between its growth and development and changes in soil microbial community structure [7]. There is a lack of research considering the impact of gravel plant communities on soil properties, and failure to evaluate whether plant cover can serve as an indicator of biotic and/or abiotic parameters for gravel colonization. To fill this research gap, Giupponi L's research aimed to determine the relationship between plant communities and soil during the colonization process of high-altitude limestone gravel [8]. The forest fungal community is formed by the interaction between the host root system and related soil conditions. Hogan J A studied how soil environment, root morphology characteristics, and root chemistry affect the root fungal communities in three tropical forest sites with different succession states in Xishuangbanna [9]. Arbuscular mycorrhizal fungi are specialized symbiotes that form mutualistic relationships with various plant species, including legumes. Gritli T conducted a study on the main soil types in the sub humid agricultural ecosystem of northern Tunisia, exploring the arbuscular mycorrhizal fungal communities associated with broad beans [10]. Although this method has shown good application prospects in some studies, its applicability and long-term effects under different terrain and climate conditions still need further verification. Therefore, this study adopts an improved simulated natural reshaping method to achieve better restoration effects in abandoned mining terrain in plateau valleys.

3. Methods

3.1 Soil Hierarchy Construction

Restoration ecology, as a branch of ecological science, has general characteristics of ecology and is closely related to the theories and methods of ecology. The concept of restoration ecology is initially proposed by Aber and Jordan from the Madison Botanical Garden at the University of Wisconsin in the United States, which is a form and method of artificial intervention in managing the environment. In 1987, Jordan defined restoration ecology as "a method for addressing ecological issues and testing ecological perspectives in the restoration and reconstruction of ecosystems." He believed that restoration ecology involves multiple disciplines, hence it is also known as integrated ecology. Aronson et al. (1993) defined restoration ecology as purposeful and conscious human modification or restoration of an ecosystem or location, in which humans mimic the dynamic changes in ecosystem structure, function, and diversity as much as possible [11-12].

Scholars, who have long been engaged in vegetation restoration practice and research in China, defined restoration ecology as a discipline that studies the causes of ecosystem degradation, technologies and methods for restoring and rebuilding degraded ecosystems, and ecological processes and mechanisms. Some scholars believed that restoration ecology is a science related to ecological restoration, aimed at studying the mechanisms and processes of ecological degradation and restoration. As a formal discipline, restoration ecology is a relatively emerging field compared to ecology, with a high degree of comprehensiveness and dual characteristics of theory and practice. Therefore, it has sparked debates in this field about theory, definition, standards, and principles. From different perspectives, people have different understandings, and the definitions of restoration ecology and ecological restoration are also different.

Meanwhile, in order to predict and evaluate the comprehensive impact of different environmental factors on ecological restoration, this study will establish a multiple regression model. This model will analyze the impact of variables such as soil nutrients, climate conditions, and plant species on the effectiveness of ecosystem restoration. The model expression is shown in formula (1):

$$Y = \beta_0 + \beta_1 X_1 + \cdots + \beta_n X_n + \epsilon \tag{1}$$

In formula (1), Y represents the effectiveness indicators of ecological restoration, $X_1,\ldots,X_n$ represents different environmental factors.

3.2 Plant Planting and Configuration

In the ecological restoration of abandoned mines, plant planting and configuration design aims to restore the structure and function of the ecosystem by selecting and arranging plant communities. Before starting planting in the mining area, the native plant species in the area are investigated. Then, species such as the white wax tree, black locust, sumac, and honeysuckle are discovered [13].

Based on the survey data, a range of suitable local trees, shrubs, and herbaceous plants are selected. For example, white wax trees are planted in areas with deeper soil to utilize their deep roots to improve soil structure and retain moisture, while herbaceous plants such as honeysuckle are used to quickly cover shallow soil and reduce soil erosion. The planting density of trees is set at 300 plants per hectare, while the planting density of shrubs and herbaceous plants is adjusted according to soil characteristics.

At the same time, intercropping planting methods are adopted to create complementarity between different plants, maximizing the overall stability of the ecosystem and the mutual benefit between species. Throughout the entire planting and configuration process, the growth stages of plants and the succession dynamics of ecosystems are closely monitored, and the performance of vegetation growth and ecological functions is regularly monitored. After more than a year of monitoring, it is found that the soil structure and water retention capacity in the restored area have improved. The survival rate of white wax trees and black locusts exceeds 80%, and the ecosystem has shown high self-sustaining ability.Formula (2) can be used to represent the increase in biomass of plants after planting:

$$B_{new} = B_{old} + r \cdot A \quad (2)$$

In formula (2), B_{new} represents the biomass after planting, B_{old} is the biomass before planting, r is the growth rate of biomass per unit area, and A represents the area of vegetation planting.

This strategy not only improves the environmental quality of abandoned mines, but also enhances the ecological resilience and resistance to external disturbances in the region by enhancing biodiversity [14].

3.3 Microenvironment Creation

Microenvironment refers to an ecological environment with specific micro terrain, soil, ground cover, and competition characteristics, and a small spatial range. At a small scale, the high variability of forest microbiota may form an asymmetric pattern of niche and competition, which leads to differentiation in the growth and distribution of regenerated seedlings. For example, microenvironmental factors such as light, soil, and litter mainly affect natural regeneration by regulating the different responses of seed germination, seedling establishment, and young tree growth to microenvironmental heterogeneity. Therefore, exploring the microbial limitations of natural regeneration in artificial forests is beneficial for improving the survival opportunities of seedlings and young trees.

The interactions between species play an important role in the natural succession of forest communities, especially neighboring competition, which is regarded as the main driving factor for community structure and dynamics. Competition drives the process of self thinning, affecting the distribution of tree size until large trees die or forest fires interfere to form gaps, promoting the renewal and growth of other species. Research has shown that there is a significant negative density dependence during the seedling stage of trees. The higher the density of adjacent plants of the same species, the lower the survival rate of seedlings, especially in the microenvironment close to the mother tree. This is because the limited resource utilization of seedlings in limited space may lead to changes in the survival and distribution of seedlings due to changes in the microenvironment.

4. Results and Discussion

4.1 Assessment of Species Diversity and Richness

In the study of plant community restoration in abandoned mines, assessing species diversity and richness is key to understanding the progress of ecological restoration. This experiment evaluates the impact of simulated natural reshaping technology on the plant community structure of abandoned mining terrain in plateau valleys. The Shannon diversity index is used in the experiment, as shown in Formula (3):

$$H' = -\sum_{i=1}^{S} P_i ln(P_i) \quad (3)$$

Among them, in Formula (3), H' represents Shannon diversity index; P_i is the relative abundance of the i-th species; S is the total number of observed species.

Figure 1 (a) shows species richness, and Figure 1 (b) shows Shannon diversity index. From Figure 1, it can be seen that the average species richness of the sample plot before restoration is 6.8, while it increases to 10.3 after restoration. In terms of Shannon diversity index indicators, the Shannon diversity index increases from an average of 1.34 before recovery to an average of 1.9 after recovery. From the data conclusion, it can be seen that the plant community in the abandoned mining terrain of the plateau valley has significantly improved in terms of species diversity and ecological functions.

 https://doi.org/10.1051/e3sconf/202456001015

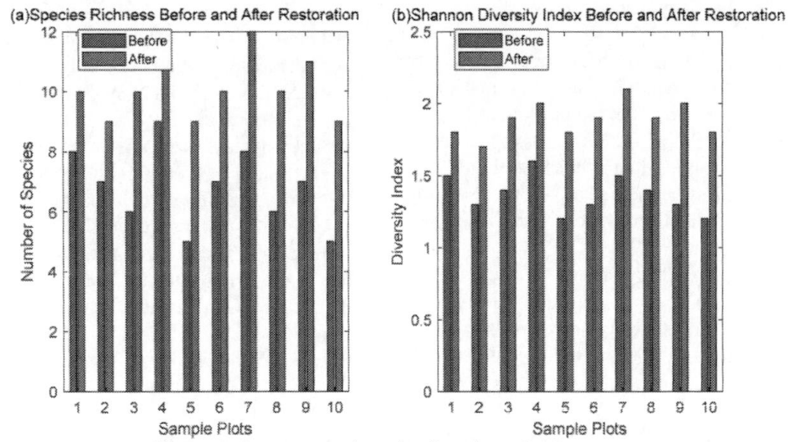

Figure 1: Assessment of species diversity and richness

4.2 Biomass Measurement Experiment

In the biomass measurement experiment, biomass data of plant communities in abandoned mining areas are collected after simulating natural remodeling. The experiment is set up in 10 fixed 1 square meter plots, and the dry weight of plants before and after recovery is measured. The data before and after recovery is plotted as a line graph. The formula for calculating biomass can be represented by Formula (4):

$$B = \frac{W_d}{A} \quad (4)$$

In Formula (4), B represents the biomass per unit area; W_d is the weight of the dried plant; A is the sampling area. The specific data situation is shown in Figure 2:

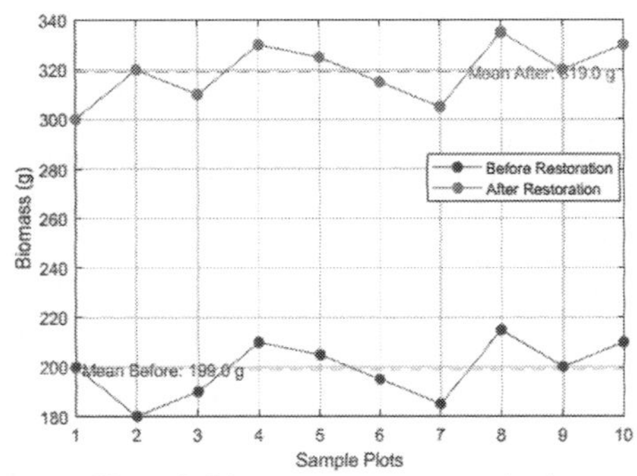

Figure 2: Biomass measurement evaluation

From Figure 2, it can be seen that the average biomass of the plant community before restoration is 199 grams, and after restoration, the average biomass increases to 319 grams. From the data conclusion, it can be seen that this significant increase in biomass indicates that ecological restoration measures have greatly improved the productivity and function of the ecosystem.

4.3 Soil Quality and Nutrient Cycle Assessment

In the experiment of evaluating soil quality and nutrient cycling, the impact of simulated natural remodeling on soil quality in abandoned mining areas is analyzed. Ten randomly selected sampling points are set up, and soil samples are collected before and after restoration to determine the pH value, total nitrogen content, available phosphorus content, and exchangeable potassium content of the soil. The nutrient content in the soil can be expressed by Formula (5):

$$C_x = \frac{m_x}{m_{soil}} \times 100\% \quad (5)$$

In Formula (5), C_x represents the concentration of nutrient x; m_x is the quality of nutrients; m_{soil} is the total mass of soil samples collected.

Figure 3 (a) shows the soil pH values before and after restoration; Figure 3 (b) shows the total nitrogen before and after recovery; Figure 3 (c) shows the available phosphorus before and after recovery; Figure 3 (d) shows the exchangeable potassium before and after recovery. From Figure 3, it can be seen that the average pH value of the soil increases from 5.63 before restoration to 6.65 after restoration; the total nitrogen content increases from 0.19% to 0.32%; the available phosphorus content increases from 0.06% to 0.16%; the exchangeable potassium content increases from 1.2% to 1.65%. From the data conclusion, it can be seen that simulated natural remodeling effectively promotes the richness of soil nutrients and the improvement of soil chemical properties, providing a good foundation for plant growth and ecosystem restoration. The specific data situation is shown in Figure 3:

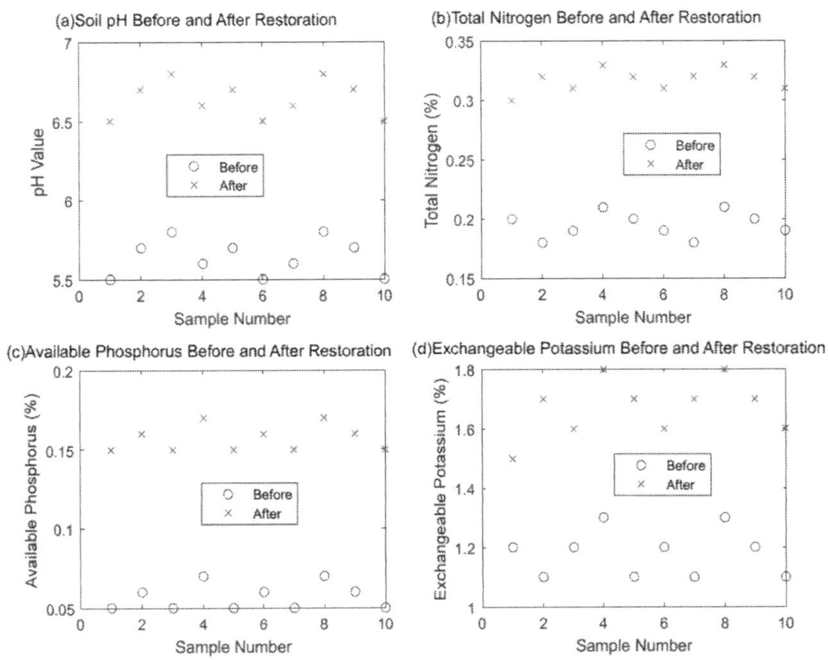

Figure 3: Soil quality and nutrient cycling assessment

4.4 Ecological Stability Experiment

In order to gain a deeper understanding of the ecological stability of simulated natural reshaping technology on a single plot, an ecological stability experiment is designed. This experiment helps to evaluate in detail the impact of ecological restoration techniques on plant resilience under specific environmental conditions. Among them, the recovery of biological coverage can be expressed by Formula (6) :

$$R(t) = R_0 \times (1 - e^{-kt}) \quad (6)$$

In Formula (6), $R(t)$ represents the plant coverage after time t month; R_0 is the coverage before interference; k is the recovery rate constant; t is the time after interference.

The specific data details are shown in Table 1:

Table 1: Ecological stability experiment

Months_After_Disturbance	Plant_Coverage
1	50%
3	65%
6	80%

From the data in Table 1, it can be seen that after experiencing interference, the plant coverage of the selected sample plot significantly recovers, increasing from 50% one month after interference to 65% three months later, and further increasing to 80% six months later. From the data changes, it can be seen that the sample plots processed by simulated natural reshaping technology have strong ecological restoration ability and stability, and can effectively restore plant community structure and function after human interference.

4.5 Time series analysis of biodiversity

In this experiment, we tracked and studied the biodiversity changes in abandoned mining areas after simulated natural reshaping over a period of one year. We measure the Shannon Diversity Index and Simpson Diversity Index monthly to evaluate changes in the structure of biological communities. The experiment mainly focuses on plants, insects, and small animals, and collects data through direct observation and sample survey methods. As shown in Figure 4:

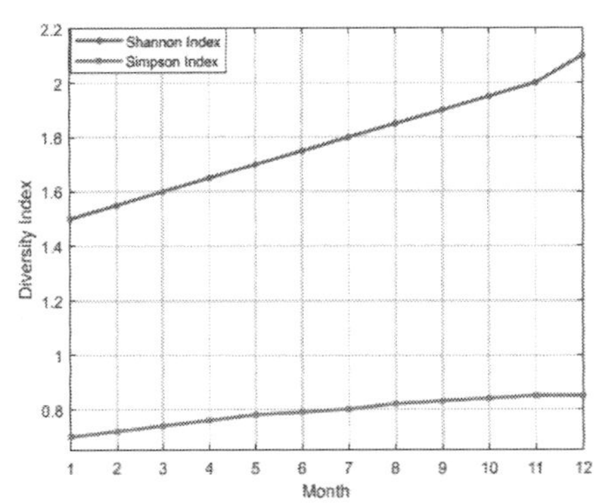

Figure 4: Assessment of changes in biological community structure

In Figure 4, the Shannon diversity index increases from 1.5 to 2.1, while the Simpson diversity index increases from 0.7 to 0.85. This indicates that after the application of simulated natural reshaping technology, biodiversity has significantly improved. Specifically, the significant increase in biomass and species richness indicates that the structure and function of the ecosystem have been effectively restored and enhanced. These data strongly support the application value of simulating natural reshaping technology in ecological restoration projects, demonstrating its significant effects in improving biodiversity and ecosystem stability.

5. Conclusion

This study comprehensively evaluated the ecological restoration of abandoned mining terrain in plateau valleys using simulated natural reshaping techniques. The results of the experimental stage showed that this technology significantly improved soil quality, increased biomass and species diversity, and enhanced ecosystem stability. Specifically, the nitrogen, phosphorus, and potassium contents in the soil have significantly increased, and biomass has also increased. Plant communities can quickly recover after being disturbed by human activities. However, there are still some limitations to the research, such as not considering the potential impact of other environmental variables on the recovery effect. Future research can further explore the effects of simulating natural remodeling under different climate and soil conditions, in order to optimize technological applications and improve the widespread applicability and persistence of ecological restoration. In addition, the combination of long-term ecological monitoring and biodiversity conservation measures is also an important direction for future research to ensure the sustainability of ecological restoration and achieve long-term ecosystem service returns.

ACKNOWLEDGMENTS

This paper is financially supported by the Engineering Research Center of Chuanxibei RHS construction at Mianyang Teachers'College of Sichuan Province.

References

1. Zhang Zeyu, Wu Xiaojing, Liang Yipeng, et al. The Near Natural Vegetation Spatial Configuration Model for Ecological Restoration of Abandoned Mines in Wula Mountain [J]. Research on Arid Areas, 2023, 40 (7): 1164-1171.
2. Lei Shaogang, Xia Jianan, Bian Zhengfu, et al. On the near natural ecological restoration of open-pit mining areas [J]. Journal of Coal Science, 2024, 49 (4): 1-10.
3. Wenying Y U , Ruipeng J I , Han X ,et al.Evaluation of the Biodiversity Conservation Function in Liaohe Delta Wetland,Northeastern China[J].Journal of Meteorology: English Edition, 2020, 34(4):8-13.
4. Dong L , Liu Z .Assessing different fire intensity on biodiversity and carbon stocks of natural Larix gmelinii forest in Great Xing'an Mountains[J].Journal of Northeast Forestry University, 2020, 48(9):45-50.
5. Kondratenko L, Gura D, Shaidullina V, et al. Restoration of vegetation around mining enterprises[J]. Saudi Journal of Biological Sciences, 2022, 29(3): 1881-1886.
6. Daskalova G N, Kamp J. Abandoning land transforms biodiversity[J]. Science, 2023, 380(6645): 581-583.
7. Xian Kanghua, Su Jiang, Fu Chuanming, et al. A study on the microbial diversity of rhizosphere soil in Huazhong Lou with different growth years [J]. Guangxi Plants, 2022, 42(12):2087-2098.
8. Giupponi L, Leoni V, Pedrali D, et al. Plant cover is related to vegetation and soil features in limestone screes colonization: A case study in the Italian Alps[J]. Plant and Soil, 2023, 483(1): 495-513.
9. Hogan J A, Jusino M A, Smith M E, et al. Root-associated fungal communities are influenced more by soils than by plant-host root traits in a Chinese tropical forest[J]. New Phytologist, 2023, 238(5): 1849-1864.
10. Gritli T, Ellouze W, Wu H, et al. Molecular characterization of arbuscular mycorrhizal communities associated with Lathyrus cicera L. grown in northern Tunisia soils[J]. Symbiosis, 2023, 90(1): 81-90.
11. Bandyopadhyay S. Plant-assisted metal remediation in mine-degraded land: a scientometric review[J]. International Journal of Environmental Science and Technology, 2022, 19(8): 8085-8112.
12. Gao Z, Liu Z, Wu Z, et al. Soil nematode communities and functions: Ecological package restoration responses in damaged desert steppe open-pit mining ecosystems[J]. Land Degradation & Development, 2022, 33(7): 1015-1028.
13. Kumari S, Maiti S K. Nitrogen recovery in reclaimed mine soil under different amendment practices in tandem with legume and non-legume revegetation: A review[J]. Soil Use and Management, 2022, 38(2): 1113-1145.
14. Alekseenko A V, Drebenstedt C, Bech J. Assessment and abatement of the eco-risk caused by mine spoils in the dry subtropical climate[J]. Environmental Geochemistry and Health, 2022, 44(5): 1581-1603.

Research and application of low oil-water ratio organoclay-free oil-based drilling fluid technology

He Shi[1,*], Xiaoxiao Ni[1], Xiaofei Wang[2], Zheng Yang[1], Jianhua Wang[1], Rongchao Cheng[1], Lili Yan[1], Jiaqi Zhang[1], Haijun Yang[1], Xiaobo Cui[1], Lining Zhang[1]

[1]CNPC Engeering Technology R&D Company Limited, Beijing 102206, China
[2]DRILLING FLUID COMPANY OF XDEC,Karamay 834099,China

Abstract. This paper introduces the formation and application of a low oil-water ratio organoclay-free oil-based drilling fluid. Shale oil and gas are important alternative energy sources. Due to the extremely complex geological conditions, oil-based drilling fluid technology is commonly used to reduce complexity. Although some achievements have been made, there are still problems such as unstable wellbore, difficult rheological control, and high cost. A stable and low-viscosity water-in-oil emulsion system was successfully prepared by using optimized fatty acid imidazolines emulsifier and alkyl alcohol ether wetting agent, which enabled the oil-based drilling fluid to have a larger solid and water phase capacity. On this basis, a special poly-fatty acid amide rheological modifier was used to assist in the regulation of rheological parameters, resulting in a low-viscosity and high-shear structure of the oil-based drilling fluid. This effectively ensured the efficiency of cuttings carrying while reducing the adverse effects of viscosity on Equivalent Circulating Density(ECD). The low oil-water ratio oil-based drilling fluid technology has been applied in a shale gas drilling platform in southern Sichuan, China, which greatly reduces the cost of oil-based drilling fluid while ensuring drilling efficiency.

1 Introduction

Oil-based drilling fluids constitute a category of shear-thinning fluids that are based on a water-in-oil emulsion[1-4]. These fluids are meticulously formulated by dispersing a range of core components, including surfactants, the aqueous phase, alkalinity regulators, oil-wet colloids, and weighting agents, within organic solvents such as diesel, white oil, synthetic oil, or ethers, which serve as the continuous phase. When compared to water-based drilling fluids, oil-based drilling fluids exhibit several notable advantages[5-9]. Firstly, they possess exceptional shale inhibition properties, which contribute significantly to maintaining wellbore stability. Secondly, their superior lubricity enhances drilling efficiency by facilitating faster penetration rates. Thirdly, they demonstrate excellent temperature resistance, making them particularly suitable for drilling ultra-deep wells[10]. Additionally, these fluids are highly resistant to contamination from sources such as cuttings and saltwater, simplifying performance maintenance. Furthermore, their recyclability allows for multiple uses, thereby reducing overall operational costs. As a result, oil-based drilling fluids find widespread applications as completion fluids, freeing fluids, workover fluids, and coring fluids, among others.

However, it is important to acknowledge the disadvantages associated with oil-based drilling fluids as well. The primary concern arises from the use of oil as the continuous phase, which elevates preparation costs. Additionally, their environmental impact is significant, raising concerns about environmental friendliness. These fluids can also interfere with geological logging activities, and the removal of filter cakes can pose challenges, potentially compromising the quality of cementing operations. Nonetheless, ongoing research and development efforts aim to mitigate these limitations and enhance the overall performance and sustainability of oil-based drilling fluids in the drilling industry.

During the exploration and development of shale oil and gas, water-based drilling fluids exhibit deficiencies in shale inhibition, temperature resistance, and resistance to acidic gas contamination, making them inadequate for safe and efficient drilling in shale formations. Conversely, oil-based drilling fluids have become the primary technical means for shale oil and gas development due to their inherent performance advantages. Despite the gradual maturation of high-performance water-based drilling fluid technology in recent years, oil-based drilling fluids still occupy a significant market share. The unit cost of oil-based drilling fluids is high, with the cost of base oil accounting for more than 70%. Conventional oil-based drilling fluid systems for shale oil and gas use organic soil as a viscosity-increasing and shear-enhancing material, which has a significant viscosity effect and high equivalent circulating density. In order to maintain reasonable rheological properties and reduce the risk of

* Corresponding author: 365441231@qq.com

lost circulation, it is necessary to maintain a high oil-water ratio, especially under high-density conditions. The oil-water ratio is often as high as 90:10, resulting in extreme cost pressure, which seriously affects the economic benefits of shale oil and gas development and has become an urgent engineering problem to be solved.

To address the aforementioned issues, this work developed an organoclay-free oil-based drilling fluids with a low oil-water ratio and conducted experimental applications in shale gas wells in southern Sichuan. By carefully selecting emulsifier and wetting agent combinations, a low oil-water ratio water-in-oil emulsion with low basic viscosity but excellent emulsification stability was formed. To ensure sufficient cutting-carrying capacity of the drilling fluid, a polyfatty acid amide rheological modifier was used instead of organo-clay for rheological control, resulting in excellent rheological parameters with low viscosity and high yield point. This system exhibited long-term high-temperature stability at an oil-water ratio as low as 60:40 and had a solid capacity not lower than that of conventional oil-based drilling fluids. Its performance remained reasonable even at a density of 2.3 g/cm^3, meeting the application requirements of drilling sites. Field application results also confirmed the superiority of this technology, which will be further promoted in the future.

2 Formation of the Drilling Fluid

2.1. Study on basic emulsion

To achieve a high-performance oil-based drilling fluid with a low oil-to-water ratio, it is necessary for the base emulsion to maintain emulsification stability comparable to that of conventional oil-based drilling fluids even at low oil-to-water ratios. Additionally, the viscosity of the base emulsion must be lower to ensure controllable rheological parameters, such as when the density is high or when there is an invasion of inferior solid phases. To achieve these two goals, a reasonable combination of emulsifiers and wetting agents is crucial, as they play a fundamental role in determining the emulsification stability and viscosity of the base emulsion. Based on numerous laboratory experiments, a mixture of imidazolines formed by the reaction of oleic acid with diethylenetriamine and hydroxyethylethylenediamine in appropriate proportions was selected as the emulsifier (named DR-EML), and alkyl alcohol ether was used as the wetting agent named DR-WEL, yielding ideal results. The emulsification stability of water-in-oil emulsions with low oil-to-water ratios based on DR-EML and DR-WEL was compared to that of conventional oil-based drilling fluid base emulsions through emulsification rate experiments.

0# diesel is used as the external phase of the emulsion. Emulsion1 has an oil-to-water ratio of 6:4, with DR-EML and DR-WEL added at 3wt% and 1wt% of the total oil-water volume, respectively. emulsion2 has an oil-to-water ratio of 8:2 and is formulated using a commonly used emulsifier combination in drilling engineering, consisting of a primary emulsifier and a secondary emulsifier, each added at 2wt% of the total oil-water volume. emulsion3 also has an oil-to-water ratio of 6:4 but uses the same type and amount of emulsifiers as emulsion2. To achieve better emulsification, 5wt% lime is added to each emulsion. After preparing the emulsions, they are aged by rolling at 150℃ for 24 hours and then left to stand at room temperature. The emulsification rate curves are obtained by comparing the ratio of emulsion volume to total oil-water volume at different times, as shown in Fig 1.

The emulsification rate of emulsion1 is higher than that of the emulsion2 in the early stage, but there is a reversal after 20 hours of standing. However, after 50 hours of standing, the emulsification rate of emulsion1 is still as high as 82%, which is not significantly different from the 87% emulsification rate of emulsion2. Therefore, it can be concluded that emulsion1 and emulsion2 have comparable emulsification stability, despite their significant difference in oil-to-water ratio. In comparison, the emulsification rate of emulsion3 is significantly lower than the other two, with only 54% after 50 hours of standing, indicating that the traditional emulsifier combination does not perform well in stabilizing emulsions with low oil-to-water ratios.

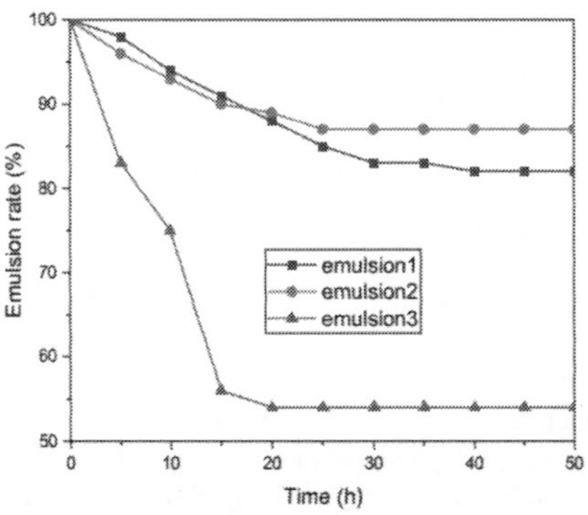

Fig. 1. Comparison of emulsification rates

The viscosity/shear rate curves under 25℃ of the above three emulsions were also compared, as shown in Fig 2. All three emulsions exhibit shear thinning behavior. The viscosity of emulsion1 is close to that of emulsion2 at 1000 1/s, but it is higher than emulsion2 in the low shear rate region. This characteristic of emulsion1 is beneficial for drilling fluid. Lower viscosity at high shear rates is conducive to raise rate of penetration（ROP）, while higher viscosity at low shear rates helps to achieve suspended cuttings and wellbore cleaning. Emulsion3 exhibits much higher viscosity than the first two in all shear rate regions, which can cause problems for drilling fluid performance control. Once formation water or solids intrude, the rheology of the drilling fluid system will quickly get out of control and become difficult to flow.

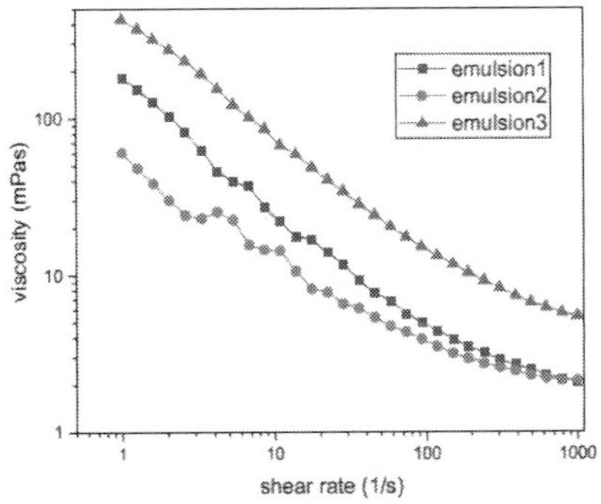

Fig. 2. Viscosity curve of emulsions

Based on the above experimental results, the emulsifier and wetting agent selected in this study demonstrated better practicality at low oil-water ratios. The emulsion with low oil-water ratio prepared by them exhibited stability and rheological properties similar to those of traditional oil-based drilling fluid emulsions, ensuring that lower oil-water ratio did not significantly affect the performance of the drilling fluid. Subsequently, an oil-based drilling fluid system will be constructed based on the low oil-water ratio emulsion stabilized by DR-EML and DR-WEL.

2.2 Study on rheology

Organoclay has long been the most commonly used rheological modifier for oil-based drilling fluids, but its usually not perfect. The common organoclay is an organophilic clay material made by cation exchange modification of montmorillonite with quaternary ammonium salts. In oil-based drilling fluids, organoclay can only influence the rheology of the emulsion by adsorbing at the oil-water interface. This leads to insufficient shear force enhancement when the organoclay dosage is small, and severe viscosity increase when the dosage is large. It is difficult to obtain the ideal flow pattern with low viscosity and high shear force, which can easily result in higher ECD and lower solid capacity, posing significant challenges to the performance maintenance of drilling fluids. The gel structure formed by organoclay recovers slowly after being destroyed, which is not conducive to improving the ROP. In addition, organoclay has the problem of slow activation. As the temperature increases, its viscosity-increasing effect gradually strengthens, easily causing high-temperature thickening of the drilling fluid. In order to avoid the adverse effects of organoclay, some special rheolocial modifiers, such as fatty acid amide polymers, are used to replace organoclay, thus the technology of clay-free oil-based drilling fluid emerges. These clay-free oil-based drilling fluids have a low-viscosity and high-shear weak gel structure, which is beneficial to the formation of low ECD while ensuring the cleanliness of the wellbore. At the same time, it enables the drilling fluid to have a larger viscosity adjustment space, which allows the drilling fluid to withstand more solid and aqueous phase pollution.

In order to maintain reasonable rheological properties of drilling fluid under high water phase content, this study also chose to construct a clay-free oil-based drilling fluid system. A polyfatty acid amide rheological modifier named DR-RM was selected, and its effect was compared with that of organoclay. DR-RM is a low molecular weight polymer produced by the reaction of dimer fatty acid, diethanolamine, and diethylenetriamine, with a number average molecular weight of 19272 Da, a weight average molecular weight of 40113 Da, a peak molecular weight of 36971 Da, and a Z average molecular weight of 66867 Da, and its molecular structural formula is shown in Fig 3.

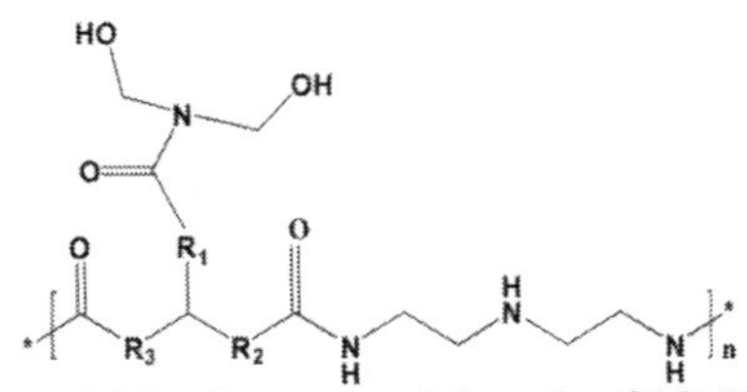

Fig. 3. Molecular structural formula of DR-RM

DR-RM has oil-philic fatty acid molecular chains as well as hydrophilic amide and alcohol amine hydroxyl groups, which means it has amphiphilic properties, and thus can adsorb on the oil-water interface just like surfactants. In a water-in-oil emulsion, DR-RM's amphiphilic properties and complex branched chain structure enable it to adsorb on the interface membranes of multiple emulsion droplets, thereby forming a network structure between the emulsion droplets and improving the gel structure strength of the water-in-oil emulsion. The principle is shown in Fig 4.

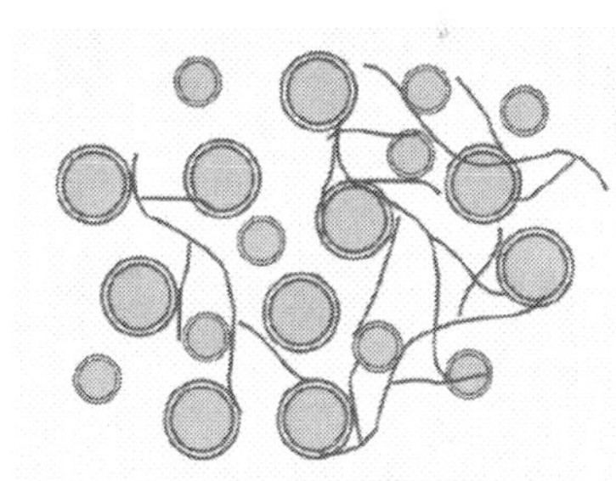

Fig. 4. The mechanism of DR-RM in water-in-oil emulsion

The selected organoclay was a commonly used commercial product on the drilling site, coded HFGEL130d. The basic formula for the evaluation experiment is shown in the Table 1.

Table 1. basic formula of drilling fluids

Component	Function	Dosage
0# diesel fuel	continuous phase	180mL
25wt% $CaCl_2$ brine	dispersed phase	120mL
DR-EML	emulsion	3wt%

DR-WEL	Wetting agent	1 wt%
CaO	alkalinity regulator	5 wt%
1000 mesh calcium carbonate	Sealing agent	5 wt%
Barite	weighting agent	350g

In the basic formula, the rheological modifier and organoclay were added to form drilling fluids. Each prepared sample underwent rolling aging at 150° C for 16 hours, and then the rheological properties were tested using a six-speed rotational viscometer at 65° C. The results are shown in Fig 5. Experimental results demonstrate that the addition of 0.5wt% DR-RM to drilling fluid achieves a comparable yield point (YP) to that obtained with 3% HFGEL130d, while maintaining a significantly lower plastic viscosity（PV）. This leads to a notably higher YP/PV ratio than the drilling fluid sample with organoclay. Increasing the concentration of DR-RM to 1wt% further elevates the yield point of the drilling fluid with minimal impact on plastic viscosity, resulting in an even higher YP/PV ratio. Compared to organophilic clay, DR-RM forms a weak gel structure that is easily broken and quickly repaired. This is beneficial for maintaining low viscosity during circulation to reduce circulating pressure loss and rapidly generating structural force to suspend cuttings and keep the wellbore clean in a static state. Such effects are highly advantageous for improving drilling efficiency. Thus, DR-RM is identified as a key additive for regulating the rheology of low oil-water ratio and soil-free oil-based drilling fluid systems.

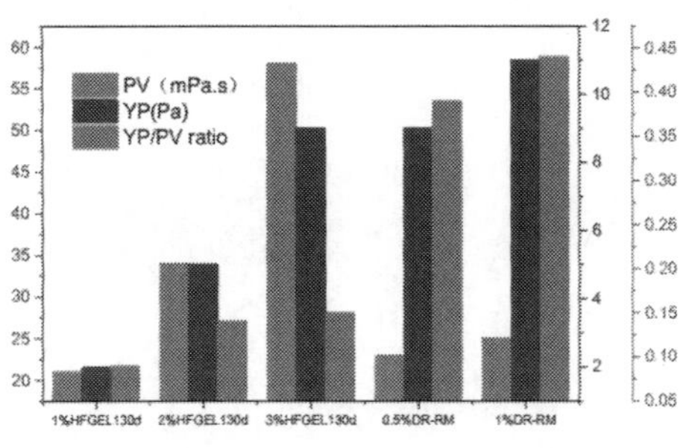

Fig. 5. The influence of DR-RM and HFGEL130d on the rheology of drilling fluid

2.3 Laboratory evaluation of the low oil-water ratio clay-free oil-based drilling fluid

Based on the experimental results, a formula for low oil-water ratio and clay-free oil-based drilling fluid was developed, as shown in Table 2.

Table 2. Formula of low oil-water ratio clay-free oil-based drilling fluid

Component	Function	Dosage
0# diesel fuel	continuous phase	180mL
25wt% $CaCl_2$ brine	dispersed phase	120mL
DR-EML	emulsion	3wt%
DR-WEL	Wetting agent	1 wt%
CaO	alkalinity regulator	5 wt%
1000 mesh calcium carbonate	Sealing agent	5 wt%
DR-RM	Rheological modifier	0.1~1wt%
Barite	weighting agent	Depends on the required density

Low oil-water ratio clay-free oil-based drilling fluids with different densities were prepared and evaluated after rolling aging at 150°C for 16 hours. The results are shown in Table 3. Within the density range of 1.5~2.3 g/cm^3, the low oil-water ratio drilling fluid system exhibits acceptable emulsion stability (Es ≥ 400V). To maintain reasonable rheological parameters for wellbore cleaning (YP between 6 and 15, Φ6 between 7 and 15), it is necessary to flexibly adjust the amount of DR-RM added based on changes in solid content in the system. At lower densities (1.5~1.7 g/cm^3), the DR-RM addition is controlled between 0.5 ~1 wt%. At higher densities (1.7~2.0 g/cm^3), the DR-RM addition is controlled between 0.3~0.5 wt%. When the density exceeds 2.0 g/cm^3, the DR-RM addition is controlled between 0.1~0.2wt%. Experimental data confirms that this regulation method is very effective.

Table 3. Performance of drilling fluids with different densities

ρ g/cm^3	$FL_{150℃}$ mL	AV/PV mPa.s	YP Pa	YP/PV	Φ6	Gel10'/10" Pa	ES V
1.5	2.4	31.5/25	6.5	0.26	7	3/4	401
1.6	2.8	33/27	6	0.22	7	3/4	414
1.7	1.8	39.5/32	7.5	0.32	10	5/6	440
1.8	2.2	58/44	14	0.32	12	6/8	556
2.0	2.2	64/50	14	0.28	11	5/7	685
2.3	1.8	74.5/62	12.5	0.2	10	5/6	868

Drilling fluid with a density of 2.0 g/cm3 was selected, and rolling aging experiments were conducted at 150°C for different durations to evaluate its long-term stability at high temperatures. The results are shown in Table 4. It can be observed that during the prolonged aging process, the rheological parameters of the drilling fluid began to decline after maintaining stability for a certain period. However, even after 96 hours, the drilling fluid still exhibited excellent performance, with an ES consistently above 500V. Additionally, despite the extended aging period, the drilling fluid sample demonstrated a minimal high-temperature high-pressure filtrate loss of just 2.2mL, along with the formation of a thin mud cake, as shown in Fig 6. This filtration performance represents an ideal state for drilling fluids. The experimental results confirm the desired long-term stability of the studied low oil-water ratio clay-free oil-based drilling fluid system at high temperatures. In practical drilling applications, such extreme conditions are rarely encountered, and timely supplementation of key additives into the drilling fluid on a daily basis will further enhance its performance stability.

Table 4. Performance of drilling fluid after long-term aging

Aging time/h	$FL_{150℃}$ mL	AV/PV mPa.s	YP Pa	YP/PV	Φ6	Gel10'/10" Pa	ES V
16	2.2	64/50	14	0.28	11	5/7	685
32		60/46	14	0.31	10	5/6	705
48		49/38	11	0.27	8	4/5	744
64		48/39	9	0.22	8	4/5	668
80		42/34	8	0.22	6	3/5	563
96		38/31	7	0.23	6	3/4	580

Fig 6. High-temperature and high-pressure filtrate loss and mud cake of drilling fluid after rolling aging for 96 hours at 150°C

Based on the characteristics of shale oil and gas wells, drilling fluid with a density of 1.8 was selected for further solid phase pollution experiments, The solid phase used is powdered drill cuttings taken from the shale oil and gas well site. The results are shown in Table 5. In this experiment, solid phase pollution did not show a negative impact on the ES, which showed an upward trend. But it had a significant effect on rheological parameters. When solid phase pollution is within 5wt%, the overall change in PV and YP is relatively small, and the YP/PV ratio remains stable. When solid phase pollution reaches 10%, the PV of the drilling fluid increases significantly, and the YP decreases, resulting in a decrease in the YP/PV ratio of the drilling fluid and the loss of the original low-viscosity, high-shear gel structure. This indicates that during application, the control of solid phase content is a key issue that needs to be focused on for low oil-water ratio clay-free oil-based drilling fluids, and strict control is required.

Table 5. Evaluation of solid phase pollution in drilling fluid

Solid pollusion wt%	$FL_{150℃}$ mL	AV/PV mPa.s	YP Pa	YP/PV	Φ6	Gel10'/10" Pa	ES V
0	2.2	58/44	14	0.32	12	6/8	556
3	1.8	61/45	15	0.33	12	6/8	573
5	1.6	77/58	19	0.32	15	7/9	632
10	1.6	91.5/83	8.5	0.10	21	11/14	740

3 Field experiment

The studied low oil-water ratio clay-free oil-based drilling fluid technology has been applied in the Longmaxi Formation of a four-open well section on a shale gas platform in the Sichuan-Chongqing region of China. The platform has a well depth of about 6600m and a horizontal section length of about 2000m. The drilling fluid density ranged from 1.6 to 2.35g/cm^3 and the oil-water ratio ranges from 70:30 to 60:40.The performance was stable and controllable throughout (Table 6). The average hole diameter expansion rate was 5.8%, and there were no falling blocks during the drilling process. As the result, the drilling cycle is shortened by more than 20 days compared with adjacent wells.Electrical measurements and casing installation were smooth, and there were zero complicated accidents caused by the drilling fluid. In addition, the unique weak gel structure of this technology achieves a balance between low ECD control, wellbore cleaning, and sedimentation stability. During the construction of nearly 2000 meters of horizontal section, the drilling fluid system maintained low-viscosity, high-shear rheological properties, and the ECD remained stable at a low value, effectively reducing downhole complexities and maintenance frequency. The average consumption of oil-based drilling fluid per meter was 0.052 cubic meters, which was 58% less than other platforms.

In addition, compared with the conventional oil-based drilling fluid system of adjacent wells, the unit cost of the low oil-water ratio clay-free oil-based drilling fluid is significantly reduced. When using 0# diesel to prepare the drilling fluid, the cost comparison per cubic meter of the two drilling fluids is shown in Fig 7. Under low-density conditions, the cost comparison between the two is most significant, and the low oil-water ratio oil-based drilling fluid can reduce costs by nearly 20%. As the density of the drilling fluid increases, the unit cost of conventional drilling fluid decreases due to the increased proportion of solids such as barite, and the gap with the unit cost of low oil-water ratio oil-based drilling fluid narrows. However, even when the density reaches as high as 2.3g/cm^3, the low oil-water ratio oil-based drilling fluid can still reduce costs by nearly 15%. The effect of low oil-water ratio oil-based drilling fluid technology in saving drilling fluid costs has reached the expected level.

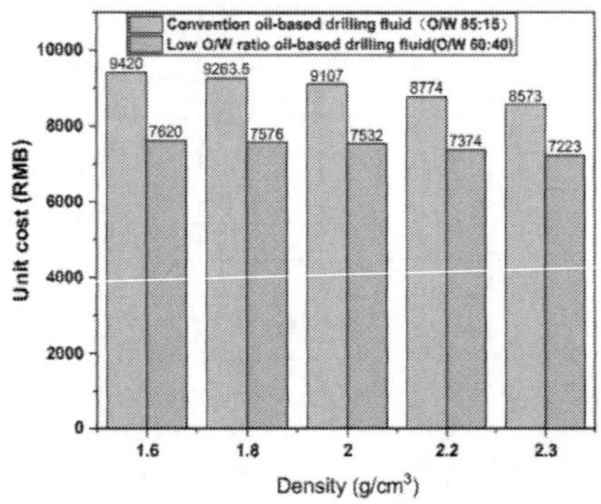

Fig 7. Comparison of unit drilling fluid costs at different densities

Table 6. Drilling fluid properties on site

ρ g/cm^3	Aging condition	AV mPa.s	PV mPa.s	YP Pa	Φ6	Gel10'/10" Pa	ES /V	FL150 /mL
1.6	Before	42	30	12	13	6/8	441	2.0
	150°C16h	43	33	10	11	5/6	454	
	150°C72h	41	31	10	10	4/5	436	
1.8	Before	51	39	12	12	6/8	444	2.2
	150°C16h	42.5	29	13.5	13	6/7	483	
	150°C72h	47	37	10	10	4.5/6	470	
2.1	Before	57	50	7	6	3/4	564	1.6
	150°C16h	61	53	8	7	3/5	558	
	150°C72h	59	51	8	7	3/5	571	
2.25	Before	73	66	7	8	3/6.5	672	1.5
	150°C16h	80	71	9	8	3/6.5	711	
	150°C72h	80	71	9	8	3/6	693	
2.35	Before	99	88	11	11	5/8	869	2.2
	150°C16h	117.5	103	14	17	9/15	887	
	150°C72h	114.5	100	14.5	15	8/12	901	

4 Conclusion

By studying the special combination of emulsifiers and wetting agents, a stable and low-viscosity water-in-oil emulsion under low oil-water ratio conditions has been successfully formed. Furthermore, by using a special polyfatty acid amide rheological modifier to regulate the rheology of the base emulsion, a weak gel structure with low viscosity and high shear characteristics has been successfully formed, which not only ensures wellbore cleaning but also improves contamination resistance and maintains a faster ROP. This technology has been successfully applied in shale gas wells in Sichuan and Chongqing, demonstrating its significant advantages in increasing speed, improving efficiency, and reducing costs. It is expected to be further promoted and applied in the future.

Acknowledgements

This work was financially supported by the scientific research and technology development project of CNPC (Grants 2020E-2803(JT)).

References

1. Amani M, Al-Jubouri M, Shadravan A. (2012) Comparative study of using oil-based mud versus water-based mud in HPHT fields. Advances in Petroleum Exploration and Development., 4: 18–27.
2. Mikhienkova E I, Lysakov S V, Neverov A L, et al. Experimental study on the influence of nanoparticles on oil-based drilling fluid properties[J]. Journal of

Petroleum Science and Engineering, 2022, 208: 109452.

3. Bui B, Saasen A, Maxey J, et al. Viscoelastic properties of oil-based drilling fluids[J]. Annu. Trans. Nord. Rheol. Soc, 2012, 20: 33-47.
4. Ghavami M, Hasanzadeh B, Zhao Q, et al. Experimental study on microstructure and rheological behavior of organobentonite/oil-based drilling fluid[J]. Journal of Molecular Liquids, 2018, 263: 147-157.
5. Oltedal V M, Werner B, Lund B, et al. Rheological properties of oil based drilling fluids and base oils[C]//International Conference on Offshore Mechanics and Arctic Engineering. American Society of Mechanical Engineers, 2015, 56581: V010T11A042.
6. Jiancheng L I, Jian G, Xiaojun W, et al. Research and application of oil-based drilling fluid technology in Block Su 53[J]. Petroleum Drilling Techniques, 2014, 42(5): 62-67.
7. Zhuang G, Zhang Z, Jaber M. Organoclays used as colloidal and rheological additives in oil-based drilling fluids: An overview[J]. Applied Clay Science, 2019, 177: 63-81.
8. Hermoso J, Martinez-Boza F, Gallegos C. Influence of viscosity modifier nature and concentration on the viscous flow behaviour of oil-based drilling fluids at high pressure[J]. Applied Clay Science, 2014, 87: 14-21.
9. Saasen A. Sag of weight materials in oil based drilling fluids[C]//IADC/SPE Asia Pacific Drilling Technology Conference and Exhibition?. SPE, 2002: SPE-77190-MS.
10. Humood M, Ghamary M H, Lan P, et al. Influence of additives on the friction and wear reduction of oil-based drilling fluid[J]. Wear, 2019, 422: 151-160.

Comprehensive Evaluation of Coal Mine Production Safety Based on Fuzzy Comprehensive Evaluation

Shuyu Shang * and Lijie Chen

(Institute of Safety Engineering, Shenyang Aerospace University,Shenyang 110136, China)

Abstract. In view of the potential safety hazards existing in the production and operation process of coal mining enterprises, through expert consultation and comprehensive investigation, sort out the relevant risk factors, establish the corresponding coal mine production risk evaluation index system, conduct fuzzy comprehensive evaluation by combining AHP and fuzzy mathematical evaluation method, and calculate the weight of each index and the risk level of coal mine production, Finally, through a case study of a coal mine in Inner Mongolia Autonomous Region, the feasibility of the risk evaluation method is verified, which provides a theoretical basis for coal mine production enterprises to formulate risk management schemes and a reference for risk management of coal mine production industry.

1 Research background

China is rich in coal resources. According to the relevant data of the National Bureau of Statistics, China National Coal Association and other official websites, China's total raw coal production in 2021 was 4.071 billion tons, accounting for 67.6% of the total primary energy production. In addition, by the end of 2021, China's coal reserves accounted for 15% of the world's total reserves, making it the world's largest coal producer. At present our country more than 95% of the coal resources rely on well work industry, and downhole operation conditions, in the production process often with gas leakage, water, transportation accident risk, coal mine accidents every year for the national economy caused more than 5 billion yuan losses, at the same time can cause employee casualties even production, the loss is numbered[1]. It can be seen that in this context, China's coal mining enterprises not only require them to adhere to the production policy of "safety first, production second" to ensure production, but also require the insurance industry to do a good job in risk assessment and insurance plan to prevent the occurrence of coal mine production accidents and reduce or transfer losses after accidents[2].

This study analyzes the risk factors existing in the production and operation of coal mine, and the corresponding risk evaluation index system is established. Finally, the risk evaluation of AHP and fuzzy mathematical evaluation methods is comprehensively used to determine the safety level of coal mine production, which provides some reference for the healthy development of coal mine production industry.

2 Coal mine risk analysis

The production environment of coal mine is complex, facing multiple risks and challenges, especially in the ventilation, safety monitoring, electrical system and other key links, there are potential threats[3]. Improper design and management of ventilation system may lead to the accumulation of gas and dust, resulting in fire or explosion[4]; if the safety monitoring system is poorly equipped or poorly maintained, it will not effectively prevent accidents; in the electrical system, if the power supply system, lightning protection lighting, cable use, electrical protection and grounding facilities do not meet the specifications, it may cause electrical accidents and seriously threaten the safety of miners.

In addition to the above technical risks, coal mine production also faces management risks. Failure to establish and improve the "six systems" of coal mine safety or not put into use will weaken the disaster relief capacity of the mine and increase the accident loss[5]. The imperfect and ineffective implementation of the safety management system will lead to the chaos of the safety management and affect the safety production. Insufficient investment in production safety, such as the lack of labor protection goods, relief materials and mechanical equipment update, will directly threaten the safety of employees[6].

In addition, the lack of safety education and training will reduce the safety awareness and skills of employees, and increase the risk of accidents. The lag of rescue team construction and the lack of professional equipment training will affect the coal mine's self-rescue and external rescue ability in emergency situations. The lack, imperfection or irregular exercise of the emergency plan will seriously affect the efficiency and effect of the

*(Corresponding author email:1022724514@qq.com)

emergency rescue and further aggravate the consequences of the accident.According to the comprehensive investigation and results of the coal mine production process and expert consultation, this study conducts risk analysis on the coal mine production industry from five aspects: power supply system, ventilation system, exploratory drainage system, roof monitoring and safety management system, and establishes a risk evaluation index system, as shown in Table 1.

Table 1. Risk evaluation index system

Coal mine production risk U	Power supply system riskU_1	grounding system U_{11}
		Overvoltage device U_{12}
		Overcurrent deviceU_{13}
		Loss of explosive deviceU_{14}
	Ventilation system riskU_2	Gas detectionU_{21}
		blowing rate U_{22}
		wind speed U_{23}
		Ventilation powerU_{24}
	Drainage exploration system riskU_3	surface water U_{31}
		confined water U_{32}
		Underground seawater layerU_{33}
		Water volume in the goafU_{34}
	Roof plate monitoring system riskU_4	lithology U_{41}
		rock pressureU_{42}
		geological structure U_{43}
		Mining field dynamic pressureU_{44}
	Safety management riskU_5	Security systemU_{51}
		Safety education and trainingU_{52}
		contingency plan U_{53}
		Safe operation managementU_{54}
		administration of the prevention and control U_{55}

3 Fuzzy comprehensive evaluation

The fuzzy comprehensive evaluation method is used to evaluate the production risk of coal mine, and AHP and fuzzy mathematics are combined with qualitative and quantitative methods to calculate the index weight and risk level, which can make the evaluation results more systematic and clear[7].

3.1 AHP method

Using AHP to determine the weight of each indicator, the steps are as follows[8]:

1) Construct the judgment matrix The nine-level scale method is used to construct the judgment matrix U.The scales and their meanings of the nine level scaling method are shown in Table 2.

Table 2. Each scale of the ninth-level scaling method and its meaning

scale	meaning
1	u_iCompared with u_j, $u_i$has the same importance as u_j
3	u_iCompared with u_j,u_i is slightly more important than u_j
5	u_iCompared with u_j, u_i is more important than u_j
7	u_iCompared with u_j, u_i is more important than u_j
9	u_iCompared with u_j, u_i is extremely important than u_j
2、4、6、8	u_iCompared with u_j,Its significance is between 1,3,5,7,9, respectively

2) Calculate the index weight. The judgment matrix U is calculated by the root method, and the formula (1) finds the eigenvector $M = (M_1, M_2, \cdots, M_n)^T$.

$$M_i = \left(\prod_{j=1}^{n} u_{ij}\right)^{1/n} \quad i = 1,2,\cdots,n \tag{1}$$

M was normalized to obtain the weights $W = (W_1, W_2, \cdots, W_n)^T$.

3) Calculate the maximum feature root. The maximum eigenvalue λ_{max} of the evaluation matrix U is obtained according to formula (2).

$$\lambda_{max} = \frac{1}{n}\sum_{i=1}^{n}\frac{(UW)_i}{W_i} \tag{2}$$

4) Conformance test Check the consistency of the judgment matrix U according to formula (3). If CR <0.1, indicate that the judgment matrix U meets the consistency requirement; if CR 0.1, the judgment matrix U must be corrected until the consistency is met.

$$CR = \frac{CI}{RI} \tag{3}$$

Where CI is obtained by formula (4), and the value of RI is checked according to Table 3.

$$CI = \frac{\lambda_{max} - n}{n-1} \tag{4}$$

Table 3. The verage random consistency index RI values of the judgment matrix.

The order of the matrix	1	2	3	4	5	6	7	8
RI price	0	0	0.52	0.89	1.12	1.26	1.36	1.41

3.2 Fuzzy mathematical evaluation method

Using the fuzzy mathematical evaluation method to determine the production risk level of coal mine, the steps are as follows[9]:

1)Establish the evaluation index set

$U = \{U_1, U_2, \cdots, U_n\}$ Standard layer index set, $U_i = \{U_{i1}, U_{i2}, \cdots, U_{in}\}$Secondary index set.

2)Establish a review set

The evaluation of coal mine production safety evaluation index system is divided into five grades,Establish a review set of

$V = \{V_1, V_2, V_3, V_4, V_5\} =$ {No risk, mild risk, moderate risk, severe risk, high − risk }[10].

3)Establish a fuzzy judgment matrix

Determine the fuzzy evaluation vector of each indicator according to formula (5).

$$R_{ij} = \frac{x_{ij}}{N} \quad (5)$$

N is the number of experts; x_{ij}Is the frequency of the indicator $U_i$being rated asV_j.

Construct a fuzzy judgment matrix： $R = \begin{bmatrix} R_1 \\ R_2 \\ \vdots \\ R_n \end{bmatrix} = \begin{bmatrix} R_{11} & R_{12} & R_{13} & R_{14} & R_{15} \\ R_{21} & R_{22} & R_{23} & R_{24} & R_{25} \\ \vdots & \vdots & \vdots & \vdots & \vdots \\ R_{n1} & R_{n2} & R_{n3} & R_{n4} & R_{n5} \end{bmatrix}$

4)Comprehensive evaluation

The fuzzy evaluation vector $B = \{b_1, b_2, \cdots, b_n\}$ is calculated according to formula (6).

$$B = (W_i R)^T \quad (6)$$

If $\sum b_i \neq 1$, normalize it and finally build a comprehensive evaluation matrix$B = [B_1, B_2, \cdots, B_n]^T$.

5)Determine the risk level

The comprehensive evaluation vector $E = WB = \{e_1, e_2, \cdots, e_n\}$is calculated in combination with the index weight of the criterion layer.

Classified the coal mine production safety risks into five grades, as shown in Table 4,

Table 4. Coal mine production risk grade evaluation table.

risk grade	I level	II level	III level	IV level	V level
degree of risk	A little bit of risk	Average risk	average risk	High risk	Extremely dangerous
Risk rating coefficient K	5	4	3	2	1

The K value is calculated according to formula (7) to determine the production risk level of coal mine.

$$K = \frac{\sum_{i=1}^{5} K_i B_i}{\sum_{i=n}^{n} e_i} \quad (7)$$

4 Case study

A coal mine of Dongxin Company of Iron Coal Group is selected as the example study object of this paper, and the production risk of the coal mine is comprehensively evaluated according to the above evaluation method.

4.1 Determine the index weight

According to the established coal mine production risk evaluation index system, the index set of evaluation objects is determined, and the target layer is U= coal mine production risk, including 5 standard layer indicators:$U = \{U_1, U_2, U_3, U_4, U_5\}$,The corresponding secondary indicators are respectively: $U_1 = \{U_{11}, U_{12}, U_{13}, U_{14}\}, U_2 = \{U_{21}, U_{22}, U_{23}, U_{24}\}, U_3 = \{U_{31}, U_{32}, U_{33}, U_{34}\}, U_4 = \{U_{41}, U_{42}, U_{43}, U_{44}\}, U_5 = \{U_{51}, U_{52}, U_{53}, U_{54}, U_{55}\}$.

Establish judgment matrices separately. The weight judgment matrix for coal mine production risk U is shown in Table 5, the weight of power supply system indicators is shown in Table 6, the weight of ventilation system indicators is shown in Table 7, the weight of exploration and drainage system indicators is shown in Table 8, the weight of roof monitoring system indicators is shown in Table 9, and the weight of safety management system indicators is shown in Table 10.

Table 5 .Judgment matrix of coal mine production risk U weight.

Concordance ratio：0.0862<1；λ_{max}：5.3862

U	Power supply systemU_1	ventilation system U_2	Exploration and drainage systemU_3	Dome monitoring systemU_4	security management system U_5
power supply system U_1	1	1/3	1/4	1/5	3
ventilation system U_2	3	1	2	3	7
Exploration and drainage systemU_3	4	1/2	1	3	9
Dome monitoring systemU_4	5	1/3	1/3	1	7
Safety management riskU_5	1/3	1/7	1/9	1/7	1

$W = (0.2481, 0.0510, 0.0589, 0.1040, 0.5380)^T$

Table 6. Power supply systemU_1 index weight.

Consistency ratio:0.0291<1；λ_{max}：4.0776

U_1	earthing deviceU_{11}	Overvoltage deviceU_{12}	Overcurrent deviceU_{13}	Hot deviceU_{14}
earthing deviceU_{11}	1	1/7	1/5	1/3
Overvoltage	7	1	3	5

deviceU_{12}				
Overcurrent deviceU_{13}	5	1/3	1	2
Hot deviceU_{14}	3	1/5	1/2	1
$W_1 = (0.5738,0.0563,0.1310,0.2388)^T$				

Table 7. U_2index weight of ventilation system.

Consistency ratio:0.0502<1; λ_{max}: 4.1341				
U_2	Gas monitoringU_{21}	blowing rateU_{22}	Wind speedU_{23}	ventilating pressureU_{24}
Gas monitoringU_{21}	1	7	5	5
blowing rateU_{22}	1/7	1	1/3	1/3
wind speedU_{23}	1/5	3	1	2
ventilating pressureU_{24}	1/5	3	1/2	1
$W_2 = (0.0510,0.5223,0.1768,0.2498)^T$				

Table 8. The U_3 index weight of the exploratory drainage system.

Consistency ratio:0.0390<1; λ_{max}: 4.1041				
U_3	surface waterU_{31}	Presswater U_{32}	Underground seawater layerU_{33}	Water volume in the goaf U_{34}
surface water U_{31}	1	5	3	5
Presswater U_{32}	1/5	1	1/3	1/2
Underground seawater layerU_{33}	1/3	3	1	3
Water volume in the goafU_{34}	1/5	2	1/3	1
$W_3 = (0.0665,0.4600,0.1486,0.3248)^T$				

Table 9. Top board monitoring system U_4 indicator weight.

Consistency ratio:0.0742<1; λ_{max}: 4.1981				
U_4	lithology U_{41}	rock pressureU_{42}	geological formationU_{43}	Mining field dynamic pressureU_{44}
lithology U_{41}	1	5	3	5
rock pressureU_{42}	1/5	1	1/5	1/3
geological formationU_{43}	1/3	5	1	3
Mining field dynamic pressureU_{44}	1/5	3	1/3	1
$W_4 = (0.0636,0.5439,0.1219,0.2706)^T$				

Table 10. Index weight of safety management system U_5.

Consistency ratio: 0.0767<1; λ_{max}: 5.3437					
U_5	Security systemU_{51}	Safety education and trainingU_{52}	contingency plan U_{53}	Safe operation managementU_{54}	Fire management U_{55}
Security systemU_{51}	1	5	7	6	1/2
Safety education and trainingU_{52}	1/5	1	3	3	1/5
Contingency plan U_{53}	1/7	1/3	1	3	1/5
Safe operation managementU_{54}	1/6	1/3	1/3	1	1/7
Fire management U_{55}	2	5	5	7	1

$$W_5 = (0.0537,0.1666,0.2846,0.4532,0.0419)^T$$

4.2 Establish a judgment matrix

A total of 10 coal mine safety experts were invited to form an evaluation expert group to establish a one-factor evaluation matrix through the method of expert scoring. The expert rating results are shown in Table 11.

Table 11. Expert evaluation results of fuzzy comprehensive evaluation in a coal mine of Iron Coal Group

	FACTOR OF EVALUATION	THE NUMBER OF EXPERTS IN EACH EVALUATION SET				
		A little bit of risk	Mild risk	Moderate risk	High risk	High-risk risk
DEVELOPMENT,MINING SYSTEMB_1	Geological environment and mining realmC_{11}	8	1	1	0	0
	Develop a wayC_{12}	1	2	4	2	1
	Bar well spacingC_{13}	1	5	3	1	0
	Number of mining working faces C_{14}	1	2	5	2	0
	supporting measureC_{15}	0	2	2	3	3
	Mine pressure monitoringC_{16}	5	3	2	0	0
LIFT、TRANSPORT SYSTEM B_2	Vertical well ascensionC_{21}	2	3	3	2	0
	Well upper and lower inclined lane track liftingC_{22}	1	1	3	3	2
	Flat lane transport C_{23}	0	2	2	3	4
	Obque lane transport C_{24}	0	0	2	3	5
	Belt conveyor C_{25}	1	3	3	2	1
	Overhead passenger device C_{26}	1	3	2	3	1
	Large mechanical operation C_{27}	2	2	3	2	1
PREVENTION AND CONTROL OF WATER SYSTEMB_3	Hydrological conditions C_{31}	3	4	2	1	0
	Water facility C_{32}	2	3	3	2	0
	Control of water management C_{33}	2	2	3	2	1
FIRE PREVENTION SYSTEMB_4	Dial the ignition source C_{41}	1	4	3	2	0
	Coal seam spontaneous combustionC_{42}	0	1	3	3	3
	Management of flammable facilities C_{43}	4	4	2	0	0
	Fire prevention measures C_{44}	3	3	2	1	1
ELECTRICAL SYSTEMB_5	Power supply and distribution system C_{51}	3	2	2	2	1
	Lightning protection and lighting: C_{52}	4	3	2	1	0
	Downhole cable: C_{53}	0	1	4	4	2
	Electrical protection, C_{54}	3	4	2	1	0
	Downhole ground ground C_{55}	3	4	2	1	0

According to the evaluation results, the fuzzy evaluation matrix of secondary indicators is constructed as follows:

$$R_1 = \begin{bmatrix} 0 & 0.1 & 0.3 & 0.4 & 0.2 \\ 0.2 & 0.4 & 0.3 & 0.1 & 0 \\ 0.1 & 0.2 & 0.5 & 0.2 & 0 \\ 0 & 0.3 & 0.5 & 0.2 & 0 \end{bmatrix}$$

$$R_2 = \begin{bmatrix} 0.2 & 0.5 & 0.2 & 0.1 & 0 \\ 0.1 & 0.3 & 0.4 & 0.2 & 0 \\ 0 & 0.2 & 0.6 & 0.2 & 0 \\ 0 & 0.2 & 0.5 & 0.2 & 0.1 \end{bmatrix}$$

$$R_3 = \begin{bmatrix} 0.2 & 0.6 & 0.2 & 0 & 0 \\ 0 & 0.2 & 0.3 & 0.4 & 0.1 \\ 0 & 0.7 & 0.2 & 0.1 & 0 \\ 0 & 0.1 & 0.7 & 0.2 & 0 \end{bmatrix}$$

$$R_4 = \begin{bmatrix} 0.2 & 0.5 & 0.2 & 0.1 & 0 \\ 0 & 0.2 & 0.3 & 0.4 & 0.1 \\ 0.2 & 0.4 & 0.2 & 0.1 & 0.1 \\ 0.1 & 0.3 & 0.4 & 0.2 & 0 \end{bmatrix}$$

$$R_5 = \begin{bmatrix} 0.2 & 0.4 & 0.2 & 0.2 & 0 \\ 0 & 0.1 & 0.3 & 0.5 & 0.1 \\ 0 & 0 & 0.3 & 0.6 & 0.1 \\ 0 & 0 & 0.4 & 0.5 & 0.1 \\ 0.2 & 0.4 & 0.2 & 0.1 & 0.1 \end{bmatrix}$$

Use formula (6) to obtain the fuzzy evaluation vector corresponding to each criterion layer index as follows:

$$B_1 = W_1R_1 = (0.0244,0.1778,0.3739,0.3091,0.1148)^T$$
$$B_2 = W_2R_2 = (0.0625,0.2675,0.4501,0.1949,0.0250)^T$$
$$B_3 = W_3R_3 = (0.0133,0.2684,0.4084,0.2639,0.0460)^T$$
$$B_4 = W_4R_4 = (0.0642,0.2705,0.3085,0.2902,0.0666)^T$$
$$B_5 = W_5R_5 = (0.0191,0.0549,0.3357,0.4957,0.0946)^T$$

The comprehensive evaluation matrix of coal mine production safety risk level is constructed as follows:

$$B = \begin{bmatrix} B_1 \\ B_2 \\ B_3 \\ B_4 \\ B_5 \end{bmatrix} = \begin{bmatrix} 0.0244 & 0.1778 & 0.3739 & 0.3091 & 0.1148 \\ 0.0625 & 0.2675 & 0.4501 & 0.1949 & 0.0250 \\ 0.0133 & 0.2684 & 0.4084 & 0.2639 & 0.0460 \\ 0.0642 & 0.2705 & 0.3085 & 0.2902 & 0.0666 \\ 0.0191 & 0.0549 & 0.3357 & 0.4957 & 0.0946 \end{bmatrix}$$

4.3 Determine the risk level

The calculated comprehensive evaluation vector E is as follows:

$$E = WB = \begin{bmatrix} 0.2481 \\ 0.0510 \\ 0.0589 \\ 0.1040 \\ 0.5380 \end{bmatrix} \begin{bmatrix} 0.0244 & 0.1778 & 0.3739 & 0.3091 & 0.1148 \\ 0.0625 & 0.2675 & 0.4501 & 0.1949 & 0.0250 \\ 0.0133 & 0.2684 & 0.4084 & 0.2639 & 0.0460 \\ 0.0642 & 0.2705 & 0.3085 & 0.2902 & 0.0666 \\ 0.0191 & 0.0549 & 0.3357 & 0.4957 & 0.0946 \end{bmatrix} = (0.0270, 0.1312, 0.3525, 0.3990, 0.0903)^T$$

According to formula (7), the production risk of the coal mine is calculated as follows:

$$K = \frac{5 \times 0.0270 + 4 \times 0.1312 + 3 \times 0.03525 + 2 \times 0.3990 + 1 \times 0.0903}{1} = 2.6056$$

According to Table 3, the production risk level of the coal mine is between level and level, and it is closer to level, that is, the production risk level of a coal mine in Inner Mongolia Autonomous Region is "grade". According to the field investigation, the coal mine belongs to the low gas area, with good geological structure and good exploration and drainage system. The main risks exist in the power supply system and safety management. It is verified that the fuzzy comprehensive evaluation results of this case are consistent with the recognition of the experts of the coal mine and in line with the actual situation of the coal mine.

5 Conclusion

1) Investigate and analyze the risks existing in the operation of the coal mining industry, and establish the risk evaluation index system of coal mine production, which can more clearly reflect the current factors affecting the production safety of coal mining enterprises.

2) using AHP model to calculate the weight of the risk factors in coal mine production, then by fuzzy mathematical comprehensive evaluation method finally determine the risk level, and select iron coal group east new company a coal mine example study, determine the risk level, provide advice for coal mine production safety, verify the feasibility of the evaluation method.

3) Provide reference for the coal mining industry to conduct risk evaluation, so that coal mining enterprises have a clearer understanding of their risk degree, and provide a theoretical basis for enterprises to reduce inherent risks, enhance safety management and formulate risk management countermeasures.

References

1. Chen Lijie. Risk Management Strategies for Coal Mining Enterprises Based on Safety Evaluation [J]. Coal Mine Safety, 2015, 46 (12): 236-237+241
2. Chen Lijie. Fuzzy Mathematics Method for Coal Mine Safety Evaluation [J]. Coal Mine Safety, 2005, 36 (12): 58-61
3. Kundan Kumar, Abhinav Sahay, Harshabardhan Vasistha. Assessment of Environment Management Plan in Tawa Coal Mines of Western Coal Fields Limited[J]. International Journal for Research in Applied Science & Engineering Technology (IJRASET), 2020, 8(10).
4. N. Sahay, A. Sinha, B. Haribabu. Dealing with open fire in an underground coal mine by ventilation control techniques[J]. Journal of the Southern African Institute of Mining and Metallurgy, 2014, 114(6):455.
5. Zhang Lan. Research and Application of Coal Mine Safety Risk Analysis System [J]. Energy and Environmental Protection, 2021, 43 (5): 8-12+38
6. Sun Chao. Research on Risk Management Countermeasures in Coal Chemical Industry [J]. Coal Mine Safety, 2018, 49 (8): 282-284
7. Su Asun. Regional Coal Mine Safety Risk Assessment Based on AHP and Fuzzy Mathematics [J]. Coal Technology, 2019, 38 (9): 124-127
8. Hu Xia. Coal Mine Safety Situation Evaluation Model Based on AHP and Entropy Weight Method [J]. Coal Mine Safety, 2021, 52 (2): 248-252
9. Hao Weihua. Research on Coal Mine Safety Evaluation Based on Entropy Weight Set Pair Analysis Method [J]. China Coal Geology, 2020, 32 (7): 73-78
10. Han Lujun. Analysis of Comprehensive Evaluation System for Coal Mine Safety Risks [J]. Energy and Energy Conservation, 2021 (6): 210-211

Screening and identification of petroleum hydrocarbon degrading bacteria in groundwater of contaminated site

Xiaorui Du [1], Haiming Li [2*], Xiao Zhang [1], Cuixia Zhang [2]

[1]Tianjin University of Science & Technology, College of Sciences, Tianjin 300457, China
[2]Tianjin University of Science & Technology, College of Marine and Environmental Sciences, Laboratory of Coastal Groundwater Utilization & Protection, Tianjin 300457, China

Abstract. In order to effectively repair groundwater pollution and remove petroleum hydrocarbons, two petroleum hydrocarbon degrading bacteria SYT-1 and SYT-2 were isolated from long-term petroleum hydrocarbon contaminated groundwater. The rescreening experiments showed that the degradation rates of SYT-1 and SYT-2 were 65.16 % and 52.52 %, respectively. SYT-1 and SYT-2 were identified as Moraxella and Acinetobacter. The degrading bacteria SYT-1 and SYT-2 were identified as Moxa and Acinetobacter by physiological and biochemical characteristics and 16S rDNA sequencing. The experimental results provide strain resources and technical support for bioremediation of petroleum pollution in groundwater.

1 Introduction

One of the necessary conditions for microbial remediation of petroleum contaminated sites is to screen petroleum hydrocarbon degrading bacteria with high degradation performance[1]. In the site polluted by petroleum for a long time, micro-organisms will constantly adjust with the change of external environment and the passage of time, so there are microorganisms with good tolerance and strong degradation in the environment polluted by petroleum hydrocarbons for a long time. Separating this dominant microorganism from the groundwater environment is the key step of microbial remediation. As early as 1975, Walker discovered a green alga (Prototheca zopfii) capable of degrading 10% and 40% of engine oil and crude oil under given conditions[2]. Varjani isolated a strain of Pseudomonas aeruginosa NCIM 5514 from petroleum contaminated soil. The degradation rate of petroleum hydrocarbons by this strain reached more than 60%, showing the broad prospects of degrading functional bacteria in the field of petrochemical industry[3]. Al-Hawash isolated fungal strains from the sludge of an oil field, and the removal rate of crude oil reached 60.3%[4]. Ravi optimized the growth conditions of Bacillus licheniformis ARMP2 and Pseudomonas aeruginosa ARMP8, and found that the degradation rates of these two strains were 88% and 73% within 48 hours[5]. A large number of studies show that functional bacteria obtained by separation and purification technology can effectively degrade target pollutants.

2 Materials and methods

2.1 Test materials

The groundwater samples used to enrich petroleum hydrocarbon degrading bacteria were collected from the groundwater of a petroleum contaminated site. The test petroleum sample is dehydrated pasty crude petroleum. The formulation of the medium used for the experiment is shown in Table 1. All the media should be sterilised in an autoclave after preparation and then be ready for use. The sterilisation condition is 120°C for 30 minutes.

Table 1. Medium Formula Table

Culture medium	Formula
Beef paste peptone medium (BPM)	beef paste, 3.0 g; peptone, 10.0 g ; NaCl, 5.0 g; pH 7.0~7.2.
Inorganic salt medium	K_2HPO_4, 1.5g; KH_2PO_4, 0.5g; $MgSO_4$, 0.6g ; $(NH_4)_2SO_4$, 1.6g; trace element solution, 10 ml; pH 7.0~7.2.
Trace element solution	$CaCl_2$, 3g; $FeSO_4 \cdot 7H_2O$, 0.5g; stored in refrigerator at 4 °C.
Inorganic salt medium containing petroleum hydrocarbon (liquid)	inorganic salt medium , 1 L; petroleum hydrocarbon 500 mg / L.
peptone medium (solid)	peptone medium (liquid), 1L ; agar, 19 g.

* Corresponding author: Lhm@tust.edu.cn

2.2 Experimental procedure

Take 10 ml groundwater samples poured into the petroleum-containing hydrocarbon-containing inorganic salt medium, and the culture medium after 7 days of incubation was inoculated at a rate of 5% in a new petroleum-hydrocarbon-containing inorganic salt medium. Repeat the above process 3 times, and gradually increase the concentration of petroleum hydrocarbons from 500 mg/L to 2000 mg/L. After dilution, 200 μL of the bacterial solution was poured into a plate for coating and then placed in a 30°C incubator. The growth of bacteria was observed, and the domesticated bacteria were separated by scribing.

The above-mentioned domesticated bacteria were inoculated and cultured on solid medium plates for screening of petroleum hydrocarbon bacteria. The bacteria with good growth were selected and inoculated in petroleum liquid medium for 7 days, and the concentration of petroleum hydrocarbons before and after the reaction was determined by ultraviolet spectrophotometry. The specific calculation formula is as follows:

$$\eta = \frac{C_1 - C_0}{C_0} * 100\%$$

In the formula, η represent the degradation rate of petroleum hydrocarbon, C_0 represent the concentration of petroleum hydrocarbon before the reaction (mg/L), C_1 represent the concentration of petroleum hydrocarbon after the reaction (mg/L).

2.3 Morphological observation

The strains were inoculated into the medium for 72 h, the pretreated dried samples were pasted on the sample stage, and the surface of the samples was coated with a metal film by ion sputtering coating, and the bacteria were observed by scanning electron microscope.

2.4 Determination of growth curve

Take 5 ml of culture solution containing bacteria and activate it in fresh culture medium for 12 hours. Suck 10 mL of activated culture medium with a pipette, inoculate it in 200 mL of beef paste peptone liquid culture medium and inorganic salt liquid culture medium, and cultivate it in a constant temperature water bath oscillator. The bacterial density (OD_{600}) was measured every 2 hours in the early stage, and every 4 hours after the bacteria grew stably, and the growth curve was drawn.

2.5 Identification of bacteria

The DNA was extracted from the bacteria, and when the DNA concentration met the test conditions, PCR amplification was performed with primers, and the primers in the amplification process were used with bacterial universal primers 27F and 1492R. The PCR products recovered from the cut were sent to BGI for sequencing. The obtained gene sequences were compared with the GenBank database, and the genera with high similarity were selected as model bacteria, and the phylogenetic tree was constructed by Maga 11 software to determine the species of the selected bacteria.

2.6 Effects of carbon sources on the growth of degrading bacteria and oil degradation

The seed liquid was inoculated into 20 ml inorganic salt liquid medium containing petroleum, and 5 % glucose, starch, sodium acetate and sodium carbonate were added respectively. The cells were cultured in a constant temperature water bath oscillator in the dark for 7 days, and the conditions were set as temperature of 30 °C and rotation speed of 150 r/min. The concentration of petroleum hydrocarbons was measured before and after culture. After the test, the degradation rate was calculated, and the best carbon source for degrading petroleum hydrocarbons was obtained in the set material.

3 Results and discussion

3.1 Morphological characteristics of degrading bacteria

By using enrichment culture method, the cultured bacterial culture solution was separated by plate streaking for many times, and two purified bacteria with degradation ability were successfully separated, which were named SYT-1 and SYT-2. The microscopic examination results of bacteria SYT-1 are shown in Figure 1.

The bacteria SYT-1 is short and wide, with no flagella and cilia, and its width is about 0.6~1.0 μm and its length is about 1.6~1.8 μm. The SYT-2 is rod-shaped, without flagella and cilia, with the width of about 0.2~0.6 μm and the length of about 0.6~1.4 μm.

Fig. 1. SEM images of degrading bacteria SYT-1 and SYT-2.

3.2 Physiological characteristics of degrading bacteria

SYT-1 and SYT-2 were identified as Gram-negative bacteria by physiological and biochemical tests. The specific measurement results are shown in Table 2. It can be seen from the table that SYT-1 and SYT-2 are Gram-negative bacteria. The results were negative. The results of V-P test, methyl red test and NO_2 reduction test of SYT-1 and SYT-2 were negative. The oxidase test of SYT-1 was positive. The gelatin liquefaction test of SYT-2 was weakly positive.

Table 2. Physiological and biochemical characteristics of bacteria

Physiological and biochemical identification	SYT-1	SYT-2
gram stain	-	-
v-p test	-	-
Methyl red determination	-	-
oxidase	+	-
Gelatin liquefaction	-	±
NO_2 reduction	-	-

3.3 Identification of degrading bacteria and construction of developmental tree

Partial sequences of SYT-1 and SYT-2 were obtained using 16S rDNA sequencing, and the sequences were uploaded to the NCBI database to screen the sequences of the model strains, and MEGA software was used to perform multiple sequence comparison, and truncated sequences were used to construct a phylogenetic tree using the NJ method.

The phylogenetic tree of the degrading bacterium SYT-1 is shown in Figure 2. From the figure, it can be seen that strain SYT-1 has high similarity with strains of the genus Moraxella sp. The homology between this strain and *Moraxella osloensis* (NR 104936.1) reached 99.14%. It has been documented that Moraxella sp. isolated by Vasudevan from an petroleum refinery unit can degrade petroleum as the sole carbon source[6]. In addition to this, Liu isolated a strain of Moraxella spp. from a contaminated site that used phenanthrene as the sole carbon source, and experiments have shown that the bacterium can remove up to 90% of phenanthrene under suitable conditions[7].

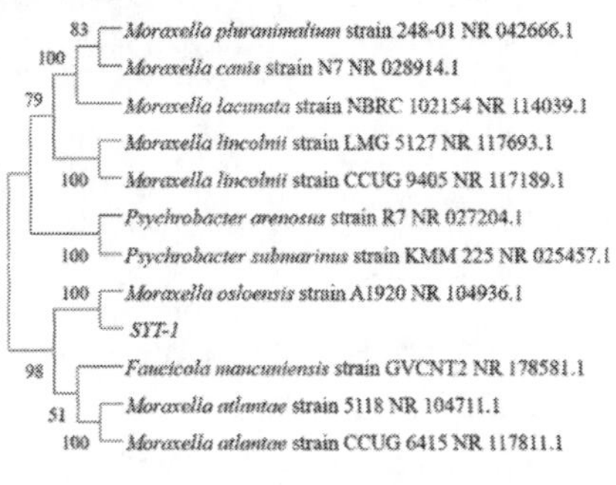

Fig. 2. Phylogenetic tree of degrading bacteria SYT-1.

The phylogenetic tree of degrading bacteria SYT-2 is shown in Figure 3. Strain SYT-2 has high similarity with Acinetobacter sp. and 98.68% homology with *Acinetobacter haemolyticus* (NR 117622.1). There have been several reported cases of Acinetobacter degrading petroleum hydrocarbons[8]. Two strains of diesel petroleum-degrading bacteria, MJ01 and MJ4, were isolated from petroleum-contaminated spetroleum in Daejeon, South Korea, and identified as Acinetobacter haemolyticus and Acinetobacter johnsonii by sequencing[9].

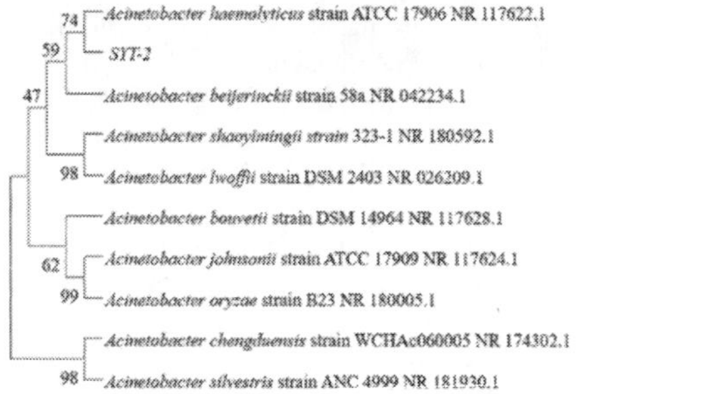

Fig. 3. Phylogenetic tree of degrading bacteria SYT-2.

Generally speaking, strains with 16S rRNA similarity less than that of ninety-seven percent will be considered as not belonging to the same species[10]. Therefore, the strains SYT-1 and SYT-2 were identified as Moraxella and Acinetobacter through the comprehensive analysis of the morphological, physiological and biochemical identification results of the strains.

3.4 Determination of growth curve of petroleum hydrocarbon degrading bacteria

The growth curve of the bacteria is shown in Figure 4. In BPM medium, the OD_{600} of SYT-1 was between 0.066~1.779, and the OD_{600} of SYT-2 was between 0.136~1.750. In the petroleum-containing inorganic salt medium, the OD_{600} of SYT-1 was between 0.045~0.246, and the OD_{600} of SYT-2 was between 0.095~0.567, and the growth of SYT-2 was better, about twice that of SYT-1. Compared with the petroleum inorganic salt medium, the two strains of bacteria proliferated faster in BPM medium.

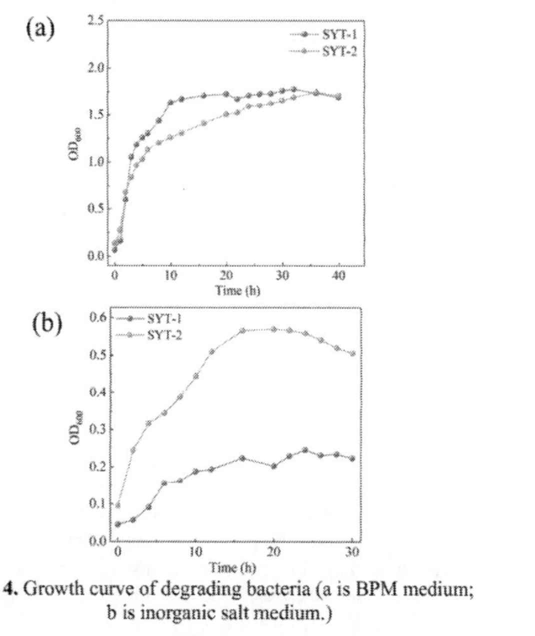

Fig. 4. Growth curve of degrading bacteria (a is BPM medium; b is inorganic salt medium.)

https://doi.org/10.1051/e3sconf/202456001018

3.5 The effect of carbon source on the growth and degradation of degrading bacteria

In the process of microbial growth and metabolism, carbon source, as an important nutrient, mainly provides the energy required for microbial life activities and constructs the carbon skeleton of cell materials. Different types of carbon sources can be decomposed by microorganisms through various pathways, and then participate in a variety of metabolic reactions to produce nutrients that can be absorbed by cells, thus affecting the growth rate and metabolic pathways of microorganisms. Therefore, in the process of carbon source selection, the characteristics of microorganisms and the conditions of culture environment should be fully considered to ensure the rationality of optimization and adjustment. In this study, glucose, starch, sodium acetate and sodium carbonate were selected as carbon sources to determine the growth and degradation performance. The results are shown in Figure 5 and Figure 6.

The strain SYT-1 grew best in sodium carbonate, and the growth density of the strain was 0.779, followed by starch, glucose, and sodium acetate. The strain had the strongest ability to degrade petroleum hydrocarbons in sodium carbonate, with a degradation rate of 65.75 %, followed by glucose and starch, and the degradation rate of petroleum hydrocarbons in sodium acetate was only 2.71 %.

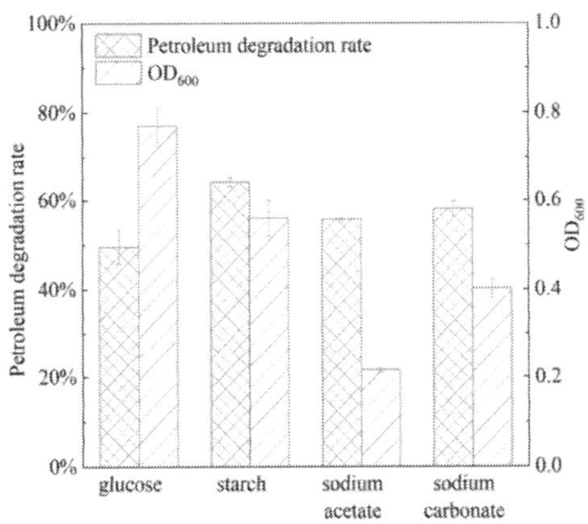

Fig. 5. Degradation and growth characteristics of SYT-1 under different carbon sources

The growth of SYT-2 strain in glucose was the best, and the OD_{600} value was 0.770, followed by starch and sodium carbonate. The growth of SYT-2 strain in sodium acetate was the slowest, and the OD_{600} value was only 0.215. Although the strain grew fastest in glucose, its ability to degrade petroleum hydrocarbons was the weakest, with a petroleum hydrocarbon degradation rate of only 49.52 %. The strongest carbon source for petroleum hydrocarbon degradation was starch, with a degradation rate of 64.27 %, followed by sodium carbonate, sodium acetate and glucose. The first three had a promoting effect on the degradation of the strain, while glucose showed a slight inhibitory effect.

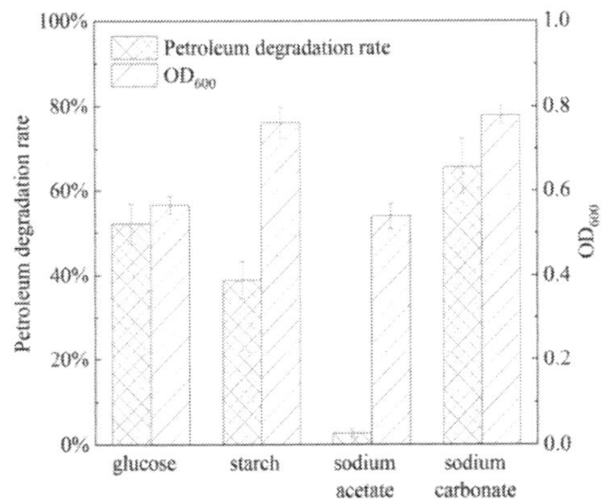

Fig. 6. Degradation and growth characteristics of SYT-2 under different carbon sources

4 Conclusions

In this study, two strains of bacteria with the ability to degrade petroleum hydrocarbons were obtained by isolating bacteria from the groundwater of an oil-hydrocarbon contaminated site, named SYT-1 and SYT-2, and identified as Moraxella sp. and Acinetobacter sp..The degradation rates of petroleum hydrocarbons by the two bacteria were 65.16% and 52.52%, respectively, which provided an experimental basis for the study of bioaugmentation of groundwater.

References

1. Liu X, He L, Zhang X, et al. Bioremediation of petroleum-contaminated saline spetroleum by Acinetobacter baumannii and Talaromyces sp. and functional potential analysis using metagenomic sequencing. Environmental Pollution, 311: 119970 (2022)
2. Walker J, Colwell R, Petrakis L. Degradation of petroleum by an alga, Prototheca zopfii. Applied Microbiology, 30(1): 79-81 (1975)
3. Varjani S J, Upasani V N. Biodegradation of petroleum hydrocarbons by oleophilic strain of Pseudomonas aeruginosa NCIM 5514. Bioresour Technol, 222: 195-201 (2016)
4. Al-Hawash A B, Zhang X, Ma F. Removal and biodegradation of different petroleum hydrocarbons using the filamentous fungus Aspergillus sp. RFC-1. Microbiologyopen, 8(1): e00619 (2019)
5. Ravi A, Ravuri M, Krishnan R, et al. Characterization of petroleum degrading bacteria and its optimization conditions on effective utilization of petroleum hydrocarbons. Microbiological Research, 265: 127184 (2022)
6. Ajona M, Vasanthi P. Bioremediation of petroleum contaminated spetroleums-A review. Materials Today: Proceedings, 45: 7117-7122 (2021)

7. Liu CY, Wang MN, Zhang JM, et al. A polycyclic aromatic hydrocarbon degrading strain and its potential of degrading phenanthrene in various enhanced systems. Chin J Biotech, 37(10): 3696-3707 (2021)
8. Kumar R, De M. Enhanced degradation of petroleum hydrocarbons by Klebsiella michiganensis RK and Acinetobacter baumannii IITG19 isolated from local spetroleum sources [J]. International Journal of Environmental Science and Technology,20(12):13387-13398 (2023)
9. Lee M, Woo S G, Ten L N. Characterization of novel diesel-degrading strains Acinetobacter haemolyticus MJ01 and Acinetobacter johnsonii MJ4 isolated from petroleum-contaminated spetroleum. World J Microbiol Biotechnol, 28(5): 2057-2067 (2012)
10. Tindall B J, Rosselló-Móra R, Busse H-J, et al. Notes on the characterization of prokaryote strains for taxonomic purposes. International journal of systematic and evolutionary microbiology, 60(1): 249-266 (2010)

The influence of N and Bi on the band structure and band offsets of InAsNBi alloys

Cheng Zhang[2,a], Ke Zhang[2,b] and Yi Duan [1,c*]

[1] Tech X Academy, Shenzhen Polytechnic University, Shenzhen, 518055, China
[2] Institute of Telecommunication and Navigation Satellites, CAST, Beijing, 100094, China

Abstract. We present the electronic properties of InAsNBi with N and Bi concentrations based on 16×16 band Hamiltonian model. This model is extended from Band Anti-crossing (BAC) and the dimension of basis states used in the Hamiltonian matrix is 16. Under the condition that the quaternary compound matches the lattice configuration of the InAs substrate, the variations of band gap (E_g), spin-orbit coupling energy (Δ_{SO}) has been examined as a dependency on the mole fractions of N and Bi. Our findings indicate the integration of N and Bi atoms in InAs gives rise to a substantial band gap energy minimization and the growth of spin-orbit coupling energy. Subsequently a crossover ($\Delta_{SO} = E_g$) is observed at a Bi concentration of 0.37%, which contributes to enhancing the effectiveness of photonic devices by suppress Auger recombination. We further reveal the band offset with different N and Bi concentrations and considered the influence of the substrate on the material deviation. The conduction band energy mislignment (ΔE_c) of InAsNBi exceeds energy misalignment in the valence bands (ΔE_v), which is required to improve thermal-insensitive characteristics of semiconductor devices. Moreover, the significant mastery over the edges of the conduction and valence bands by adjusting the concentration of N and Bi will increase the flexibility to design InAsNBi/InAs structures.

1 Introduction

In recent years, InAs has drawn widespread attention owing to its potential application of amplification of a single carrier while maintaining low noise excess [1] and dark current levels photodiodes [2,3,4]. However, its low cut-off wavelength of 3.5μm makes it unsuitable for the use as mid-infrared (IR) or even far-IR device. The bandgap reduction of III-V alloys can be realized with the integration of a slight concentration of nitrogen or bismuth within the matrix of the III-V lattice and this may be used to resolve the problem of low cut-off wavelength for photodetector. With the introduction of a trace amount of nitrogen, InAsN exhibits a significant bandgap bowing effect, which is attributed to the notable disparity in atomic sizes and electronegativity between nitrogen and As in InAsN alloy. The conduction band anti-crossing (CBAC) model can explain this process very well, where the substitution of a trace amount of N atoms will establish a localized state of N-impurities proximal to the conduction band edge of InAs and this impurity level interact with host conduction band. More recently, the content of N in InAsN can reach to 2.5% experimentally and its wavelength of emission has reach to 4.5 μm [5].

Furthermore, introducing bismuth in III-V alloys induce a significant decreasing in the bandgap and a substantial augmentation in the spin-orbit energy separation. Since the radius of N atom is smaller than that of Sb atom, Bi atom is larger than Sb atom, the corporation of N and Bi atoms in InAs lattice will give rise to strain compensation, which has been applied in experimental growing of GaAsNBi [6,7]. Liang [8] used the density-functional theory (DFT) and local-density approximation (LDA)to study the character of GaAsNBi and InGNBi respectively, it has been concluded that the compensation between the Bi and N atoms reduces the formation energy of quaternary alloy. The growth of an InAsNBi layer on an InAs substrate may provide comparable benefits to those of GaAsNBi material, thereby making it compelling prospect for the development of mid-infrared optoelectronic devices. Different epitaxial techniques [9,10] have been used to grow InAsN and InAsBi alloys successfully, but to our knowledge, InAsNBi materials system has not been grown experimentally. To examine the electronic properties of InAsNBi materials, we take a detailed theoretical investigation of the promising InAsNBi/InAs materials system with 16×16 k·p Hamiltonian model.

2 Mathematical model

The energy band configuration of bulk $InAs_{1-x-y}N_xBi_y$ have been determined through the application of a 16×16 Hamiltonian model for calculations. This 16th order matrix is derived by the (8×8) k·p matrix denoted ($H_{8\times8}$) which is used to describe the band structure of binary alloy system. $H_{8\times8}$ matrix incorporated an eight-band Kane model of the second order, founded upon the

* Corresponding author: [c]dyi0206@szpu.edu.cn
[a]zhangcheng4@126.com
[b]13439757136@139.com

Löwdin perturbation theory, to accurately capture the electronic structure of the system [11], which has been utilized for computing the electronic properties of zinc blende structures. When N, Bi atoms are incorporated, the model needs to add corrections to the diagonal elements of $H_{8\times8}$ (ΔE_{CBM}, ΔE_{VBM} and ΔE_{SO}). The modified matrix is denoted by $H_{8\times8}^{mod}$. ΔE_{CBM}, ΔE_{VBM}, and ΔE_{SO} respectively signify the linear variations in the band energy of the CBM, VBM, and spin-orbit separated band during the transition of InAsNBi from InAs to InBi. In the process of combining BAC (Band Anti-crossing) model and 8×8 matrix $H_{8\times8}^{mod}$, within the host matrix, two localized s-like states of N atoms and six localized p-like states of Bi atoms, along with their respective fundamental functions, have been incorporated. These energy levels are captured by an augmented diagonal (8×8) matrix, designated as $H_{N,Bi(8\times8)}$. $H_{N,Bi(8\times8)}$ incorporates the impurity level position of N (E_N), representing the energy levels of nitrogen impurities. It also includes the heavy hole states and light hole states of Bi (E_{Bi}), reflecting the energy states of bismuth's valence band, as well as the associated spin-orbit splitting level ($E_{Bi\text{-}so}$), which accounts for the splitting due to spin-orbit coupling. According to BAC model, the interaction between the impurity state and the intrinsic semiconductor energy state is described by matrix $V_{8\times8}=\mathrm{diag}[V_N(y),V_N(y),V_{Bi}(x),...,V_{Bi}(x)]$, where the $V_N(y)=C_N\sqrt{y}$ represents the interaction between localized N impurity states and conduction band energy levels and $V_{Bi}(x)=C_{Bi}\sqrt{x}$ represents the interplay between the impurity-induced energy levels of Bi and the valence band structure of the host substrate. $C_{N/Bi}$ is fitting parameter, indicating the strength of coupling among the host band and impurity energy levels. After these steps, the resulting 16×16 Hamiltonian can be expressed as follows

$$H_{16\times16}=\begin{pmatrix} H_{8\times8}^{mod} & V_{6\times6} \\ CC & H_{N,Bi(8\times8)} \end{pmatrix} \quad (1)$$

Since the 8×8 matrix contains conduction band, the Luttinger parameters (γ_1 , γ_2 , γ_3) used in $H_{8\times8}$ are different from what is used in 6×6 Hamiltonian. The relation between the VBAC related Luttinger parameters (γ_1^{VB} , γ_2^{VB} , γ_3^{VB}) used by Alberi et al. [12] and the adjusted Luttinger parameters (γ_1 , γ_2 , γ_3) is given as [13] $\gamma_1=\gamma_1^{VB}-\frac{E_P}{3E_g}$, $\gamma_2=\gamma_2^{VB}-\frac{E_P}{6E_g}$, $\gamma_3=\gamma_3^{VB}-\frac{E_P}{6E_g}$, P is the matrix element in Kane model and is normally written as [14] $E_P=\frac{2m_0}{\hbar^2}P^2$. Some related parameters required in our calculation have listed in Table 1. S. A. Barnett has predicted the energy gap size of the InBi alloy as -1.62 eV employing quantum dielectric theory. In the binaries composed of group III-V elements that include indium (In), since the valence band edge changes linearly with the lattice constant [15], the valence band misalignment between InAs and InBi is obtained as 1.30 eV. The energy separation due to spin-orbit coupling in InBi is 2.2 eV .

Table 1. BAC parameters used in our work.

Parameter (eV)	E_N	E_{Bi}	$E_{Bi\text{-}so}$	C_N	C_{Bi}
--	1.44	-0.4	-1.9	2.2	1.0
ΔE_{CBM}^{InAsN}	ΔE_{CBM}^{InAsBi}	ΔE_{VBM}^{InAsN}	ΔE_{VBM}^{InAsBi}	ΔE_{SO}^{InAsN}	ΔE_{SO}^{InAsBi}
-0.27	-1.03	-1.79	0.94	-1.4	0.87

The lattice incompatibility between quaternary alloy layers and the substrate material results in strain effects, which in turn leads to a separation in the energy levels in the conduction and valence bands. In order to study the effect of InAs substrate on the band offset of InAsNBi, some calculations with strain effects have also been taken into our calculation. The lattice constant of InAs1-x-yNxBiy can be ascertained through a proportional estimation technique, based on the lattice constants of the constituent end compounds. [16].

$$a_{InAs_{1-x-y}N_xBi_y}=(1-x-y)a_{InAs}+xa_{InN}+ya_{InBi} \quad (2)$$

The reported value for a_{InAs} , a_{InBi} and a_{InN} are 5.869 Å, 6.686 Å and 4.980 Å [16] respectively. According to Eq. (2), when the proportion of N and Bi impurity concentrations (y/x) equals to 0.92, $InAs_{1-x-y}N_xBi_y$ epitaxial layer can be designed to achieve lattice matching with an InAs substrate. The stress that causes the semiconductor band structure to shift was divided into two types, one is hydrostatic component of strain (ΔE^{hy}) and another is shear component of strain (ΔE^{sh}). ΔE^{hy} get conduction band shifted while both ΔE^{hy} and ΔE^{sh} influence the valence band's energy alteration. With shear component of strain, the degenerate state of the valence band peak is removed and splits into two sub-bands $E_{VB\text{-}HH}$ and $E_{VB\text{-}LH}$. For a completely strained structure, mathematical formula for band offsets caused by strain can be given as:

$$E_{CB}(\varepsilon)=E_{CB}(\varepsilon=0)+\Delta E_{CB}^{hy} \quad (3)$$

$$E_{VB\text{-}HH}(\varepsilon)=E_{VB}(\varepsilon=0)+\Delta E_{VB}^{hy}+\Delta E_{HH}^{sh} \quad (4)$$

$$E_{VB\text{-}HH}(\varepsilon)=E_{VB}(\varepsilon=0)+\Delta E_{VB}^{hy}+\Delta E_{LH}^{sh} \quad (5)$$

$$\Delta E_{HH}^{sh}=-\frac{1}{2}\Delta E_S \quad (6)$$

$$\Delta E_{LH}^{sh}=-\frac{1}{2}\Delta_0+\frac{1}{4}\Delta E_S+\frac{1}{2}\left[\Delta_0^2+\Delta_0\cdot\Delta E_S+\frac{9}{4}(\Delta E_S)^2\right]^{\frac{1}{2}} \quad (7)$$

$\Delta E_{CB}^{hy}=2a_C\left(1-\frac{C_{12}}{C_{11}}\right)\varepsilon$ and $\Delta E_{VB}^{hy}=2a_V\left(1-\frac{C_{12}}{C_{11}}\right)\varepsilon$ quantify the shifts in the conduction band and valence band induced by hydrostatic strain. $\Delta E_S=2b\left(1+\frac{C_{12}}{C_{11}}\right)\varepsilon$ is the influence of shear strain on the valence band. a_C and

a_V represent the hydrostatic strain potentials, each corresponding to the conduction band and valence band, respectively. b represents the axial strain potential. Lattice mismatch between $InAs1_{-x-y}N_xBi_y$ and InAs is denoted as ε and Δ_0 represents the energy separation due to spin-orbit interaction.

3 Result and discussion

We study the electronic band structures of InAs and $InAs_{1-x-y}N_xBi_y$ alloys utilizing the eight-band Kane model of second order and 16×16 band BAC Hamiltonian, respectively. In Fig. 1(a), E_N, E_{Bi} and $E_{Bi\text{-}so}$ indicate the energy levels pertaining to N and Bi impurities. The symbols accompanied by the indices CB, HH, LH, and SO represent the conduction band, heavy-hole band, light-hole band, and spin-orbit splitting band, respectively. $InAs_{0.863}N_{0.04}Bi_{0.097}$ is lattice-matched with InAs substrate and the strain effects can be neglected in the electronic band structure calculation. The electronic band structure of $InAs_{0.863}N_{0.04}Bi_{0.097}$ near Γ point along Δ and Λ directions which was defined in cubic Brillouin zone is shown Fig. 1(b). The solution of 16-band BAC model yields eight doubly generate eigenvalues, which are denoted by $E_{CB\pm}$, $E_{HH\pm}$, $E_{LH\pm}$ and $E_{SO\pm}$. The sign in the corner indicates two different spin directions. The band gap of $InAs_{0.863}N_{0.04}Bi_{0.097}$ is calculated as 0.22eV. Compared with the intrinsic InAs, the conduction band minimum of $InAs_{0.863}N_{0.04}Bi_{0.097}$ is reduced by 0.197eV and the valence band maximum of $InAs_{0.863}N_{0.04}Bi_{0.097}$ is increased by 0.006eV, which leads to the band reduction of 0.213eV. It is obvious that E_{CB} is sub-divided into upper (denotes E_{CB+}) and lower sub-bands (denotes E_{CB-}) after introducing N and Bi atoms, which is due to the coupling effect between E_N and conduction band. The same kind of splitting also arises in the valence and spin-orbit splitting bands due to the interplay between the valence bands and the impurity states associated with Bi. It should be noted that despite the incorporation of N, Bi causes the valence band to shift upwards but still does not eliminate the coincidence of energy levels between the heavy-hole and light-hole sub-bands of InAsNBi at k=0. The spin-orbit splitting energy (denoted Δ_{SO+}) of $InAs_{0.863}N_{0.04}Bi_{0.097}$ is defined as the difference between E_{HH+} and E_{SO+} and the value has been increased to 0.45eV.

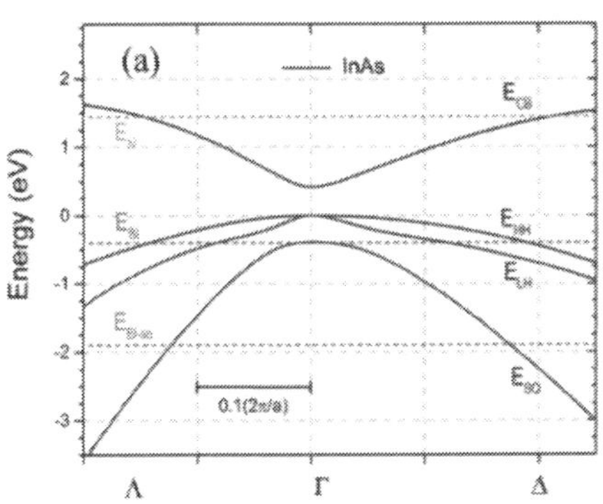

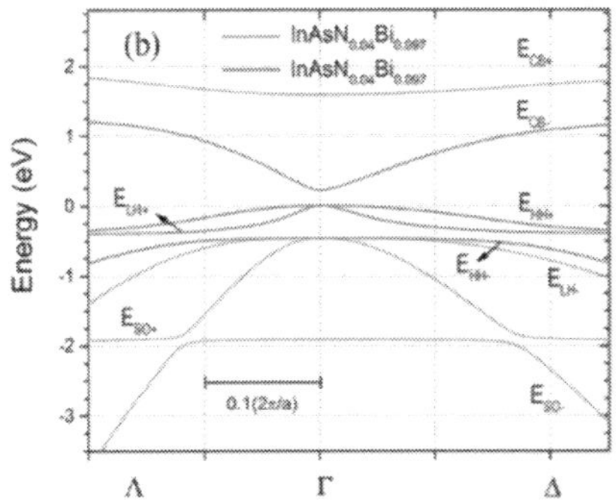

Fig. 1. (a) Electronic band of InAs, the blue solid line represents CB, HH, LH and SO band respectively. E_N and E_{Bi} and $E_{Bi\text{-}so}$ denoted by dotted lines represents the impurity levels. (b) Band structure details of $InAsN_{0.04}Bi_{0.097}$ (lattice-compatible with InAs substrate).

Fig. 2 shows the variation of relative sub-band states of the CB, HH/LH and SO band edge with Bi mole fraction at the Γ point. The values of $E_{CB\pm}$, $E_{HH/LH\pm}$ and $E_{SO\pm}$ are calculated by resolving the 16-band Hamiltonian under lattice-matched condition. It has been observed that the interplay between the native bands and the energy states imparted by N and Bi impurities manifests as an anti-crossing phenomenon, resulting in a distinct separation of the energy levels within the conduction band (CB), heavy-hole and light-hole bands (HH/LH), as well as the spin-orbit splitting band (SO). The incorporation of N and Bi in InAs results in a decrease of E_{CB-} by 35.4meV/%Bi and an increase of $E_{HH/LH-}$ by 19.9meV/%Bi, leading to the narrowing of the band gap of ~55.3 meV/%Bi. Compared with the upward shift of the valence band edge, the conduction band edge undergoes a notable decrease, accounting for approximately 64% of the overall reduction in the band gap. Regarding the overall reduction in the E_{CB-} value, the contribution from Bi 6s states are 10.3meV/%Bi. The positioning of $E_{HH/LH-}$ and E_{SO-} energy states is dictated by the presence of Bismuth-induced impurity energy spectra and it is also related to the value of V_{Bi}. Similarly, the positioning of E_{CB-} level is contingent upon the value of E_N.

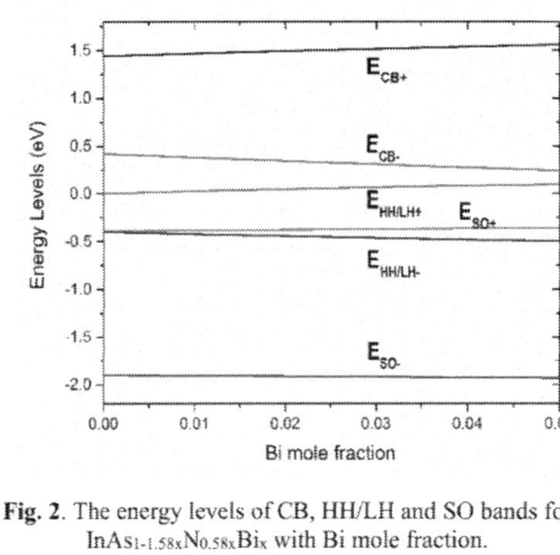

Fig. 2. The energy levels of CB, HH/LH and SO bands for $InAs_{1-1.58x}N_{0.58x}Bi_x$ with Bi mole fraction.

Variations in the energy gap and spin-orbit coupling-induced energy separation. for $InAs_{1-1.58x}N_{0.58x}Bi_x$ as a function of Bi mole fraction has been presented in Fig. 3. Two models were implemented to ascertain the energy value of the bandgap of $InAs_{1-1.58x}N_{0.58x}Bi_x$, one of which is 16-band k·p model, and the other model combines VBAC and CBAC models. The band gap in the combined VBAC and CBAC model is characterized by the energy gap from the peak of the valence band obtained through the VBAC model and the minimum value of the conduction band derived from the CBAC model. The result is expressed as:

$$E_g(x, y) = E_C(y) - E_V(y) \qquad (8)$$

$E_C(y)$ denotes the Conduction Band Minimum (CBM) energy, while $E_V(y)$ signifies the Valence Band Maximum (VBM) energy. The band gap values from two models are basically equal in the range of low Bi concentration as shown in Fig. 3, while there are deviations in the case of high concentrations of Bi and N compositions. This can be attributed to the fact that the combined VBAC and CBAC model solely takes into account the interplay between the N impurity level and the conduction band, as well as the interaction between Bi-related impurity levels and the valence band, neglecting other potential interactions. Therefore, when the Bi concentration reaches 5%, the band gap calculated using the 16-band k·p model yields a higher value of 276meV, whereas the combination of the VBAC and CBAC models results in a lower band gap value of 222meV. A crossover between Δ_{SO} and E_g is observed at a Bi concentration of 0.37% and the $\Delta_{SO} > E_g$ regime in InAsNBi can be realized with Bi concentration greater than 0.37%. This methodology exhibits the capability to alleviate the detrimental impacts of Auger recombination and Inter-Valence Band Absorption (IVBA) mechanisms, ultimately leading to an improvement in the performance of semiconductor lasers and photodetectors.

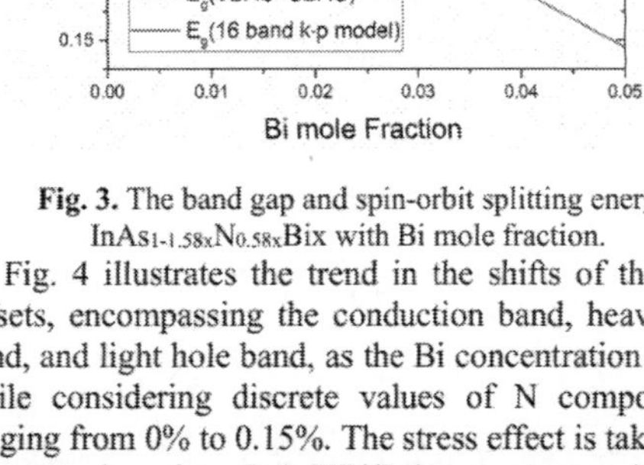

Fig. 3. The band gap and spin-orbit splitting energy of $InAs_{1-1.58x}N_{0.58x}Bix$ with Bi mole fraction.

Fig. 4 illustrates the trend in the shifts of the band offsets, encompassing the conduction band, heavy hole band, and light hole band, as the Bi concentration varies, while considering discrete values of N compositions ranging from 0% to 0.15%. The stress effect is taken into account in the InAsNBi/InAs system. As the concentrations of N and Bi rise, the energy level of the conduction band edge exhibits an upward shift, whereas the valence band edge moves in the opposite direction, downward, thereby suggesting a type-I band alignment configuration in the InAsNBi/InAs material system. Furthermore, the effect of N on the conduction band offset is more significant than The impact of Bi on the shift in the valence band energy level. At N concentration of 1%, the ΔE_C, ΔE_{HH} and ΔE_{LH} are basically equal as the Bi content up to 5%. With the concentration of Bi beyond 5%, ΔE_C exceeds the given values of ΔE_{HH} and ΔE_{LH} and their values are 0.215eV, 0.104eV, 0.153eV. As the concentration of N continues to rise, the band offset of the conduction band significantly surpasses the band offset of the valence band. Larger band offset will help to improve the thermal sensitivity of semiconductor devices through providing greater carrier confinement.

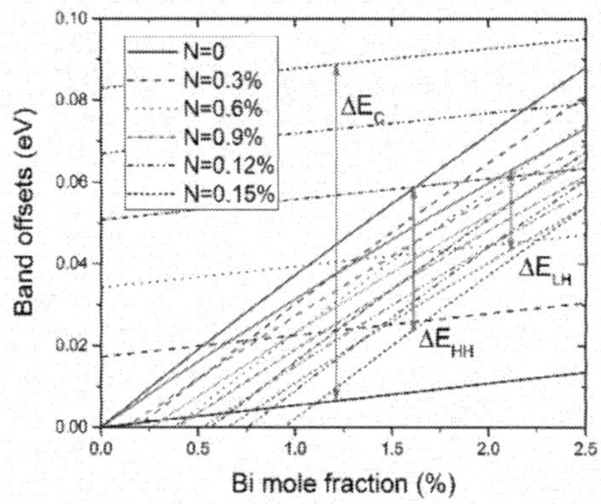

Fig. 4. Predicted band alignments of the CB (ΔE_C), HH (ΔE_{HH}), and LH (ΔE_{LH}) bands in InAsNBi/InAs junctions varying with Bi content at multiple N compositions at 10 K.

4 Conclusions

The electronic band structure of $InAs_{1-x-y}N_xBi_y$/InAs material system is measured by using a 16 × 16 Hamiltonian which combines the CABC and VBAC model. The stress effect resulting from InAs substrate is also taken into account in our work. The CBAC and VBAC effects between related impurity levels and host bands result in a substantial narrowing of the band gap of 0.213 eV for $InAsN_{0.04}Bi_{0.09}$, which has the potential to significantly extend the transition wavelengths for near-infrared optoelectronic devices. As Bi mole fraction increasing, the band gap value of InAsNBi decreases, while spin-orbit energy separation increases. A crossover ($\Delta_{SO} = E_g$) is observed at a Bi enrichment of 0.37% and the $\Delta_{SO} > E_g$ regime in InAsNBi can be realized with Bi concentration greater than 0.37%, which might help to facilitate the mitigation of Auger recombination and optical deficiencies in mid-infrared devices and LEDs. Apart from the concurrent integration of Bi and N into InAs, this methodology empowers InAsNBi with the flexibility to finely adjust the band edge positions, thereby positioning it as a versatile platform, capable of operating under tensile, compressive, or unstrained conditions, tailored to meet diverse application. When juxtaposed with traditional ternary compound semiconductors, this distinctive feature unlocks novel pathways for the development of advanced optoelectronic device designs.

References

1. Marshall A R J, Vines P, Ker P J, et al. (2011) Avalanche multiplication and excess noise in InAs electron avalanche photodiodes at 77 K[J]. IEEE Journal of Quantum Electronics, 47(6): 858-864.
2. Jain N , Mal I , Samajdar D P ,et al. (2022) Theoretical exploration of the optoelectronic properties of InAsNBi/InAs heterostructures for infrared applications: A multi-band k·p approach[J].Materials science in semiconductor processing, 2022:148.
3. Ajnef N , Habchi M M , Rebey A .(2022)The band structure calculation of tensile strained GaNAsBi/GaAs quantum well heterostructure[J].Micro and nanostructures, (Apr.):164.
4. Wang X, Cui Y, Li T, et al. (2019) Recent advances in the functional 2D photonic and optoelectronic devices[J]. Advanced Optical Materials, 7(3): 1801274.
5. De la Mare M, Zhuang Q, Krier A, et al. (2009) Growth and characterization of InAsN/GaAs dilute nitride semiconductor alloys for the midinfrared spectral range[J]. Applied Physics Letters, 95(3).
6. Ruzhevich M.(2021)Investigation of Radiation Recombination Channels in Long-Wavelength InAs/InAsSb/InAsSbP LED Heterostructures[J]. Reviews on advanced materials and technologies.DOI:10.17586/2687-0568-2021-3-4-24-28.
7. Lin S, Wang H, Wu F, et al. (2019) Room-temperature production of silver-nanofiber film for large-area, transparent and flexible surface electromagnetic interference shielding[J]. npj Flexible Electronics, 3(1): 6.
8. Liang D, Zhu P, Han L, et al. (2019) Composition dependence of structural and electronic properties of quaternary InGaNBi[J]. Nanoscale Research Letters, 14: 1-6.
9. Zhang J, Liu G, Hu H, et al. (2019) Graphene-like carbon-nitrogen materials as anode materials for Li-ion and Mg-ion batteries[J]. Applied Surface Science, 487: 1026-1032.
10. Bi K, Yang D, Chen J, et al. (2019) Experimental demonstration of ultra-large-scale terahertz all-dielectric metamaterials[J]. Photonics Research, 7(4): 457-463.
11. Voon L C L Y, Willatzen M. (2009) The kp method: electronic properties of semiconductors[M]. Springer Science & Business Media.
12. Alberi K, Wu J, Walukiewicz W, et al. (2007) Valence-band anticrossing in mismatched III-V semiconductor alloys[J]. Physical Review B, 75(4): 045203.
13. Ng S T, Fan W J, Dang Y X, et al. (2005) Comparison of electronic band structure and optical transparency conditions of In x $Ga_{1-x}As_{1-y}N_y$/GaAs quantum wells calculated by 10-band, 8-band, and 6-band k· p models[J]. Physical Review B, 72(11): 115341.
14. Bahder T B. (1990) Eight-band k· p model of strained zinc-blende crystals[J]. Physical Review B, 41(17): 11992.
15. Samajdar D P, Dhar S. (2014) Valence band structure of and alloy semiconductors calculated using valence band anticrossing model[J]. The Scientific World Journal, 2014.
16. Wu L, Wu K, Liu D, et al. (2018) Largely enhanced energy storage density of poly (vinylidene fluoride) nanocomposites based on surface hydroxylation of boron nitride nanosheets[J]. Journal of Materials Chemistry A, 6(17): 7573-7584.

Research on aviation carbon reduction using aviation biofuels

Yao Meng, Huahua Liu*, Jianqiang Xuan, Yan Han and Fangyuan Wen

Aerospace System Engineering Shanghai , 201109, Shanghai, China

Abstract. In order to achieve sustainable aviation, countries are actively implementing actions to achieve plans to reduce the emissions of greenhouse gas and develop renewable energy. With the rapid development of the aviation industry, the impact of traditional aviation fuel combustion on climate change is significant, and the airline industry is facing severe challenges in reducing greenhouse gas emissions. Aviation biofuel is one of the promising alternative aviation fuels in the future. It has attracted increasing interest as it can effectively reduce greenhouse gas emissions as a clean and sustainable energy source. Various studies have been conducted to develop and evaluate biofuels. This report introduces the history and development of biofuels, discusses the preparation process, analyzes the greenhouse gas emission reduction effects, and discusses the prospects of aviation biofuels.

1 Introduction

The greenhouse gases (GHG) emitted from the fossil fuels become the major cause of the exacerbation of the greenhouse effect as well as global climate change. The commercial aviation' emissions account for approximately 2% of the whole CO_2 emissions caused by humankind [1]. How to reduce the GHG emissions in the airline industry has become a common problem faced by airlines in various countries. The International Air Transport Association (IATA), representing the entire airlines, proposed a goal to the International Civil Aviation Organization (ICAO): carbon emissions in 2050 are 50% lower than in 2005 [2]. To achieve these goals, the industry has made tremendous efforts. Improvements in aircraft and engine technology can help reduce greenhouse gas emissions and fuel consumption. However, with the improvement of aircraft combustion efficiency and airline operational efficiency, it is still difficult to fulfill IATA's commitment to global climate change and fundamentally reduce aviation greenhouse gas emissions.

At present, global airlines, including China, are actively seeking solutions to cope with the pressure of fuel price fluctuations and aviation carbon reduction. The European Emissions Trading System (EUETS) has attracted global attention to aviation fuels' emissions and the development of aviation biofuels. With the rapid development of the airline industry, the impact of traditional airline fuel combustion on the climate change is significant, and the aviation industry is facing severe challenges in reducing greenhouse gas emissions. In recent years, the development and the application of low-carbon aviation biofuels and renewable raw materials have received widespread attention in the airline industry to achieve the goal of reducing aviation greenhouse gas emissions.

As shown in Figure 1, according to IATA's prediction, the main contributors to the reduction will be alternative aviation fuels and related technologies. Therefore, exploring clean aviation fuel has become the main direction for airlines at present.

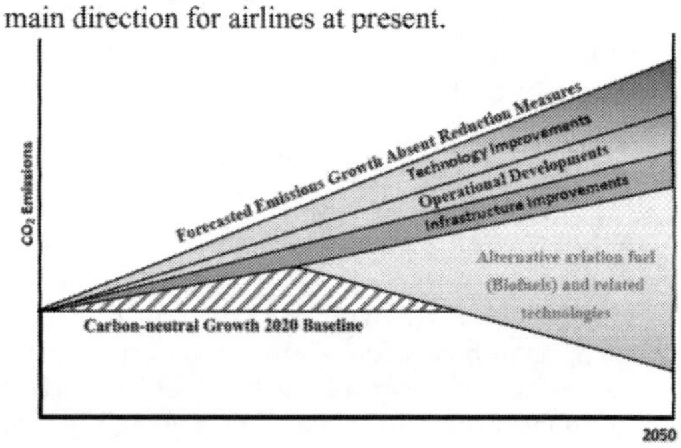

Fig. 1. Potential ways to reduce GHG emissions.

Biofuels have become a research hotspot due to their renewability, wide range of raw materials, low cost, low emissions, and the concept of circular economy. The use of biofuels not only reduces carbon emissions, but also gradually alleviates the cost pressure brought about by the rise in crude oil prices, bringing billions of business opportunities. The development of biofuels has brought new dawn to the international aviation industry in response to climate change.

2 Production and development of aviation biofuel

2.1. Alternative aviation fuel

Aviation fuel is a type of aviation fuels specifically designed for airline.Faced with the dual pressures of

* Corresponding author: m18646756369@163.com

severe greenhouse gas emission reduction challenges and resource shortages, research and exploration of bio aviation fuels are imperative. The aviation industry has identified biofuel development as one of the major means of reducing greenhouse gas emissions. Biofuels provide partial substitution for the aviation industry and bring the possibility of completely replacing carbon intensive petroleum fuels in the future [3].

2.2. Biofuel

Biofuels is a kind of fuels that extracted or composed from the biomass[4]. In most cases, biofuels refer to liquid biofuels like: biodiesel, methanol, as well as bioethanol.

Bioethanol is a fuel alcohol obtained from various biomass through microbial fermentation [5]. It can be used alone or mixed with gasoline to produce ethanol gasoline as a car fuel. Bioethanol is the most mature and widely used bioenergy to date. Bioethanol is developed from the first generation of biofuel technology based on food and edible oil. It has developed to the third generation and gradually shifted its raw materials from food to non primary foods, such as straw, animal and plant fats, seaweed, etc., making it an ideal choice for humans to prepare biofuels.

Biodiesel refers to renewable diesel fuel produced through ester exchange reactions using oils from aquatic plants such as oil crops, wild oil crops, and engineered microalgae, as well as animal fats and kitchen waste. These mixtures are mainly organic compounds with high molecular weight, almost including various oxygen-containing organic compounds such as ethers, aldehydes, ketones, phenols, organic acids, alcohols, etc. Plant oil, waste engine oil, and refinery by-products with catalysts. Biodiesel is a typical "green energy" with characteristics such as good environmental performance, good fuel performance, wide range of raw materials, and renewability. Can be used as a substitute for petrochemical diesel. The development of biodiesel has important strategic significance for sustainable economic development, promoting energy substitution, reducing environmental pressure, and controlling urban air pollution.

Because liquid biofuels are considered the best alternative or supplement to gasoline as well as petrochemical products[5]. Biofuels have the advantages of thewide range of raw materials, renewability, easy availability, low cost, low emissions, and circular economy.Compared with other forms of fuel, although still facing technological bottlenecks, the development of aviation bioenergy is still considered a necessary path to achieve carbon reduction or even zero carbon in the aviation transportation industry.

2.3. Production and development of biofuels

Since the 1970s, many countries have attached increasing importance to the development of biofuels due to limited world oil resources, rising oil prices, environmental awareness, and global climate change, and have achieved significant results. Figure 2 summarized the progress of raw materials, production methods, and fuel products in the production of aviation fuels[5-7].

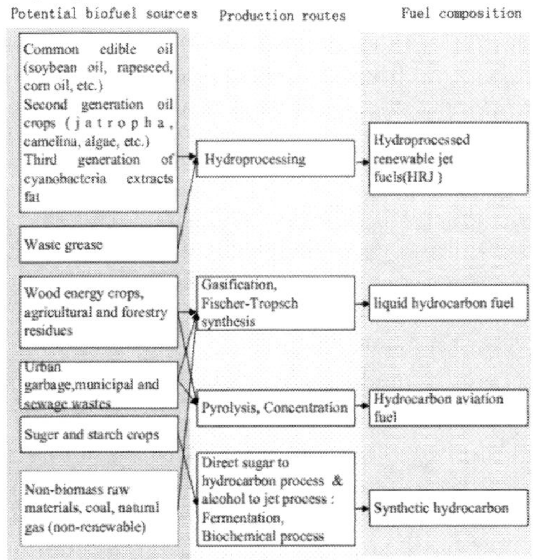

Fig. 2. Production and Development of Biofuel.

Most of the biofuels currently used in the market are secondary biofuels. They are mainly divided into three categories by generation. The first generation biofuel was extracted from crops like corn, soybeans, and sugarcane through a fermentation process. These fuels are not suitable for commercial aircraft as they don't possess the safety qualities as well as high performance of aviation fuel [8]. The second generation mainly uses alcohol made from materials containing cellulose, such as corn stems and leaves, wood, etc. The third generation uses raw materials such as leprosy trees, flaxseed, algae, and other different wastes to produce biodiesel and other fuels [9]. The second and third generation fuel can be made from the inedible resources, therefore they will not affect the food chain[9]. The main purpose of second-generation and third-generation fuels is to increase the production of biofuels, reduce dependence on fossil fuels, and not disrupt ecological balance.

There are different preparation routes. The ultimate goal is to convert raw materials into products with the same performance and structure as traditional fuels to meet the combustion conditions of existing aircraft engines. Different preparation routes can affect the performance, composition, and cost of the product.

2.4. Aviation biofuel performance analysis

Aviation alternative fuels prepared using biomaterials can effectively reduce the GHG emissions. The entire process of aviation biofuels from growth to final combustion is related to greenhouse gas emissions, and Life-cycle Analysis(LCA) analysis is usually used to fully analyze the characteristics of biofuels.

However, currently researchers are working to improve computational models, like GREETI_2014 model and GCAM analysis model[10-11], the main difference lies in the role of bio jet fuels, rather than a detailed comparative evaluation of the potential of specific technologies. Based on the analysis of the global energy system, related data from 32 different geographical locations in Europe are analyzed [12]. The tools used by Ou et al. It is the Well to Wheels (WTW) module of the Tsinghua-CA3EM model, which discusses the characteristics of biofuels in China based on its current situation. This LCA study on biofuels in China is unique in geography and differs greatly from the analysis process and conclusions of countries like the United States, the European Union, as well as Brazil [13].A research shows that biofuels can reduce LCA carbon dioxide emissions by an average of 20% [6]. The LCA greenhouse gas emissions of the biofuel can be calculated as follows:

$$E_{\text{totalGHG}} = E_{\text{iGHG}} + E_{\text{dGHG}} \quad (1)$$

Where, E_{iGHG} represents the indirect biofuel GHG emissions, E_{dGHG} represents the direct biofuel GHG emissions.

$$E_{\text{iGHG}} = \sum\nolimits_{ab} m_a C_b \quad (2)$$

Where, m represents the biofuels' emission coefficients of raw materials and equipment, a represents the related materials, C represents the biofuels' LCA input of raw materials and equipment, b represents the related stages.

$$E_{\text{dGHG}} = \sum \lambda E \varepsilon \quad (3)$$

Where, λ represents the biofuels' emission coefficients, E represents the biofuels' consumption, ε represents the biofuels'oxidation rate.

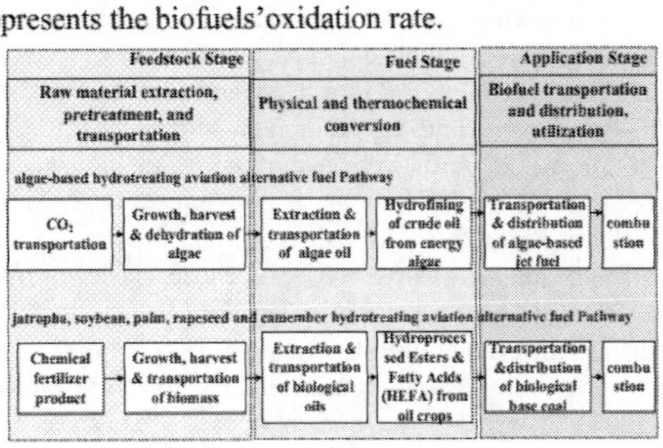

Fig. 3. Stages of biofuels' LCA.

Figure 3 shows the stages of biofuels' LCA. Coal based Fischer Tropsch synthetic aviation alternative fuels have the lowest PM2.5 emissions during their lifecycle, accounting for 35.7% of petroleum based aviation fuel emissions, and PM10 emissions accounting for 75.3% of petroleum based aviation fuel emissions. The production of space based alternative biofuels using energy algae accounts for 71.4% and 74.7% of petroleum based aviation fuels and 70% and 67.5% of soybean based fuels, respectively, during the lifetime of PM10 and PM2.5 emissions[13]. Therefore, using biomaterials to prepare aviation alternative fuels has advantages in air pollution prevention and control.

However, not all biofuels have lower greenhouse gas emissions than fossil fuels. O'Connor et al. also studied the greenhouse gas emissions of several biofuels from different resources [14]. Due to the increase in N2O, plants such as palm plantations can even produce biofuels with higher greenhouse gas emissions. They calculated and compared the carbon dioxide equivalent emissions of sunflowers, soybeans, and rapeseed, and the results were similar to European research.

More importantly, other factors can also have an impact on greenhouse gas emissions, such as the saturation of fatty acids in oil, soil composition, isomerization, additional transportation, etc. For example, peatlands can significantly increase greenhouse gas emissions from the production of biofuels[13]. In addition, different types of resources are a key factor. Generally speaking, compared to the best performing oil crop pathway in greenhouse gas emissions, the lignocellulosic pathway performs much better[14].

Generally speaking, the method of producing fuel can affect fuels' composition, performance, economic, usability, as well as their environmental influence. Industrial development is achieved through the adoption of new production technologies as well as rational utilization of the locally potential raw materials to improve production processes.

3 Application and Policy of Aviation Biofuels

Over the past 100 years, the extensive use of fossil fuels has caused serious environmental problems, especially in developing countries. Therefore, the public's desire for sustainable development is becoming increasingly strong. In the energy sector, biofuels are highly anticipated.

A biofuel powered flight test was conducted on the G-VWOW Boeing 747-400 of Virgin Atlantic Airways in the UK. Although aviation biofuels only provided 20% of the engine's kinetic energy during the test flight, they are still considered a breakthrough in environmental protection technology in the aviation industry. From then on, the prelude to aviation biofuel testing and application began globally.

In recent decades, countries have continuously strengthened cooperation and mutual supervision, actively achieving the goal of energy conservation and greenhouse gas emissions reduction, especially in the field of transportation. The government and authorities have issued many relevant policies, either supporting the use of aviation biofuels or restricting the production of certain types of biofuels.

The International Civil Aviation Organization is exploring ways to reduce aviation emissions under the guidance of the Kyoto Protocol. To achieve this goal, US-RFS, as well as EU-RED have developed about 15 biofuel production and process standards[14], which is crucial for increasing the utilize of alternative aviation fuels, and these studies will help create a more environmentally friendly environment.

The EU ETS underwent revisions between 2013 and 2016, and legislation was also revised. In 2016, the International Aviation Carbon Emissions Establishment and Reduction Plan (CORSIA) adopted by the ICAO on the impact of incorporating aviation into the carbon trading system is considered successful. In 2021, the International Energy Agency (IEA) released a report stating the goal of creating an energy system with net zero emissions by 2050.[15].

4 Concerns and prospects for aviation biofuels

Since Virgin Air successfully conducted its first flight test of aviation biofuels in February 2008, several demonstration flights have been conducted by organizations including New Zealand, the United States, the country, Japan, France, Germany, the Netherlands, Brazil, Thailand, the United Arab Emirates, Mexico, Chile, Australia, and other countries and regions to verify the feasibility of aviation biofuels as a new alternative energy source. While biofuels are rapidly developing, governments and scholars around the world are also highly concerned about the potential impact of their development. For alternative fuels, many challenges need to be overcome, such as manufacturing process, economic factors, feasibility of raw materials, environmental impact, and compatibility with traditional fuels [15].

For one thing, developing avaition biofuel can alleviate resource crisis, reduce avaition carbon emissions, enhance agriculture's economic position so as to increase local income, For another hand, the development of biofuel is likely to threaten world foods' security, as they may need to expropriate a huge amount of arable land, consume a large amount of water resources, as well as drive up prices of agricultural products. The energy demand of the rich in the world will take away the food demand of the poor. Bringing about social justice issues. Therefore, balancing the relationship among environmental security, source security, and food security makes great significance.

Although research on aviation biofuels has achieved promising results, the availability and price competitiveness of commercial fuels widely used in aviation biofuels have become important concerns for airlines [16]. This requires breakthroughs in the production of aviation biofuels, which are competitive in price compared to traditional fuels, and finding a stable supply of raw materials to ensure the sustainable supply of biofuels as aviation fuel.

Due to the use of aviation kerosene in most civil aircraft, bio kerosene can become a key biofuel for aviation. Although the current process can convert petroleum into bio aviation kerosene, there are still some issues that need to be addressed. One is that the process route is long and the yield of bio aviation kerosene is low. Secondly, there are many cracking products during the reaction process, and the content of iso alkanes in the products is relatively low, which affects the low-temperature flow performance of biological aviation. The third is the catalyst used in the reaction process. The lifespan is not long. The active ingredients currently used are mostly precious metals, but most have a lifespan of only a few hundred hours, and their toxic effects are inactivated. The cost of producing aviation bio kerosene is more than that of traditional aviation kerosene, which hinders the development of biofuel aviation kerosene. Large scale industrial production. Therefore, improving the production process route of bio aviation kerosene and developing long-life, low-cost bio aviation kerosene catalysts are the key development directions of aviation bio kerosene.

5 Conclusion

With the increasing pressure of high oil prices and greenhouse gas emissions reduction, the combination of aviation biofuels and renewable raw materials is the fundamental way for the global aviation industry to address emission reduction challenges and achieve sustainable development.

Research has confirmed that biofuels as an alternative to aviation fuels can achieve better greenhouse gas emissions reduction than that of tradition airline fuels. Palm, energy algae, soybeans, and other raw materials have enormous potential. In addition to controlling greenhouse gas emissions, the use of biofuels can also significantly reduce PM2.5 and PM10 emissions and improve air quality.

At present, although the development and application of aviation biofuels have made gratifying progress, they still face some important problems that need to be solved urgently: firstly, further reduce the production cost of aviation biofuels and improve market competitiveness; Secondly, expand the sources of aviation biofuel raw materials to ensure sufficient supply of raw materials; The third is to expand the production scale of aviation biofuels, further expand production, and replace more petroleum resources. The fourth is to strengthen the promotion and application of aviation biofuels to ensure the achievement of emission reduction targets.

Aviation biofuels have gone through stages such as technological research and development, flight testing, etc. Currently, their development has begun to take effect, but they still need to reach the stage of large-scale commercialization in order to experience a long-term downturn. In order to overcome this downturn, taking measures to optimize the collection technology, efficiency, and economy of raw materials can help enhance the competitiveness of biofuels from the root. In promoting the sustainable development of biofuels, social organizations such as governments, enterprises, and research institutions should play different roles to jointly guide and promote the orderly and healthy development of the biofuel industry. The raw material collection.

References

1. Arias A ,Nika E C ,Vasilaki V , et al., Assessing the future prospects of emerging technologies for

shipping and aviation biofuels: A critical review, Renewable and Sustainable Energy Reviews, 197,114427 (2024).

2. Ahlström Johan,Jafri Yawer,Wetterlund Elisabeth & Furusjö Erik. Sustainable aviation fuels Options for negative emissions and high carbon efficiency.International Journal of Greenhouse Gas Control,125 (2023).
3. Nygren, E., Aleklett, K., and Hook, M., Aviation fuel and future oil production scenarios, Energy Policy, vol. 37, pp. 4003 4010 (2009).
4. Yilmaz, N., and Atmanli, A., Sustainable alternative fuels in aviation, Energy, vol. 140, pp. 1378–1386 (2017).
5. Kandaramath, T., Yaakob, Z., and Binitha, N. N., Aviation biofuel from renewable resources : Routes , opportunities and challenges, Renewable and Sustainable Energy Reviews, vol. 42, pp. 1234–1244 (2015).
6. Wang Z., Wang Q.,W., Wang C., B., et al., Comparative study of carbon emissions from green aircraft in China based on mixed life cycle assessment, Propulsion Technology, 45.03: 26-36 (2024).
7. Hu, X., Qi, P., Fu, X., He, H., Huang, G., and Li, X., Technology development background and application status on aviation biofuel, Chemical Industry and Engineering Progress, vol. 31, pp. 1625–1630 (2012).
8. Kousoulidou, M., and Lonza, L., Biofuels in aviation : Fuel demand and CO2 emissions evolution in Europe toward 2030, Transportation Research Part D, vol. 46, pp. 166–181 (2016).
9. Marangon B B ,Castro S D J ,Assemany P P , et al., Wastewater-grown microalgae biomass as a source of sustainable aviation fuel: Life cycle assessment comparing hydrothermal routes, Journal of environmental management, 360, 121164-121164 (2024).
10. Zhao, J., Guo, F., ALS, and Yang, X., Research of sustainable feedstock for future alternative aviation fuels, Journal of Beijing University of Aeronautics and Astronautics, vol. 42, pp. 2378 2385 (2016).
11. Wise, M., Muratori, M., and Kyle, P., Biojet fuels and emissions mitigation in aviation : An integrated assessment modeling analysis, Transportation Research Part D, vol. 52, pp. 244–253 (2017).
12. Ou, X., Zhang, X., Chang, S., and Guo, Q., "Energy consumption and GHG emissions of six biofuel pathways by LCA in (the) People s Republic of China, Applied Energy, vol. 86, pp. S197 S208 (2009).
13. O Connell, A., Kousoulidou, M., Lonza, L., and Weindorf, W., Considerations on GHG emissions and energy balances of promising aviation biofuel pathways Committee on Aviation Environmental Protection, Renewable and Sustainable Energy Reviews, vol. 101, pp. 504 515 (2019).
14. Ozgur B ,Nesrin C ,Hakan C ., Aviation, energy, exergy, sustainability, exergoenvironmental and thermoeconomic analyses of a turbojet engine fueled with jet fuel and biofuel used on a pilot trainer aircraft, Energy, 263(PD) (2023).
15. Anderson Brian J., et al."Social Science Applications in Sustainable Aviation Biofuels Research: Opportunities, Challenges, and Advancements."Frontiers in Energy Research 9.(2022).
16. Thanikasalam K ,K. G B ,Zuhdi M M , et al., Exploring socioeconomic and political feasibility of aviation biofuel production and usage in Malaysia: A thematic analysis approach using expert opinion from aviation industry, Green Energy and Intelligent Transportation,2(5) (2023).

Analysis of uneven density of oil-based drilling fluid caused by shale respiration

Xiaoxiao Ni[1,a], Jian Li[1,b], Jianhua Wang*[1,c], Haijun Yang[1,d], Rentong Liu[1,e], He Shi[1,f], Lili Yan[1,g], Shan Gao[1,h], Xiaofei Wang[2,i], Die Zhang[1,j],

[1]CNPC Engineering Technology R&D Company Limited, China
[2]Drilling Fluid Company of CNPC Xibu Drilling Engineering Company Limited, China

Abstract: Longmaxi shale formation has developed beddings and fractures in southern Sichuan, which is prone to well leakage and wellbore instability. In the drilling process of horizontal section of Ning 209H58-2 well, drilling fluid would loss during circulation and return after pump stopped in the same section, which was the obviously respiration effect. The larger amount of leakage and pumping back, the greater the difference of drilling fluid density would be, accompanied by the increasing value of the gas. In this paper, shale was determined to have "screen effect" by assuming methods, the drilling fluid would loss as the equivalent circulating density was greater than the crack opening pressure, and the crack width was just in the size range of barite particle, then the larger size barite cannot enter in shale cracks end gathered together, and low density drilling fluid would penetrated into the formation, and then the drilling fluid density would be high in the first and low then after circulation again. The equivalent circulating density would be controlled by reducing the drilling fluid density, the effects of microfracture opening and respiration would be avoided, and the drilling fluid density would return to normal after circulation. This paper illustrates that the shale screen effect was the main reason for the uneven density of drilling fluid, which was expected to provide technical support for the development of horizontal shale gas Wells.

1 Introduction

The uneven density of drilling fluid caused by shale respiration is a phenomenon that has been paid more and more attention in the process of shale gas drilling in Sichuan. Improper handling will lead to more time and capital consumption[1-3]. The strata in the Sichuan Basin have strong heterogeneity, and the anisotropy of the strata is very prominent. The shale of Longmaxi Formation in the southern Sichuan area is dominated by black silty shale and carbonaceous shale, and the shale bedding fractures are developed, and lost circulation occurs from time to time[4-6]. When the external force exceeds the load pressure of the shale, the micro-fractures of the shale will be opened, then the drilling fluid in the wellbore would be pushed into the formation under the the pressure difference. It leads to uneven density at the outlet, which in turn affects the drilling process[6]. After the pump is stopped, the drilling fluid in the fractures is pushed back into the wellbore due to the decrease of the equivalent density (ECD) at the bottom hole, and then the overflows would be occurred. Since the late 1980s, many researchers have carried out research on the mechanism of respiration effect and research on natural fractured strata[7-14]. But there is still no systematic conclusion. Therefore, the description and mechanism analysis of respiration were discussed in this paper.

2 Background of Well Ning 209H58-2

Well Ning 209H58-2 is a horizontal well located in East District of Daba, Yibin, Sichuan. The reservoir is buried at a depth of 2531-2783m. The production layer is mainly a tight sandstone pore gas reservoir with micro-fractures. The horizontal section is designed to be 2000m length. In depth of 4502m, there occurred drilling fluid loss without return, the maximum leakage rate was $46m^3/h$ and there loss about 23 m^3 drilling fluid. At that time, the density of the oil-based drilling fluid was $1.81g/cm^3$, the funnel viscosity was 73s, the displacement was 30L/s, and the horizontal section is 1550m long. After the lost circulation, one plugging while drilling and two bridge plugging were operated unsuccessful due to the "respiration effect". The drilling fluid continued to leak, and then returned to the site after pump stopped. It prepared to drill about 3-5m to expose more leakage layers by reducing the density of the drilling fluid and increasing the displacement, which would beneficial to consolidate and plug the leakage.

[a]18810906075@163.com, [b]253537929@qq.com, *[c]wjhdri@cnpc.com.cn, [d]yanghj_dri@cnpc.com.cn, [e]liurtdr@cnpc.com.cn, [f]shihedr@cnpc.com.cn, [g]Yanlilidr@cnpc.com.cn, [h]gaosdr@cnpc.com.cn, [i]15276203168@163.com, [j]Zhangdiedr@cnpc.com.cn

After drilling about 0.65m, the overflow stopped drilling, they prepared to adjust the performance of the drilling fluid to kill the well. After turning on the pump again, the bottom hole ECD increases, and the leakage and return volume increase accordingly with the uneven density of the drilling fluid and high gas measurement value. The density of the returned drilling fluid is ranging from 1.68g/cm³ to 1.89g/cm³. In the process, the cumulative loss of oil-based drilling fluid was about 187.9m³. And some conclusions were conducted, (1) The returned volume of drilling fluid is large, which is always smaller than the leakage volume. It is an important symbol to distinguish breathing effect from overflow. (2) As the density of the drilling fluid increases, the leakage increases, and the return flow increases accordingly after the pump stopped. (3) As the drilling fluid density increases, the gas measurement value increases gradually, and the duration of over 35% is longer.

3 Analysis of uneven density of drilling fluid

(1) Drilling fluid is affected by formation water or fracturing fluid from adjacent wells, which will result in uneven density at the outlet.

The oil-water ratio of drilling fluids with densities of 1.76g/cm³ and 1.69g/cm³ were measured, the results of both tests were the same as the oil-water ratio of 80/20 as the mixing drilling fluid (Fig 1). The results show that the drilling fluid would not be affected by the invasion of formation water or fracturing fluid from adjacent wells. Therefore, the hypothesis does not hold.

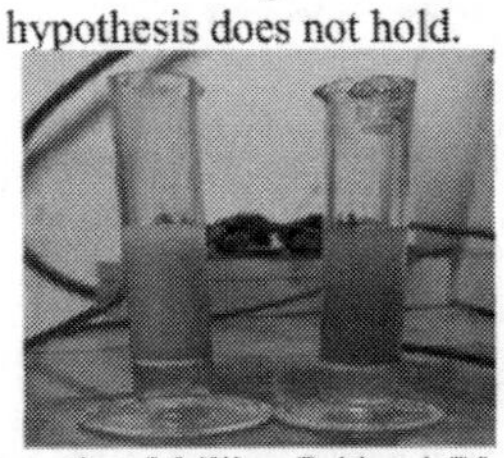

Fig.1 oil-water ratio of drilling fluid at 1.76g/cm³ (left) and 1.69g/cm³ (right)

(2) Drilling fluid settles after standing for a long time.

The wellbore is divided into two parts by assuming subsidence through the analysis of the wellbore structure. The first is the wellhead to A point: the high-density at the wellhead to A point should be concentrated near the inclined well section of A point due to gravity. The low-density should be returned first, followed by the high-density drilling fluid. Horizontal section: The high-density drilling fluid will be concentrated at the bottom of the wellbore due to gravity, and the low-density drilling fluid will be at the upper part. When the drilling tool is stirring, the drilling fluid will be stirred evenly, and the density of the drilling fluid should not fluctuate during the upward returning.

To verify the above assumption, the drilling fluid with a density of 1.74g/cm³ was placed in the aging tank at 90°C for 8 hours, the density of the upper and lower layers were both 1.74g/cm³ (Fig 2). The density of 1.77g/cm³ was measured after standing at room temperature for 10 hours. The the upper and lower layers were both 1.77g/cm³. Above results showed that the drilling fluid does not settle at the bottom hole temperature and normal temperature, so the assumption does not hold.

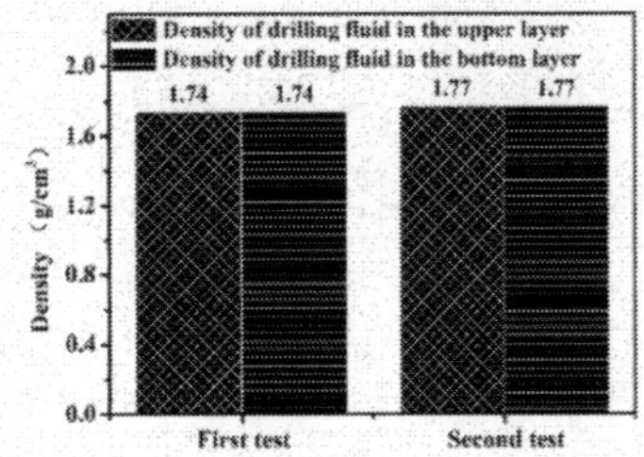

Fig. 2 Difference of drilling fluid density before and after aging

(3) Drilling fluid settles under dynamic conditions.

Place the drilling fluid in the API drilling fluid rheology measuring cup, keep the temperature at 90 °C, adjust the rotational speed of 100 r/min, and use a syringe to extract the drilling fluid in the bottom of the cup. Then the sample density was measured to about 1.77g/cm³. After 30 minutes, the density was still 1.77g/cm³ (Fig 3). It verified that the drilling fluid would not settle under dynamic conditions, and then the assumption does not hold.

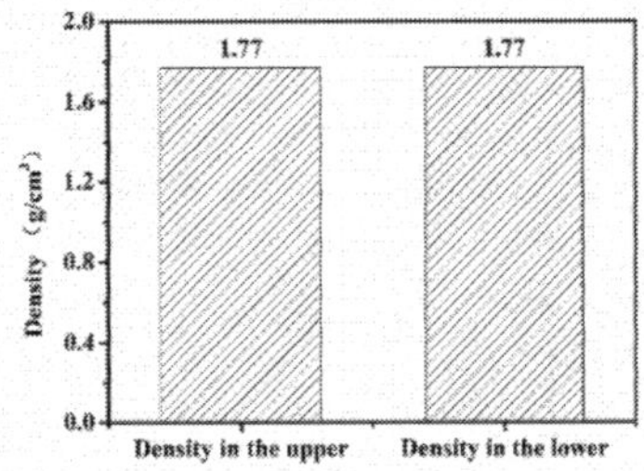

Fig. 3 Drilling fluid dynamic settlement

(4) The shale has screen effect and the respiration of shale results in uneven density.

It was assumed that increasing the bottom hole ECD would cause the circulating equivalent density to be larger than the cracking pressure of the formation fractures, resulting in the opening of the fractures and the leakage in the wellbore. As the size of barite is larger than the fracture width, the barite will stay in the leakage end face to form a high-density drilling fluid, and the low-density formed by barite with a particle size smaller than the fracture width will enter into the formation through the fracture. After the circulation was stopped, the circulation pressure loss was disappear, the equivalent density of the bottom-hole is less than the closing pressure of the fracture, so that the bottom-hole fractures would be closed, then the high-density in the wellbore would be returned out first, the low-density and shale gas would be spit out, and the gas transported at the bottom of the hole. After removal, the high-density drilling fluid would return, resulting in uneven density and high gas measurement values. In order to verify this hypothesis, an indoor experiment was designed. Oil-based drilling fluid with the density of 1.81g/cm³ was filtered with a 325 mesh (45μm) sieve cloth, and the density after sieving and the remaining drilling fluid were measured respectively. Figure 4 show that the density after sieving is 1.58g/cm³, and the density of the remaining drilling fluid is 1.99g/cm³, resulting in a serious density difference and no filter cake on the screen surface. It is speculated that the width of the open cracks in the bottom hole shale is range from 40 to 60 μm, and the large particle of barite remain on the surface of the shale cracks, and the drilling fluid has a serious density difference when it was regurgitated.

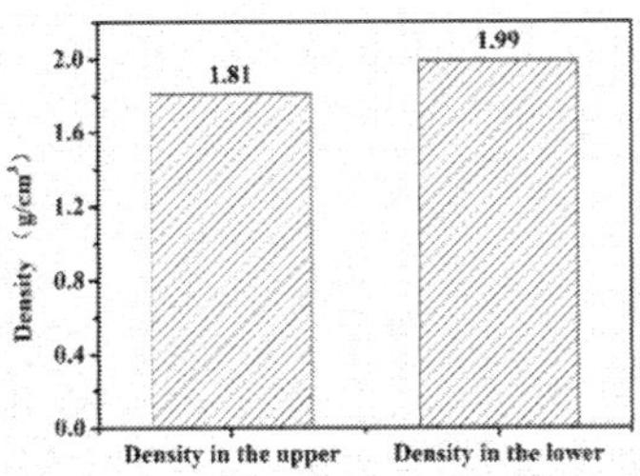

Fig. 4 Density of the upper and lower parts of the drilling fluid after filtration with sieve

4 Method to eliminate respiration

The uneven density at the outlet was eliminated by reducing the shale respiration. Calculating the bottom hole ECD, determining the closing pressure and hole leakage pressure at bottom to control the leakage. Reduce the pump displacement cycle to make the density circulate evenly in the whole well, and reduce the uneven circulating density and high gas measurement value.Then adjusting the density, controlling the bottom hole equivalent density would make it safe to continue drilling.

Pump pressure transfer relationship was showed in the following equation:

$$P_s = \triangle P_g + \triangle P_{st} + \triangle P_a + \triangle P_b$$

P_s: Pump pressure, ΔP_g: Ground manifold pressure loss, ΔP_{st}: Pressure loss in drill string, ΔP_a: Annular pressure loss, ΔP_b: Bit pressure drop.

When a lost circulation occurs, the surface manifold pressure loss, the pressure loss in the drill string, and the bit pressure drop would do not change, only the annular pressure loss decreases, so it can be determined that the change in pump pressure was equivalent to the change in annular pressure loss. The annular pressure loss was equal to the sum of the annular pressure loss of the tool string, the annular pressure loss from the tool string to the casing shoe, and the annular pressure loss above the casing shoe.

The above experimental results can be summarized as a coefficient for evaluating the severity of formation respiration effect, namely the formation respiration effect coefficient. The calculation formula is as follows:

$$\alpha = \frac{P_d - P_c}{P_c}$$

α: breathing effect coefficient , P_d : circulating pressure of the drilling fluid, P_c: fracture closed pressure.

As the breathing effect coefficient decreases, the throughput of the breathing effect of the fracture continuously decreases. Therefore, the size of the formation breathing effect can be adjusted by adjusting the circulating pressure of the drilling fluid, thereby reducing the impact of the breathing effect.

Table 1 Logging data

Time	Depth (m)	Drill position (m)	Pump pressure (MPa)
22:28:18	4502.05	4499.74	27.08
22:28:49	4502.05	4499.74	27
22:29:20	4502.05	4499.74	26.65
22:29:51	4502.05	4499.94	24.71
22:30:21	4502.05	4499.94	24.25
22:30:52	4502.05	4499.94	24.04
22:31:23	4502.05	4500.18	23.89
22:31:53	4502.05	4500.18	23.91
22:32:24	4502.05	4500.18	24.03
22:32:55	4502.05	4500.18	24.17
22:33:26	4502.05	4500.18	24.18

It can be seen from Table 1 that the total pool volume starts to decrease significantly at 22:29:20, and the pump pressure decreases by 0.35MPa. The initial stage was imbibition. According to the logging data, the pump pressure fluctuation range from 2.3MPa to 2.75MPa. Assuming that leakage occurs at a depth of 4502m, total annular pressure loss calculated to about 2.4MPa. Among them, the annular pressure loss of the bottom instrument combination is 0.18MPa, the annular pressure loss of drill pipe was 1.44MPa, and the annular pressure loss of drill pipe was 0.78MPa. So it can be preliminarily determined that the leakage layer is about 4502m, and respiration of the leakage layer leads to the uneven density.

5 Conclusions

(1) Respiration is caused by the opening and closing of shale micro-fractures under pressure.

(2) Drilling fluid density increases, the bottom hole ECD increases, and the respiration will occure.

(3) Shale has a "screen effect", large size barite accumulates on the leakage end face of the fracture, and the leakage and respiration lead to the uneven density.

(4) Reducing the bottom hole ECD, the shale respiration will be weakened, the problems of uneven density and increased gas value will be avoided.

Acknowledgments

This work was supported by scientific research and technology development project of CNPC (Grants 2024ZS49).

References

1. Luo W, Yang Y C. Identification of formation breathing effect and its impact on cemeting quality[J]. Petrochemical industry technology, 2024,31(05): 286-288.
2. Zou C N. Advances, challenges, and

countermeasures in shale gas exploration of underexplored plays, sequences and new types in China[J]. Oil & Gas Geology,2024,45（2）: 309-326.

3. Ozdemirtas M, Babadagli T, Kuru E. Numerical Modelling of Borehole Ballooning/Breathing-Effect of Fracture Roughness[J].Petroleum Society of Canada, 2007, PETSOC-2007-038.
4. Zhang J H. Shale gas accumulation patterns in China[J]. Natural Gas Industry, 2022,42(8): 78-95.
5. Zhang H H. Source rock characteristics and geological significance of Upper Permian Linxi Formationin Northern Songliao Basin[J]. Journal of Northeast Petroleum University, 2023,47(4):1-13.
6. Wang W. Study on the mechanism of fiber fracturing fluid controlling pulverized coal transportation[J]. Energies, 2022, 15(6): 2165.
7. Shahri M P, Zeyghami M, Majidi R. Investigation of Fracture Ballooning and Breathing in Naturally Fractured Reservoirs: Effect of Fracture Deformation Law[C]. Nigeria Annual International Conference and Exhibition, 2011, SPE-150817-MS.
8. Chen T. Progress of polymer application in coated proppant and ultra-low density proppant[J]. Polymers, 2022, 14(24): 5534. .
9. Xiang K M. Reservoir characteristics and main controlling factors of gas content of deep shale gas in Southern Sichuan Basin[J]. Journal of Northeast Petroleum University, 2023, 47(03): 44-55+8-9.
10. Guo J C. Efficient placement technology of proppants based on structural stabilizers[J].

    Petroleum Exploration and Development, 2024, 51(3): 616-623.
11. Yue W H. Study and field application of fiber slickwater proppant carrying capacity[J]. Drilling & Production Technology, 2023, 46(5): 151-156.
12. MA X H. Prospects and challenges of shale gas development in China［J］. Petroleum Science Bulletin, 2023, 8(4):491-501.
13. Guo X H. Mechanisms for lacustrine shale oil enrichment in Chinese sedimentary basins[J]. Oil & Gas Geology, 2023, 44（6）: 1333-1349.
14. Lavrov A, Tronvoll J. Mechanics of Borehole Ballooning in Naturally-Fractured Formations[C]. SPE Middle East Oil and Gas Show and Conference, 2005, 10.2523/93747-MS.

 https://doi.org/10.1051/e3sconf/202456001022

Study on the Development of Distributed Energy in Northern China of Rural Energy Transformation

Shaohua Liu[1,2,*], Yingyuan Zhang[1,2], Shu Zheng [1,2]

[1] NARI Group Corporation (State Grid Electric Power Research Institute), Nanjing, Jiangsu, 211106, China
[2] NARI Technology Nanjing Control Systems Co., Ltd., Nanjing, Jiangsu, 211106, China

Abstract—Distributed energy is a crucial breakthrough and focus for the green and low-carbon transformation of rural energy in China. It has experienced rapid development in recent years. However, mismatches between rural energy production and consumption capacities, and between the unregulated construction of distributed energy systems and the capacity of rural power grids, have become increasingly severe. These issues are hindering the establishment of a new energy system and the modernization of rural areas. This paper addresses the problems and challenges faced in the development of rural energy in China. Through an investigation and analysis of the experiences and shortcomings in the development and utilization of distributed energy in Northern China, the paper proposes measures and suggestions to promote the scientific development of distributed energy in rural China.

1. Introduction

Green energy transformation is an important direction for rural development and a key measure to implement the comprehensive promotion of rural revitalization and modernization of agricultural countryside. In 2023, to promote the largescale and high-quality development of renewable energy, the National Energy Administration accelerated the planning and construction of a new energy system and guided local governments to comprehensively initiate the construction of pilot counties for the rural energy revolution, while consolidating and enhancing the quality of rural power grid projects. In March, the National Energy Administration, the Ministry of Ecology and Environment, the Ministry of Agriculture and Rural Affairs, and the National Rural Revitalization Administration formulated the "Construction Plan for Pilot Counties of the Rural Energy Revolution" [1]. In July, the National Development and Reform Commission, the National Energy Administration, and the National Rural Revitalization Administration issued guidelines on implementing the Rural Power Grid Consolidation and Enhancement Project [2].

The rapid development of renewable energy has played a positive role in rural energy supply and rural revitalization [3–5]. At the same time, the continuous rapid growth of new energy has gradually exceeded the capacity of the power grid, putting significant pressure on rural power grids. In some areas, the conflict between the development of new energy and the grid's capacity for integration and absorption has become prominent. In some rural regions, there have been instances of wind and solar power being curtailed, and transformer overloads caused by reverse power flow, threatening grid stability and safety [6–11].

Based on the levelized cost of electricity (LCOE) model and the levelized revenue per electricity (LROE) model, this paper considers the policy, technology, society, economy and other influencing factors to improve the levelized social cost of electricity (LSCOE) model and the levelized social benefit of electricity (LSBOE) model to evaluate the social investment income of distributed photovoltaics, and proposes a distributed energy consumption distribution microgrid topology that is suitable for rural areas. It also puts forward measures to promote the scientific development of distributed energy in rural areas in northern China to assist in the transformation of rural green energy[14].

2. Construction Status And Business Model Of Distributed Photovoltaics

According to the National Energy Administration's Notice on matters relevant to the construction of wind and photovoltaic power generation projects in 2019 (National Energy Ad- ministration [2019] No. 49), distributed photovoltaics generally refer to photovoltaic power generation projects with a single-point grid-connected installed capacity of less than 6 MW. Distributed photovoltaics have characteristics such as flexible scale, strong regional adaptability, low energy loss, and minimal environmental impact, making them suitable for promotion and application in rural areas. Before 2020, the installed capacity of distributed photovoltaics was consistently lower than that of centralized ground-mounted stations. Influenced by subsidy policies, the installed capacity of distributed photovoltaics saw a brief

*Corresponding author's e-mail: liushaohua2@sgepri.sgcc.com.cn

decline in 2019 but has since maintained a rapid upward trend. As the degree of marketization in the domestic energy sector continues to increase, the advantages of distributed photovoltaics have gradually become apparent, standing out among various energy types, with new installations in 2021 surpassing those of centralized ground mounted stations for the first time[15].

Currently, rooftop distributed photovoltaic power generation projects in rural areas are mainly divided into three types based on investment and construction methods: self-funded construction, photovoltaic loan-funded construction, and rooftop leasing. Power consumption modes are categorized into three types: full feed-in, self-consumption, and surplus feed-in.

Affected by national policy factors such as the "comprehensive list of counties with distributed photovoltaic" project, distributed photovoltaic overall continues to maintain a rapid development trend. However, from the perspective of operation within transformer areas, about 50% of low voltage transformer areas with distributed photovoltaic access experience reverse power flow. Among these, some areas face severe issues of reverse heavy load and reverse overload.

3. Evaluation Model Of Social And Economic Benefits Of Rural Photovoltaic Systems

3.1. Levelized Social Cost of Electricity (LSCOE) Model

The Levelized Cost of Electricity (LCOE) model is currently the most widely used model for measuring the power generation cost of photovoltaic projects internationally. The LCOE model is a cost analysis method that discounts all costs incurred over the lifecycle of a photovoltaic project to their present value using a certain discount rate. Currently, research on the LCOE model, both domestically and internationally, primarily focuses on the cost measurement of energy generation projects and does not consider loan costs, nor the investments made by the grid to upgrade the distribution network to fulfill the social responsibility of fully accommodating distributed energy.

$$LSCOE = \frac{C_{project} + C_{power\_grid} + C_{loan} + R_{land\_rent}}{\sum_{t=0}^{N} \frac{E}{(1+d)^t}} \tag{1}$$

$$C_{project} = I_{project} + \sum_{t=0}^{N} \frac{C_{operation}}{(1+d)^n} - \frac{V}{(1+d)^N} \tag{2}$$

$$C_{power\_grid} = k_{conversion} \times k_{apportion} \times \frac{I_{power\_grid}}{N_{power\_grid}} \tag{3}$$

$$LSBOE = \frac{\sum_{t=0}^{N} \frac{T \times E_{surplus}}{(1+d)^t}}{\sum_{t=0}^{N} \frac{E}{(1+d)^t}} \tag{4}$$

In these equations:

- $LSCOE$ is the Levelized Social Cost of Electricity, in CNY/kWh.
- $I_{project}$ is the total investment of the project, in ten thousand CNY.
- $C_{operation}$ is the total operating cost of the project in year t, in ten thousand CNY.
- V is the residual value of the photovoltaic system, in ten thousand CNY.
- E is the electricity generation of the photovoltaic system in year t, in kWh.
- N is the lifespan of the project, in years.
- d is the discount rate.

Due to the scale of distributed photovoltaic individual projects, the construction period is short, and the actual operation and maintenance costs are very low.

- $C_{power\ grid}$ represents the investment by the grid company to expand the grid assets to accommodate distributed energy. This includes new construction and renovation of substations, lines, and primary and secondary equipment in the distribution network. This article only considers the annual depreciation cost of the initial investment

$I_{power\ grid}$ in the grid, without considering the operating and maintenance costs of the grid equipment. Since grid investments provide universal benefits to society, they are converted based on the relevance of the investment $k_{conversion}$ and allocated according to the scale of distributed photovoltaic access $k_{apportion}$.

- C_{loan} is the cost paid by photovoltaic operators and individuals to the bank for construction and operation loans, where P is the total loan amount, r is the interest rate, and n is the total repayment period (annualized conversion).
- $R_{land\ rent}$ is the cost for distributed photovoltaic operators to lease rooftops and land from farmers for construction.

3.2. Levelized Social Benefit of Electricity (LSBOE) Model

The income of photovoltaic projects mainly comes from the revenue from selling generated electricity and photovoltaic subsidies. In practice, the amount of electricity used for household consumption by farmers is very low. The daily routines of farmers have an inverse characteristic compared to the peak and trough of photovoltaic power generation, resulting in relatively low 'self-consumption' of rooftop photovoltaic electricity. Additionally, photovoltaic subsidies have gradually been reduced or cancelled as the cost of photovoltaic modules decreases. Therefore, the direct social benefit of photovoltaic income mainly refers to the profit from 'the rest of the generation to the grid.'

$$LSBOE = \frac{\sum_{t=0}^{N} \frac{T \times E_{surplus}}{(1+d)^t}}{\sum_{t=0}^{N} \frac{E}{(1+d)^t}} \tag{5}$$

- T is the direct social benefit generated from surplus electricity (feed-in tariff) per unit of electricity from the photovoltaic project, measured in CNY/kWh.

- $E_{surplus}$ is the amount of surplus electricity fed into the grid by the user in year t, measured in kW.
- E is the total electricity generated by the user in year t, measured in kWh.

For an individual photovoltaic project, the levelized cost of electricity (LCOE) revenue should exceed the LCOE. However, considering that the weak grid capacity leads to the inability to fully absorb the photovoltaic electricity, resulting in decreased revenue from surplus electricity fed into the grid, and the increased LCOE due to subsequent grid investments, the levelized social benefit of electricity might be less than the LCOE in that year.

Additionally, photovoltaic loans are a model that enables users to utilize financial products to support green energy production and investment. However, many photovoltaic operators exploit this farmer-friendly policy by taking out loans in the name of farmers, thereby burdening them with debt. The only income farmers receive is from the rental of their land.

Therefore, the orderly development of distributed energy, a reasonable construction of grid capacity, and an effective increase in farmers' income are key to the rural energy transition.

4. The Main Problems And Challenges Facing The Development Of Distributed Energy

4.1 The disorderly expansion of distributed photovoltaics seriously restricts the healthy development of new energy

As the cost of photovoltaic construction decreases and the economic benefits are significantly improved, many places have strictly approved centralized photovoltaics and required them to be equipped with energy storage. Some photovoltaic operators that do not meet the approval requirements have "broken up into pieces" such that they built distributed residential photovoltaics through renting sites and buying and selling procedures. At the same time, the qualifications of some photovoltaic operators are uneven, and the construction quality is difficult to supervise and guarantee, which poses a risk of endangering the safety of farmers and damaging their interests.

Taking Chaoyang as an example, the current maximum carrying capacity of new energy in the power grid is about 3.8 million kilowatts, and the actual installed capacity of new energy has reached 4.3278 million kilowatts; in 2024, the power grid will increase its maximum carrying capacity of new energy to 7.5 million kilowatts through transformation and expansion and other reinforcement methods, basically meeting the grid connection needs of the new energy indicators (7.22 million kilowatts) that have been strived for, but the overall carrying capacity of the power grid is still saturated. The disorderly growth and uncontrollable access to distributed photovoltaics will squeeze the output of centralized development projects and affect the healthy development of the regional new energy industry.

4.2 Distributed power sources are difficult to consume locally, and grid access capacity is seriously insufficient

North China is a typical large power supply and medium load system. Most rural areas are underdeveloped areas with small loads and slow growth. On the one hand, the unit construction capacity of household distributed photovoltaic projects ranges from 10 to 50 kilowatts, and the user load is only about 2 to 5 kilowatts. The load growth rate is far less than that of distributed photovoltaics. On the other hand, the peak electricity consumption in rural areas is often at night, and the photovoltaic output and load characteristics are misaligned, this has led to a severe shortage in the distribution grid's absorption capacity, and issues have already arisen with local and nearby balance.

According to the national "Guidelines for the Assessment of the Carrying Capacity of Distributed Power Sources Connected to the Grid" [12], the current grid-connected capacity of distributed power sources in Chaoyang is 569,300 kilowatts, and the carrying capacity is only 280,000 kilowatts; the gridconnected capacity of distributed power sources in Weichang is 113,600 kilowatts, and the carrying capacity is only 25,000 kilowatts.

4.3 Power quality issues are prominent, threatening the safe and stable operation of large power grids

On the one hand, the electricity load in underdeveloped areas is small, and the seasonal peak-to-valley difference is large. After a large number of distributed power sources are connected, the distribution network has changed from a 'passive' network to an 'active' network, and heavy overload is common in some periods of the substation area; on the other hand, the simultaneous rate of distributed photovoltaic power sources is relatively high (the highest area reaches more than 95%), and it is difficult to achieve remote control. The power generation increases sharply after sunrise every day, and the power generation decreases sharply when there is cloud cover and other situations. The flow direction changes frequently, and the voltage exceeds the limit and other power quality problems are prominent, which seriously threatens the safe and stable operation of the large power grid.

For example, the "Mulan" 220 kV substation in Weichang has encountered the problem of distributed power supply reverse transmission, with 38 of 10 kV reverse transmission lines, and all 1,101 substations with photovoltaic installations have photovoltaic reverse transmission to the Internet. Some substations in Chaoyang have experienced voltage exceeding the limit many times. There are a total 20 of 220 kV substations in the city, of which 11 of 220 kV substations and 29 of 66 kV substations have reverse transmission problems and overload problems caused by reverse transmission. Some lines no longer meet the N−1 requirements (the minimum standard to ensure the safety of the power grid), which seriously threatens the safety of the power grid.

4.4 Low level of informatization and increasing professional operation and maintenance issues

Firstly, distributed new energy, adjustable loads and other resources have the characteristics of large quantity, small scale and wide distribution. Due to the insufficient wireless public network coverage in rural areas, line corridors, distributed photovoltaic power stations and remote areas generally have poor signal communication quality and other problems. The coverage rate of the rural power grid is low, and the real-time collection of access equipment has not yet been achieved. It is widely facing the problems of "unobservable, unmeasurable and uncontrollable," which restricts the development of intelligence and informatization in the rural power grid system.

Secondly, the power supply area of the rural power grid is generally large, the power supply station is short of personnel, and the methods for routine maintenance are relatively outdated. Distributed photovoltaic and energy storage equipment require professional operation and maintenance personnel. In addition, the natural environment in rural areas is complex, and it is necessary to simultaneously consider constraints related to supply security, safety, and fire prevention. The difficulty of professional operation and maintenance is becoming increasingly prominent.

Thirdly, due to the volatility and randomness of distributed power output, microgrids often need to be equipped with a large amount of energy storage to meet the requirements of power supply reliability, resulting in the disadvantages of high construction and operation and maintenance costs and difficult operation and maintenance technology in microgrids. In particular, microgrids in rural areas are often not equipped with energy management systems to control construction costs and reduce the difficulty of operation and maintenance, and the level of intelligence is relatively low.

5. Measures Taken To Address The Energy Transformation And The Achievements Obtained

Although the development of distributed energy in the region faces difficulties in access and consumption as a whole, the government and power companies are still actively responding and adhering to the national green energy transformation policy. The power company conscientiously implements national policy decisions and adheres to systematic and holistic thinking. It integrates various aspects of distributed photovoltaic business with grid safety, energy transformation, and industry development to maximize the "three simultaneous assurances" of industry development, orderly grid integration, and grid safety[16-17].

5.1 Seeking policy support from local governments in various ways

Grid-side control of distributed photovoltaic access. In September 2022, the Hebei Provincial Development and Reform Commission issued the "Notice on Strengthening the Management of Rooftop Distributed Photovoltaic Power Generation" [13], clearly outlining the requirements for the grid connection and operation of distributed photovoltaic for the first time. It specifically states "to ensure that the photovoltaic capacity connected to each level of the main transformer (distribution transformer) does not exceed 80% of the equipment's rated capacity," and "the rooftop distributed photovoltaic capacity that can be connected to the county power grid is calculated in principle based on not reverse-power to the 220 kV and above power grid". This means following two connection standards: "substation area penetration rate + openable capacity", to achieve orderly integration of distributed photovoltaic.

In 2022, Liaoning introduced new policies on photovoltaic power generation, further regulating photovoltaic integration and power generation criteria. Projects must meet national energy conservation, emission reduction, and industrial structure upgrading needs, with a minimum installed capacity of 300 kilowatts per project to prevent disordered photovoltaic grid connection. The owner must ensure the long-term operation of the project, with the power generation system's operational period not less than 20 years, and ensure residential use. The government also standardized the photovoltaic application process, requiring the energy authority and local grid companies to approve applications and complete bidding procedures accordingly.

5.2 Scientifically matching green and clean power generation, grid, load and storage construction

In terms of new energy development, power companies coordinate the development of centralized projects and distributed projects. Distributed projects have great difficulty in supporting the transformation of power grids, while centralized projects can quickly achieve the strategy of strengthening the city with clean energy. In terms of the current situation, Beipiao in Chaoyang actively promotes centralized photovoltaic projects, and plans to install a total installed capacity of more than 80,000 kilowatts of centralized photovoltaics; scientifically promote distributed photovoltaics, explore and promote the integrated development of livestock and photovoltaic complementary industries, and plan to install a total installed capacity of 1.244 million kilowatts of distributed photovoltaics. At the same time, vigorously promote the construction of centralized and distributed wind power industries, with a total installed capacity of 950,000 kilowatts of centralized wind power and a total installed capacity of 80,000 kilowatts of distributed wind power, adopting the "self use first and the rest of the generation to the grid" mode to promote the nearby consumption and utilization of wind power.

In terms of energy storage, Weichang Power Supply Branch actively promotes three types of electrochemical energy storage methods based on the actual situation in the region, including energy storage with centralized new energy, independent energy storage on the grid side, and independent energy storage on the power supply side; in addition to electrochemical energy storage, Weichang area also plans a 1 million kilowatt pumped storage power station, which is expected to be completed and put into operation in 2027.

In terms of load, Chaoyang District has gradually implemented green electricity substitution in all industries and fields. In combination with local characteristic industries, the development of intelligent agricultural tools such as electric irrigation and drainage, electric drying, and all-electric greenhouses in agricultural parks has been accelerated to achieve agricultural electrification production; centralized, heat storage, solar thermal compensation electric boilers (heat pumps) and other electric heating technologies and new kitchen appliances have been gradually popularized to improve the quality of rural electrification life; encourage the acceleration of public transportation electrification, promote smart household new energy electric vehicles, and strengthen the construction of charging and swapping infrastructure; create a low-carbon industrial park with a full industrial chain of "source, network, load, storage, and sales", and gradually promote the electrification, automation, digitization and intelligence of industrial production.

5.3 Innovative microgrid reliable power supply mode

To solve the power supply problem in remote mountainous areas, in 2011, the distributed generation/energy storage and microgrid access control pilot project implemented in Yudaokou Village, Weichang, was the first project in China to connect and control distributed clean energy generation to the grid. It was also the first microgrid project in China in the field of intelligent rural grid construction.

In recent years, with the continuous improvement of the ecological environment in Weichang and the increase in the greening rate year by year, the contradiction between how to ensure the quality of power supply to farmers in remote areas and forest and grassland fire prevention has become increasingly prominent. Weichang Power Supply Branch has innovatively adopted a power supply solution that mainly builds microgrids in remote areas and uses distribution networks as cold standby.

Pilot the "distribution-microgrid collaboration + field-grid collaboration" model to improve the reliability of power supply in remote areas,as shown in Fig.1. Fuxingdi Village was selected as the pilot project location. The village is about 23.8 kilometres away from the 35 kV substation and is at the end of the 10 kV line. In combination with the load characteristics of the village, "rooftop photovoltaic + energy storage" is used for self-use first and the rest of the generation to the grid. In order to meet the village's heating electricity needs, wind farm fans are used as backup power sources to provide the village with a second reliable power source. A 250 kW/564 kWh energy storage device is also installed to achieve full green electricity operation for the entire village.

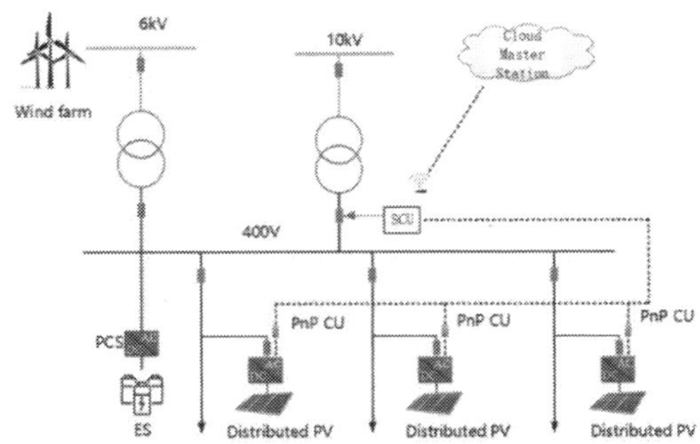

Fig. 1. "distribution-microgrid collaboration + field-grid collaboration" model diagram

Zhangjiawan Village was selected to build a small-scale photovoltaic storage "off-grid + distribution network cold standby" model,as shown in Fig.2. The project is located in a remote mountainous area with only 5 households using electricity. The maximum load is about 15 kilowatts and the maximum daily power consumption is 57 kWh. Under normal conditions,

photovoltaics can meet the daily electricity consumption of residents during the day, and the remaining electricity is used to charge the energy storage system to meet the electricity needs of users at night, on rainy days, and after photovoltaic power generation is stopped. The power supply branch lines that cross the forest farm and have fire hazards are converted to cold standby. In extremely severe weather or when the energy storage system fails, the user's load can be switched to the grid power supply mode and then switched to the off-grid mode after the system resumes normal operation, effectively improving the reliability of power supply in remote areas while reducing the probability of forest and grassland fire risks.

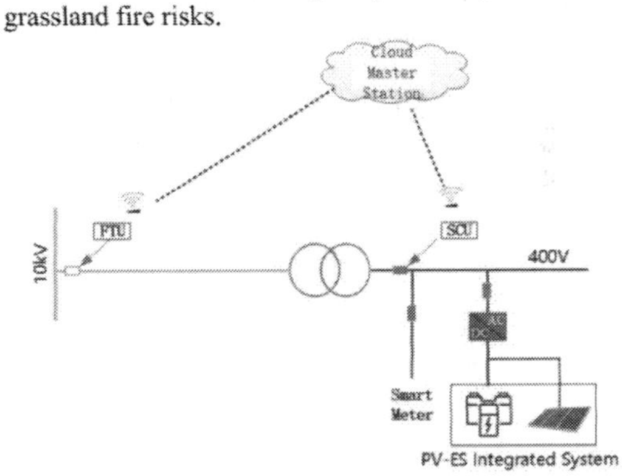

Fig. 2. "off-grid + distribution network cold standby" model diagram

5.4. Ensure the steady improvement of intelligent operation and maintenance level in remote areas

To improve the communication quality of rural power grid equipment and ensure intelligent operation and maintenance of the grid, the Weichang Power Supply Branch has collaborated with industry, academia, and research institutes to create an exemplary model of informatized communication through multi-party cooperation. District Communications primarily relies on

https://doi.org/10.1051/e3sconf/202456001022

mobile 4G and HPLC. Newly added transformers are equipped with remote signalling and remote measurement TTU that use the 4G network for communication, while areas with no signal use the 2G network. In the operation and maintenance of the distribution network, drone inspections for the distribution network and photovoltaic systems have been initiated. Currently, 17 drones are in use for the distribution network, significantly improving inspection efficiency.

6. Recommendations For Sustainable Development Of Distributed Energy

6.1 Continuously standardizing industry management to promote the healthy development of distributed new energy

Firstly, standardize project filing management and reasonably distinguish between household photovoltaic projects for natural persons and non-natural person construction photovoltaic projects. Enterprises cannot conduct distributed power projects under the guise of household photovoltaics, especially to prevent the safety risks of a large number of distributed photovoltaics disorderly connecting to rural low-voltage grids. When enterprises carry out distributed photovoltaic projects, it is advisable to conduct project construction in the form of 10 kV photovoltaic power stations, while strictly controlling the trading of new energy project procedures.

Secondly, strengthen government-enterprise coordination, establish a photovoltaic operator evaluation mechanism and a household photovoltaic project access mechanism, and provide financial and technical assistance to residents who carry out household photovoltaic projects.

Thirdly, correctly guide villagers to understand the costs and benefits of photovoltaic projects, encourage villagers to use their own residences to build household photovoltaic projects in a certain proportion, recommend the adoption of the "self use first and the rest of the generation to the grid" mode, reasonably utilize rural unused land, village committee public idle land, or building's rooftops, to construct photovoltaic poverty alleviation power stations, increase collective and villagers' income, and effectively implement policies benefiting the people such as photovoltaic poverty alleviation and rural revitalization.

6.2 Strengthening the integration of source and grid planning and construction to support the orderly development of renewable energy in rural areas.

Actively seek government understanding and support, with the guarantee of grid safety and reliable power supply as a prerequisite, to coordinate the coordinated development of centralized power plants and distributed power sources with the grid. Distributed project grid supporting transformation faces significant difficulties, while centralized projects are safer and more controllable. It is recommended to prioritize the development of centralized projects following the principle of 'first centralized, then distributed'. Conduct an analysis of the rural grid's capacity for distributed energy, establish a graded map of distributed renewable energy access capacity in the region, scientifically assess regional open capacity, actively communicate with the government, and guide the development of distributed energy in an orderly manner.

According to the "Guidelines for Distributed Power

Generation Access Network Capacity Assessment" (DL/T 2041—2019), develop distributed projects in green areas (with abundant access capacity) in an orderly manner, limit the development of distributed projects in yellow areas (with insufficient access capacity), and suspend the development of distributed projects in red areas (with saturated access capacity). No new distributed photovoltaic projects are allowed to be connected until the grid-carrying capacity is effectively improved. Coordinate the development order and utilization methods of rural energy development, promote the steady advancement of distributed energy development through coordinated planning, and resolve conflicts of interest among various hosts such as the government, grid enterprises, distributed power enterprises, and farmers.

6.3 Strengthen rural power grids and coordinate the development of power sources and loads

Reasonably plan and layout power points, scientifically guide the deployment of distributed power sources to areas with concentrated loads, prioritize local and nearby absorption, reduce energy transmission distances and conversion processes, and improve the comprehensive utilization efficiency of rural energy resources. Taking into account the characteristics of local rural production, energy use for living, resources, and living arrangements, actively promote rural electrification alternatives, such as promoting coal-to-electricity heating, fully electric kitchens, and greenhouse construction in northern regions, transforming the energy use structure in rural areas, and increasing the electrification level and renewable energy utilization rate of rural production and living.

Based on the thorough exploration of absorption potential, timely promote the construction, transformation, and upgrading of rural power grids, strengthen the weaknesses of rural power grid infrastructure, increase per capita distribution capacity, and adapt to the growing demand for rural grid loads. Strengthen the construction of load connection channels, gradually solve the problem of weak connections between rural power grids in remote areas and the main grid, tailor economic and easy-to-maintain rural microgrid applications according to local conditions, support both grid-connected and off-grid operations, pay special attention to the observability and measurability of important parameters of rural microgrids in remote areas, and the operation and maintenance issues of energy storage batteries in extremely cold climates in northern regions, to prevent the shortening of the lifecycle of rural microgrids due to operational maintenance issues after construction.

6.4 Strengthen the construction of rural power grid digitalization capabilities and improve the level of lean operation and maintenance

Coordinate with network operators to fill the gaps in rural power grid communication infrastructure, ensuring the observability, measurability, adjustability, and controllability of lines, substations, distributed energy, microgrids, etc. Explore coordinated control technologies for main-distribution-micro (substation) systems and optimize the energy management mechanisms of substations (microgrids) for autonomy, hierarchical mutual assistance, and regional balance. Establish a lean operation and maintenance management system adapted to rural grids, enhance the perception of grid status and the intelligence of fault diagnosis, strengthen professional operation and maintenance training, improve the technical support capability of new equipment, and utilize drones, mobile terminals, etc., to enhance inspection efficiency and meet the diversified operation and maintenance requirements of rural areas.

6.5 Explore the construction of distributed energy storage to meet the needs of flexible regulation of distribution networks under large-scale distributed power access

Explore distributed energy storage to support the massive integration of distributed power sources into distribution networks. In terms of policy and business models, clarify the relationships and responsibilities among multiple stakeholders such as government, enterprises (power generation, energy storage, and grid), storage-access principles, access and regulatory mechanisms. Explore operation models such as shared energy storage and virtual power plants, study the revenue models of shared energy storage, and encourage distributed energy storage to participate in market-oriented transactions such as power auxiliary services and electricity spot markets. In terms of technology, conduct research on key technologies for optimizing the configuration and safe integration of distributed energy storage, explore applications such as supporting the efficient absorption of new energy, reactive power adaptive adjustment, dynamic load balancing, proactive management of power quality, and rapid self-healing of the grid, to enhance the distribution network's ability to flexibly regulate under the massive integration of distributed power sources.

7. Conclusion

This paper summarizes the current status of rural green energy transition, proposes a levelized cost of electricity (LCOE) and revenue model for distributed photovoltaic considering social factors, explains and analyzes the issues in the development of distributed energy, and further suggests strategies for the sustainable development of county-level distributed energy. It also proposes a model for distributed energy consumption adapted to rural microgrids, exploring a healthy development model that combines market-based operations with government policy support. It is hoped that the government, enterprises, and all sectors of society can work together to increase support for the rural green energy transformation, achieving sustainable development and a green ecological environment in rural areas.

Acknowledgment

This work was supported by the State Grid Corporation of China Science and Research Program under Grant 5400202440179A-1-1-ZN

References

1. Notice of the National Energy Administration, Ministry of Ecology and Environment, Ministry of Agriculture and Rural Affairs, and National Rural Revitalization Bureau on Organizing the Construction of Rural Energy Revolution Pilot Counties [EB/OL]. National Energy Administration〔2023〕No. 23, March 15, 2023. [Online]. Available: https://www.gov.cn/zhengce/zhengceku/2023-03/24/content_5748120.htm
2. Guiding Opinions on Implementing Rural Power Grid Consolidation and Improvement Project [EB/OL]. Development and Reform Energy Regulation〔2023〕No. 920, July 4, 2023. [Online]. Available: https://www.gov.cn/zhengce/zhengceku/202307/content_6891875.htm
3. Du Xiangwan. Research on China's Rural Energy Revolution and Distributed Low-carbon Energy Development Strategy[M]. Beijing: Science Press, 2019.
4. Zhu Y, Xu S M, Ding X H, et al. Background situation, strategic suggestions and future prospects of construction of new energy system[J]. Bulletin of Chinese Academy of Sciences, 2023, 38(8): 1187-1196, doi: 10.16418/j.issn.1000-3045.20230320002.
5. Feng K., Li Q., Huang B., Yan H., Zhang H. Key Issues on Rural Energy Development in China[J]. China Electric Power, 2022,55(6):1-8 DOI: 10.11930/j.issn.1004-9649.202202016
6. TIAN Yishui. Research on the development path of deeply coupling utilization of coal and new energy[J]. Energy of China, 2020, 42(5): 25–30.
7. Charlie Dou, Zuo Chong, Jia Yan, Jiang Lu, Zheng Lina. Analysis on development potential of rooftop distributed photovoltaic power generation of rural residential dwellings in China[J]. Solar Energy, 2023 345(1): 5-16
8. LUO Guoliang, XU Xuan, ZHANG Jiaxin. An empirical analysis on the changes and influencing factors of rural energy consumption in Beijing, Tianjin and Hebei[J]. Journal of North China Electric Power University (Social Sciences), 2021(2): 20–30.

9. Yan Xiaohui, Gao Dan, Li Yanjie Thoughts and Countermeasures on Promoting Energy Revolution in the Beijing–Tianjin– Hebei Region[J].Chinese Journal of Engineering Science, 2021 23(1):24-31
10. SHI Jianxun, SHI Hanwen , MENG Ting , YANG Shu The Impact of Distributed Photovoltaics on China's Energy Market[J]. Gas and New Energy, 2024,36(1):22-27
11. Ye Zongxiu THE dilemmas and solutions of rooftop distributed photovoltaic promotion in rural areas of northern China from the multisubjects perspective[D] Master of public administration, North China Electric Power University 2023
12. National Energy Administration. DL/T 2041-2019 "Guidelines for the Assessment of the Carrying Capacity of Distributed Generation Access to the Grid" [S]. National Energy Administration, 2019-06-04.
13. Hebei Provincial Development and Reform Commission. Notice on Strengthening the Management of Rooftop Distributed Photovoltaic Power Generation [EB/OL]. Hebei Development and Reform Energy [2022] No. 1243, September 27, 2022. [Online]. Available: https://hbdrc.hebei.gov.cn/xxgk_2232/zc/wzcwj/202309/t20230907_92136.html
14. XU Wei, LIU Zhenling, WANG Huazhi, LIU Susu. Evaluation of socio-economic benefits of rural PV systems based on LCOE model: An example of rooftop distributed PV projects in Beijing suburbsJJ.Journal of China Agricultural University,2023,28(09):208-217.
15. Xu Zuxiang, Deng Zhaoyi, Liu Ruifeng, Shi Xu.Discussion on construction of roof distrubuted PV power station based on low-carbon development in urban and rural areas[J]. Solar Energy,2024,359(3):23-26.
16. Zhang Yingyuan, Liu Shaohua,Zheng Shu.Rural Green Transformation of Energy in a Demonstration County in China[C].2023 IEEE the 7th Conference on Energy Internet and Energy System Integration(EI2 2023 hangzhou,China):4529-4533
17. Liu Xiaolong , Ge Qin , Jiang Lingling , Huhetaoli , Cui Leilei , Li Bin , Du Xiangwan .Typical Paradigms and Implementation Paths for Ecological Civilization Construction Based on Rural Energy Revolution[J].Strategic Study of CAE,2019,21(5): 106-112

Research on Ecological Management Strategies for Urban River Water Environment

Liqin Du *

University of Birmingham, Birmingham, UK

Abstract. The ecological management of urban river water environments holds significant importance for sustaining city development and safeguarding ecological balance. Human activities have led to severe pollution in urban rivers, diminishing their self-healing capacity and disrupting their essential functions. This article highlights the significance of ecological management, principles guiding such management, challenges posed by water shortage and pollution in urban rivers, and strategies to address these issues. Respect for nature, restoration of ecosystem functions, and comprehensive governance emerge as key principles for effective ecological management. Water scarcity and pollution issues are explored, including the impact of urbanization and industrialization on water quality. The article suggests strategies such as thoughtful channel design, sewage recycling projects, and scientifically managed revetment works to restore and maintain healthy urban river ecosystems.

1 Introduction

Urban rivers play a pivotal role in the sustainable development of cities and the preservation of ecological balance. However, rampant human activities have led to severe pollution in these vital water bodies, undermining their self-regeneration capacity and essential functions. This degradation not only disrupts the well-being of aquatic organisms but also impairs the daily lives and productivity of urban populations. As water pollution in urban rivers escalates, downstream ecological disturbances exacerbate, posing a threat to the survival of diverse riverine species. To address these challenges and safeguard both ecology and economic growth, it becomes imperative to shift towards responsible urban river development that prioritizes water environment purification from its root causes [1]. Embracing ecological management practices for urban river water environments holds substantial promise in fostering a robust urban ecology. This article delves into the significance of such practices and outlines key principles and strategies essential for restoring and maintaining the health of urban river ecosystems. The discussion encompasses respecting nature, restoring ecological functions, embracing comprehensive governance, and addressing the pressing issue of water scarcity in urban rivers [2]. Furthermore, it explores the menace of urban water pollution, including rainwater contamination, river siltation, and deteriorating water quality, underscoring the urgency of ecological intervention [3]. With a focus on sound ecological management, this article advocates for scientific cross-sectional design, sewage recycling projects, and mindful revetment works as crucial steps toward rejuvenating urban river water environments [4]. By adopting holistic strategies, cities can pave the way for thriving urban ecosystems that balance ecological health and sustainable urban development.

2 The significance of ecological management of urban river water environment

Engineering Urban rivers are of great significance to the sustainable development of cities and the protection of the ecological environment. The long-term unreasonable behavior of people has caused serious pollution to the river ecosystem, which will not only reduce the self-healing ability of the water environment but also affect the basic functions of the river.

Seriously affects people's normal production and life. When the water pollution of urban rivers becomes more and more serious, the ecological pollution downstream will also increase, affecting the survival of various organisms in the rivers. Therefore, whether it is from the perspective of ecology or economic development, it is necessary to avoid the actual.

During the development period, it will bring pollution to the water environment of the river, purify the water environment of the urban river to the greatest extent, and curb the problem of water pollution from the root [1]. On the whole, the ecological management of the urban river water environment can provide great power support for the healthy development of the urban ecological environment, thereby providing a more suitable living environment for biological survival. In addition, the implementation of

* 2482516799@qq.com

ecological governance can also improve the purification capacity of the river's water environment and give full play to the role of the river's flood discharge and drought relief.

3 Principles of Ecological Governance of Urban River Water Environment

3.1 Respect nature

At present, people have higher requirements for the ecological environment, and pay more attention to the harmony between man and nature. In the process of urban river management, it is necessary to comprehensively consider, and restore the original appearance of nature as much as possible, focus on ecological protection, implement governance strategies in an all-round way, reasonably carry out ecological restoration, maintain biodiversity, and finally form an ecological balance between man and nature.

3.2 Recovery function

During the ecological management of river water environment, it is necessary to fully restore the ecological environment of the city as a whole, to provide strong support for the balance of the ecosystem, and to fundamentally restore the flood and waterlogging prevention functions of the river. As an important component of the urban ecological landscape system, river channels play a role in transportation, flood control and drainage,It has multiple functions such as reducing pollution and maintaining ecological balance [2]. In addition, carrying out this ecological governance work can also better promote urban ecological environmental protection, provide guarantee for urban economic development, and achieve the purpose of improving social and economic benefits.

3.3 Comprehensive governance

When carrying out ecological management of the urban river water environment, it is necessary to ensure that the ecological management work is carried out in an all-round way. It is necessary to know whether the cross-sectional shape matches the actual application requirements, and at the same time fully consider the width between the banks of the embankment. In addition, the landscape function of the river should also be brought into play. As an important component of the urban landscape ecosystem, the river plays an important role in urban ecological construction and beautification of the urban environment. Therefore, this aspect cannot be ignored in the governance work. It is necessary to deploy various tasks scientifically and rationally, formulate an effective governance plan, and finally achieve the goal of ecological governance.

4 Water shortage in urban rivers

4.1 Water shortage in urban rivers

(1) Functional water shortage in the city. The lack of an urban water environment is mainly manifested in the relative shortage of domestic water resources. Many cities will build and develop around water sources in the actual development process. River water is an extremely important freshwater resource in the life and production process of the city. The ecological adjustment ability of urban rivers is relatively weak. During the actual construction process, the original ecological environment has been damaged to a certain extent. In addition, domestic wastewater, industrial wastewater, and agricultural wastewater are continuously discharged into the river, resulting in a decline in the stability of the urban surface water environment[1]. During the development of modern society, the number of river water environmental problems has increased significantly. In the actual development process, the water resources of many urban rivers have been unable to meet the requirements of domestic water, and even the standards for industrial production water, resulting in the city's functional permanent water shortage has seriously disturbed the daily life of residents and restricted urban development.

(2) Insufficient water supply in the river course. The water in the corresponding river courses in the city is mainly the actual stagnation of the plain water network, its own water flow is relatively small, and the water circulation speed is relatively slow. Urban rivers also have a strong seasonality. In the dry season, the amount of water is not enough, which leads to the obstruction of water circulation; in the rainy season, there is sufficient rainwater, but it also brings rainwater pollution problems, and finally gathers in urban rivers to pollute the river water. Regardless of the dry season or the rainy season, the clean water resources available in urban waterways are in short supply. Secondly, the city's surface water and groundwater are over-exploited, the water level of the river drops sharply, the groundwater replenishes it insufficiently, the river course is silted up, and the self-cleaning ability of the water environment is poor, which leads to the problem of urban water resource pollution control.

4.2 urban water pollution

(1) Urban rainwater pollution. Urban rivers are often regarded as important channels for urban sewage discharge. The rapid development of cities, the significant increase in population, and the further increase in the degree of industrialization have not only improved people's living standards, but also laid hidden dangers for environmental pollution. If a city encounters a sudden heavy rain, when the rainwater flows on the surface, the garbage scattered or piled on the roadside will be washed into the urban pipe network; in many cities, due to industrial development, a surge in the number of cars, winter heating, coal-fired power generation and other reasons, lead to serious air pollution. In addition, due to human urban activities, ground and roof subsidence and contamination

There are many pollutants polluted by rainwater, and these pollutants flow into rainwater pipes or urban sewer network under the influence of rainwater flushing. Surface runoff generated during urban rainfall, or rainwater flowing into rainwater pipes and sewage pipes, will eventually converge in urban rivers, causing serious visual and environmental pollution to rivers. Rainwater pollution also occurs in agricultural and industrial areas around cities and along urban rail lines. These will lead to problems such as water quality imbalance and eutrophication in cities, resulting in the reduction of economic and ecological functions of water bodies[2].

(2) River siltation. Among the water environment problems of urban rivers, the silt and pollutants at the bottom of the rivers are also very prominent, resulting in blocked river flooding, which is not good for urban flood control and flood fighting. Secondly, the bottom mud accumulated in the river channel with excessive pollution will emit extremely strong corrosive stench in summer, breed a large number of mosquitoes, and affect the image of the city; it is a breeding ground for various pathogenic bacteria, spreading diseases through the river water, and threatening human health; it is River water quality continues to be polluted. One of the sources, leading to a serious imbalance in the water environment, water eutrophication and other environmental problems[7].

(3) The water quality of the river deteriorates. Some cities blindly pursue economic speed Rapid growth, no emphasis on environmental value, no focus on water pollution.

Pollution problems, and lax supervision of sewage discharge, resulting in failure to meet discharge standards. The standard waste water is discharged into the river, and even surface water pollution leads to groundwater pollution, threatening the water safety of the city.

5 Ecological management strategy of urban river water environment

5.1 Scientifically and rationally design the cross-section of the channel.

In the process of ecological management of urban river water environment, when encountering cross-section or longitudinal section, special attention should be paid to the safety performance of reservoir flood discharge, to maintain the original shape of the river to the greatest extent, and scientifically and rationally design the cross-section plan of the river. Ensure that the width of the river is coordinated and the depth is integrated. When designing the cross-section scheme of the river, it is necessary to ensure that the basic elements of the river will not suffer from various hazards. At the same time, on the basis of ensuring that the ecology of the urban river will not be damaged, the design of the river can fully reflect the development needs of urban intelligence. Thus creating a more beautiful ecological water landscape [3].

5.2 Construction of sewage recycling projects

The total population of cities in our country is increasing, and the discharge of a large amount of domestic sewage is not up to the standard. The indiscriminate discharge of large areas of sewage will not only damage the urban environment, but also seriously pollute the urban soil and water sources. People's lives and health are seriously threatened. under construction

In the process of establishing the urban ecological environment, relevant personnel should do a good job in the construction of sewage recycling projects, reasonably and effectively control the phenomenon of sewage discharge, and lay a good foundation for the healthy development of the urban ecological environment. For the construction of ecological cities, when treating sewage, it is necessary to effective combination of treatment and decentralized treatment to form a complete set of sewage treatment cycle mechanisms to realize the effective utilization of urban sewage. For example, in order to better recycle sewage, a city has effectively classified and treated the sewage in the city. In the city, schools, residential areas, commercial areas

There are not many toxic substances in the sewage, and a large amount of sewage can be reused with simple treatment. Therefore, in response to this phenomenon, the city clearly stipulates that school sewage can be used for urbanization after effective treatment, urban greening consumes a lot of water, and sewage recycling can meet the needs of green[6].

It can also improve the utilization rate of water resources, realize the optimal allocation of resources, and save water resources, which is consistent with the concept of green development [4]. However, this method is not suitable for sewage discharge in urban industrial areas. For sewage containing more toxic and harmful impurities, relevant

Personnel should formulate a special sewage treatment plan and corresponding discharge channels, do a good job in centralized treatment, and prevent the random discharge of industrial sewage from polluting the overall urban environment.

5.3 Scientific management and control of revetment works.

From During the construction of revetment works, it is necessary to understand the principles of water ecological management. According to the actual demand, according to the natural conditions in the embankment and the specific situation of the water surface line, Choose the construction scheme that has the least impact on biological survival. for urban waterways. In terms of water environment management, the selection of river protection structure and river embankment type. It is necessary to consider comprehensively all aspects to ensure that the construction technology is consistent with the actual plan. On the basis of meeting the design requirements, it is necessary to choose the most economical. The construction plan can improve the stability and safety of revetment works. At present, the ecological environment of waters in many areas of my country's cities is still in an unreasonable state, the engineering design should be consistent with the natural development of the river. In line with the law, the use of concrete should be reduced as much as possible in the construction of the project, so that the river course can maintain its original appearance. During the construction of revetment works, it is necessary to focus on widening the channel blockage and straightening the river channel, among which widening the channel blockage will continuously increase the degree of downstream inundation loss, while intercepting

If the bend is taken straight, the water will continue to scour the river channel and cut down.timing should be compared from the perspective of discharge flow and economic loss, choosing the most cost-effective construction plan improves the economic benefits of the project. in the river in the process of Tao governance, we must adhere to the scientific concept of development and the principle of sustainable development.

To improve the level of urban river management [5].

6 Conclusion

In conclusion, the ecological management of urban river water environments holds immense significance for the sustainable development of cities and the protection of the ecosystem. The improper actions of humans have led to severe pollution in these river systems, which not only diminishes their ability to self-heal but also disrupts their vital functions. This, in turn, affects the well-being of both residents and the environment. To address these challenges, ecological governance strategies must be implemented. the joint development of a multi-ownership economy. Adopting effective ecological management strategies becomes imperative[7]. This includes scientifically designing channel sections, constructing sewage recycling projects, and implementing proper revetment works management. By striking a balance between engineering solutions and ecological preservation, cities can enhance water quality, alleviate water scarcity, and promote a healthier environment. The ecological management of urban river water environments is not only a pressing need for environmental protection but also a vital aspect of ensuring the well-being and prosperity of urban communities. By following sound ecological principles and implementing targeted strategies, cities can rejuvenate their river ecosystems, mitigate water shortages, and foster a harmonious coexistence between urban development and nature.

References

1. Xu, Zuxin, et al. "Urban river pollution control in developing countries." Nature Sustainability 2.3 (2019): 158-160.
2. Wang, Yibo, et al. "Analysis and prevention of Urban River pollution." Journal of Physics: Conference Series. Vol. 1549. No. 2. IOP Publishing, 2020.
3. Honingh, Dorien, et al. "Urban river water level increase through plastic waste accumulation at a rack structure." Frontiers in earth science 8 (2020): 28.
4. Garrick, Dustin, et al. "Rural water for thirsty cities: A systematic review of water reallocation from rural to urban regions." Environmental Research Letters 14.4 (2019): 043003.
5. Mihai, Florin-Constantin, et al. "Plastic pollution, waste management issues, and circular economy opportunities in rural communities." Sustainability 14.1 (2021): 20.
6. Jamieson, R., et al. "Assessing microbial pollution of rural surface waters: A review of current watershed scale modeling approaches." Agricultural water management 70.1 (2004): 1-17.
7. Hu, Ran. "Pollution control and remediation of rural water resource based on urbanization perspective." Environmental Technology & Innovation 20 (2020): 101136.

Advancing Water Sustainability: Innovative Technologies in Hydrology and Water Resources Engineering

Xu Liu *

Harbin Engineering University, Harbin, China

Abstract. In the evolving field of water environment, hydrology, and water resources engineering, advanced water treatment techniques have emerged as pivotal elements in addressing contemporary challenges and ensuring sustainable water management. This academic article delves into a thorough examination of these advanced techniques, offering a comprehensive review that bridges the gap between cutting-edge technology and practical applications. It systematically explores a variety of key methods, including membrane technologies, advanced oxidation processes, and biological treatment systems, highlighting their effectiveness in various scenarios. The article further assesses the implications of these technologies on sustainable water management, focusing on their environmental impact and contributions to hydrological balance. It discusses how these methods aid in mitigating water pollution, enhancing water quality, and ensuring the efficient use of water resources. Additionally, the study sheds light on the integration of these technologies into current water management practices, examining their role in addressing the challenges posed by urbanization, industrialization, and climate change. Emphasizing the interdisciplinary nature of water resources engineering, the article also explores the interplay between these advanced water treatment methods and broader environmental and hydrological systems. It underscores the necessity for an integrated approach that combines technological innovation with sustainable practices, policy-making, and community involvement. By highlighting the importance of these technologies in contemporary water management strategies, the study aims to inform and inspire ongoing research, development, and implementation in this crucial field.

1 Introduction

In an era marked by escalating environmental challenges and burgeoning global populations, securing clean and sustainable water resources is imperative. Advanced water treatment technologies have emerged as pivotal instruments in this quest for sustainable water management. This article systematically examines advanced water treatment methods, focusing on Membrane Filtration Technologies and Advanced Oxidation Processes (AOPs). These technologies play crucial roles in enhancing water quality, combating pollution, and securing water sustainability. Moreover, the integration of advanced water treatment technologies with hydrology is reshaping water resource management. Sustainable water supply, ecosystem protection, and the mitigation of water-related extremes are transforming hydrological practices and environmental stewardship. In the environmental context, advanced water treatment fosters responsible environmental practices, ecological well-being, biodiversity conservation, and the promotion of sustainability [1]. This article serves as a beacon, illuminating the critical role that advanced water treatment technologies play in addressing contemporary water challenges. Through the fusion of science and engineering, these technologies offer promising avenues for securing clean water resources, protecting ecosystems, and ensuring a resilient hydrological landscape.

2 Advanced Water Treatment Technologies

Advanced water treatment technologies encompass a diverse spectrum of methods meticulously engineered to elevate water quality, bolster its availability, and ensure long-term sustainability. In this section, we delve deeper into two prominent categories: Membrane Filtration Technologies and Advanced Oxidation Processes (AOPs), elucidating their pivotal roles in the quest for cleaner and more sustainable water resources.

2.1 Membrane Filtration Technologies

Microfiltration. Microfiltration stands as a stalwart guardian of water quality, functioning as a physical barrier that sieves out impurities. Employing a porous membrane with pore sizes typically in the range of 0.1 to 10 micrometers, this method adeptly captures suspended

*2482516799@qq.com

solids, bacteria, and even some viruses. Microfiltration's versatility renders it indispensable in various applications, from wastewater treatment to the production of potable water [2].

Ultrafiltration. Ultrafiltration takes the filtration process a step further by employing even finer membranes, often with pore sizes between 0.001 and 0.1 micrometers. This technology excels at removing colloidal particles, macromolecules, and certain dissolved contaminants. It finds widespread use in purifying drinking water, industrial processes, and treating effluents laden with challenging pollutants.

Nanofiltration and Reverse Osmosis. Nanofiltration and Reverse Osmosis represent the pinnacle of membrane-based filtration technologies, utilizing membranes with nanopores and molecular sieving capabilities. While nanofiltration selectively removes divalent ions, organic compounds, and colorants, reverse osmosis goes a step further by tackling monovalent ions and virtually all solutes. These technologies play pivotal roles in desalination, the treatment of brackish water, and producing high-purity water for critical applications like pharmaceuticals and electronics manufacturing.

2.2 Advanced Oxidation Processes (AOPs)

Photocatalysis. Photocatalysis harnesses the power of light to trigger chemical reactions that degrade organic pollutants. Typically, a photocatalyst, often titanium dioxide (TiO_2), is irradiated with ultraviolet or visible light, generating highly reactive radicals that break down contaminants into harmless byproducts. This eco-friendly technique finds applications in purifying both air and water, offering a sustainable solution to mitigate environmental pollution.

Ozonation. Ozonation is a robust oxidative process that employs ozone (O_3) to dismantle recalcitrant organic compounds [3]. Ozone, a potent oxidant, readily reacts with a wide range of contaminants, breaking down their chemical structures into simpler, less harmful constituents. Widely adopted in wastewater treatment plants, ozonation enhances the removal of persistent organic pollutants and mitigates the release of harmful substances into the environment.

In conclusion, these advanced water treatment technologies, membrane filtration, and advanced oxidation processes exemplify the scientific and engineering prowess dedicated to ensuring water quality, safeguarding ecological balance, and promoting sustainable water resource management. Their multifaceted applications continue to evolve, making them indispensable tools in the quest for a cleaner, healthier, and more resilient aquatic environment.

2.3 Descriptive Statistics

Desalination technologies, crucial for addressing water scarcity, involve converting seawater or brackish water into fresh water. These technologies are vital in arid regions and areas where freshwater resources are limited. Seawater desalination mainly focuses on removing salts from ocean water, making it suitable for drinking, agriculture, and industry. Techniques like reverse osmosis, which filters water through semi-permeable membranes, and thermal distillation, which involves heating and condensing seawater, are commonly used. Brackish water desalination treats water with lower salinity, often found in estuaries or underground. It's less energy-intensive due to lower salt content and crucial for augmenting freshwater supplies in areas with limited surface water [4]. While desalination offers a solution to water scarcity, it faces challenges like high energy demands and managing saline by-products. Continuous advancements aim to enhance efficiency and reduce environmental impact, making desalination a key component in modern water resource management.

3 Applications in Hydrology

The incorporation of advanced water treatment techniques into hydrological studies and water resource management is nothing short of transformative, ushering in a new era of sustainable water practices and enhanced environmental protection.

3.1 Sustainable Water Supply

Ensuring Resilient Water Sources. The utilization of advanced water treatment technologies stands as a linchpin in guaranteeing a sustainable and secure water supply. By purifying alternative water sources such as treated wastewater and brackish water, communities reduce their reliance on conventional freshwater reserves. This diversification of water sources bolsters resilience against droughts, climatic fluctuations, and burgeoning water demands.

Fostering Water Conservation. Moreover, advanced treatment methods actively promote water conservation. By maximizing the efficiency of water purification, less water is wasted during treatment processes [5]. This virtuous cycle not only conserves water but also reduces the energy required for treatment, contributing to a more sustainable and eco-friendly water supply.

3.2 Ecosystem Protection

Minimizing Adverse Effects. One of the most significant ramifications of integrating advanced treatment techniques is the protection of aquatic ecosystems. Untreated water discharges can unleash a barrage of contaminants and pollutants into rivers, lakes, and oceans, causing harm to delicate ecosystems. Advanced treatment methods serve as a robust defense, minimizing the adverse effects of such discharges and promoting healthier hydrological environments.

Fostering Ecological Balance

Furthermore, these techniques foster ecological balance within aquatic ecosystems. By reducing the influx of harmful substances, they create conditions more conducive to the flourishing of indigenous species. This, in turn, supports fisheries, wetlands, and other vital ecosystems that rely on clean and well-maintained aquatic habitats.

3.3 Flood and Drought Mitigation

Managing Floodwaters. The role of advanced water treatment technologies extends beyond water purification. They are instrumental in managing floodwaters effectively. By treating and storing excess stormwater, urban areas can mitigate the destructive impacts of floods. This proactive approach not only safeguards communities but also prevents pollutants from being swept into water bodies during intense rainfall events.

Mitigating Drought Effects. In times of drought, advanced water treatment technologies can be deployed to augment water supplies [6]. By treating and reusing wastewater, communities can stretch their available water resources further. This strategy enhances overall hydrological resilience, ensuring that water remains accessible even in the face of prolonged drought conditions.

4 Environmental Considerations

Effective water treatment measures are intrinsically tied to responsible environmental management, with profound implications for ecological well-being.

4.1 Water Quality Improvement

Mitigating Ecological Footprint. Perhaps the most immediate benefit of advanced water treatment is the substantial improvement in water quality. By removing contaminants, nutrients, and pathogens, these methods reduce the ecological footprint of polluted water bodies. This, in turn, allows for the recovery and rejuvenation of aquatic ecosystems, benefiting both flora and fauna.

Safeguarding Aquatic Life. Enhanced water quality safeguards aquatic life, providing the necessary conditions for thriving ecosystems. Fish, invertebrates, and other species dependent on clean water benefit from the restoration of their natural habitats [7]. This ripple effect contributes to the conservation of biodiversity within freshwater and marine environments.

4.2 Sustainable Environmental Practices

Promoting Responsible Water Use. The incorporation of advanced water treatment aligns seamlessly with sustainable environmental practices. It champions responsible water use by emphasizing the importance of treating and conserving this finite resource. This approach resonates with broader sustainability goals, encouraging the prudent management of water resources.

4.3 Biodiversity Conservation

Holistic Ecosystem Support. Ultimately, enhanced water quality resulting from advanced treatment methods bolsters biodiversity conservation efforts. Healthy aquatic ecosystems, nurtured by clean water, provide crucial habitats and sustenance for a diverse range of species. The conservation of these ecosystems, in turn, safeguards the delicate balance of nature and the intricate web of life within water bodies.

In summation, the integration of advanced water treatment technologies into hydrological studies and water resource management transcends mere technological innovation. It represents a pivotal shift towards more sustainable water practices, ecosystem protection, and environmental stewardship. These advanced methods are, without a doubt, instrumental in shaping a more resilient, cleaner, and healthier hydrological landscape for generations to come.

5 Discussion

In this discussion, we will explore the integration of advanced water treatment technologies, specifically the combination of Membrane Bioreactors (MBRs) and Advanced Oxidation Processes (AOPs), in wastewater treatment. This integration represents an innovative approach within the field of water resources engineering, providing powerful tools to address contemporary water environmental challenges and ensure sustainable water resource management.

Membrane Bioreactors (MBRs) are effective wastewater treatment technologies that employ porous membranes to separate suspended solids, microorganisms, and organic matter, thereby improving water quality. MBRs efficiently remove contaminants from wastewater and prevent the backflow of microorganisms, enhancing the efficiency and quality of wastewater treatment. However, MBRs face challenges such as membrane fouling and energy consumption. Advanced Oxidation Processes (AOPs) encompass a range of techniques that utilize highly reactive oxidants to degrade recalcitrant organic pollutants. These techniques include photocatalysis, ozonation, and advanced peroxidation. AOPs are particularly effective in breaking down persistent organic compounds into simpler, less harmful byproducts. Nevertheless, some AOPs may require high energy input, necessitating careful consideration in practical applications. The combined application of MBRs and AOPs holds significant promise in wastewater treatment. MBRs efficiently remove biodegradable pollutants and microorganisms from wastewater, while AOPs can further target non-biodegradable and recalcitrant organic compounds. This synergistic approach enhances the overall efficiency of wastewater treatment and facilitates the removal of challenging contaminants. Additionally, the integration can contribute to reduced energy consumption compared to using AOPs alone, as MBRs pre-treat the wastewater. It is important to note that the successful implementation of MBR-AOP systems requires careful design and operation to ensure system stability and efficiency. Factors such as cost and energy consumption should be comprehensively considered in practical applications. Despite these challenges, the integration of MBRs and AOPs represents a cutting-edge advancement in water resources engineering, offering a more sustainable solution to wastewater treatment and environmental protection.

6 Conclusion

Advanced water treatment technologies are vital components in addressing modern water challenges and achieving sustainable water management. This review has extensively examined these technologies, focusing on Membrane Filtration Technologies and Advanced Oxidation Processes (AOPs), showcasing their effectiveness in improving water quality and mitigating pollution. The integration of advanced water treatment methods into hydrology has ushered in a transformative era of sustainable water practices, offering solutions for sustainable water supply, ecosystem protection, and flood and drought mitigation. In the realm of environmental considerations, these technologies have the potential to significantly improve water quality, protect aquatic life, promote responsible water use, and support biodiversity conservation. However, there is still much work to be done. Future research should focus on optimizing these technologies, reducing energy consumption, and enhancing cost-effectiveness. Collaborative efforts among scientists, engineers, policymakers, and communities are essential for effective implementation. In conclusion, advanced water treatment technologies represent a promising path toward a cleaner and more resilient hydrological landscape. Continued research and innovation are crucial to harness their full potential and ensure a sustainable water future for generations to come.

References

1. La Rosa, Giuseppina, et al. "Coronavirus in water environments: Occurrence, persistence and concentration methods-A scoping review." Water research 179 (2020): 115899.
2. Rathi, B. Senthil, P. Senthil Kumar, and Dai-Viet N. Vo. "Critical review on hazardous pollutants in water environment: Occurrence, monitoring, fate, removal technologies and risk assessment." Science of The Total Environment 797 (2021): 149134.
3. Amarasiri, Mohan, Daisuke Sano, and Satoru Suzuki. "Understanding human health risks caused by antibiotic resistant bacteria (ARB) and antibiotic resistance genes (ARG) in water environments: Current knowledge and questions to be answered." Critical Reviews in Environmental Science and Technology 50.19 (2020): 2016-2059.
4. Cheng, Dongle, et al. "A critical review on antibiotics and hormones in swine wastewater: Water pollution problems and control approaches." Journal of hazardous materials 387 (2020): 121682.
5. Sun, Alexander Y., and Bridget R. Scanlon. "How can Big Data and machine learning benefit environment and water management: a survey of methods, applications, and future directions." Environmental Research Letters 14.7 (2019): 073001.
6. Zounemat-Kermani, Mohammad, et al. "Ensemble machine learning paradigms in hydrology: A review." Journal of Hydrology 598 (2021): 126266.
7. Vélez-Nicolás, Mercedes, et al. "Applications of unmanned aerial systems (UASs) in hydrology: A review." Remote Sensing 13.7 (2021): 1359.

Effects of corncob carbon on peanut growth and soil properties

Haoyuan Tang[a], Zheng Wang[b*]

College of Geology and Environment, Xi'an University of Science and Technology, Xi'an 710054, China

Abstract. In this paper, corncob biochar was prepared by pyrolysis, and the basic physicochemical properties of the carbon were analysed by characterisation, and potting tests were used to study the effects of different carbon application rates on peanut growth indexes and soil conventional physicochemical properties. The results showed that the corncob carbon at 500°C for 30 min was suitable for soil improvement; the carbon had a pH of 9.98 and could regulate soil pH; it had a rich pore structure with a specific surface area of 11.6 $m^2 \cdot g^{-1}$, which could improve soil pore density; and its N、P、K contents were 5.8、8.3、49.5 $g \cdot kg^{-1}$, respectively, which could provide exogenous nutrients to the soil. The higher the amount of carbon applied, the better the effect on soil physicochemical properties and yield promotion. When the amount of carbon applied was 7%, the soil capacity was reduced by 8.8%, the plant height was reduced by 5.5%, the stem thickness was increased by 18.4%; soil pH increased by 3%, CEC by 20.2 %, and organic matter by 52.03%, resulting in a 40.8% increase in peanut yield.. This corncob carbon has good physicochemical properties and is an ideal soil conditioner.

1. Introduction

According to statistics[1], China's corn production in 2022 reaches 277.2 million tonnes, an increase of 1.7% over last year's production, and it is estimated that corn kernel production could reach 58.2 million tonnes. Pyrolysis is an economical and effective way to treat agricultural waste, and its derivative biochar has special physicochemical properties that are very suitable for soil improvement. The properties and efficacy of biochar vary greatly depending on the charring process conditions, raw materials and preparation methods, and the results of related studies are difficult to compare effectively and in common, and even some conclusions are contradictory, which is of limited significance and value in guiding agricultural production[2]. Therefore, it is especially necessary to carry out research on the characteristics and application of biochar of different materials and carbonisation processes[3]. In this paper, corncob carbon was prepared by pyrolysis carbonisation of corncobs as raw material, and the effect of corncob carbon on soil improvement was further explored on the basis of clarifying the changes in the basic physicochemical properties before and after carbonisation.

2. Materials and Methods

2.1. Experimental materials

The test materials, corncobs, were taken from Langfang, Hebei Province, naturally air-dried and then crushed to 1~2 cm particles with a crusher.The test peanut variety was Hanyu 19; the test soil was taken from Langfang, Hebei Province, with the following basic physicochemical properties: pH 7.9, CEC content of 12.55 $cmol \cdot kg^{-1}$, organic matter content of 12.01 $g \cdot kg^{-1}$, alkaline dissolved nitrogen content of 91.3 mg kg^{-1}, effective phosphorus content of 12.24 mg kg^{-1}, and quick-acting potassium content of 74.2 mg kg^{-1}. Corncob carbon was taken from the subject group.

2.2 Preparation of corncob carbon

The test was carried out using the pyrolysis furnace developed by the group. Weighing the appropriate amount of corn heart particles placed in the pyrolysis furnace after the passage of N_2 blowing for 5 min, to maintain an oxygen-free atmosphere in the furnace, with a temperature increase rate of 10°C/min were raised to 400, 500, 600°C, and stayed at the target temperature time of 30 min; 10°C/min temperature increase rate was raised to 500°C, and stayed at the target temperature for 15, 30, 45 min, respectively. After the end of pyrolysis, continue to pass N2 to cool the corncob carbon to room temperature and then remove it for storage, labelled as CCB-400, CCB-500, CCB-600, CCB-15, CCB-30 (CCB-500), CCB-45, respectively.

2.3. Characterisation methods

The main elements were determined by elemental analyser (vario MACRO cube), ash, volatile matter and fixed

* Corresponding author: [b]8795463@qq.com
[a]t2675492895@163.com

carbon content were determined by reference to the standard test method for charcoal (GB/T 17664-1999), calorific value was determined by high-precision microcomputer full-automatic calorimeter (ZDHW-8), specific surface area was determined using a physical adsorption apparatus (AutoSorb-iQ), microstructure was determined by scanning electron microscope (Merlin), trace elements were determined by X-ray fluorescence spectrometer (PANalytical B.V.AXIOS), and pH was determined by pH meter (HQ40d).

2.4. Soil improvement test

Corncob carbon was homogeneously mixed with soil (10 kg) at mass ratios of 0, 3%, 5% and 7% respectively and placed in pots. The experiment was set up with four treatment groups, which were recorded as CK-P, CCB3-P, CCB5-P, CCB7-P according to the amount of charcoal applied to each, and three replicate experiments for each treatment group, totalling 12 pots. Each pot had 3 peanuts of each kind, and after the seedlings grew to 8~10 cm, two seedlings with the same growth were selected and retained; routine management such as water application, pest control and weeding were unified.

2.5. Sampling and measurement methods

When the peanuts are mature, collect the plants of each treatment group, measure the plant height and stem thickness indexes, and harvest the peanuts and soybeans of each treatment group to calculate the yield; take 1 kg of soil samples from 0~20 cm of each treatment group into a self-sealing bag, and then air-dry and sieve them and store them in wide-mouthed bottles for the determination of soil chemical properties.

The soil bulk weight was measured by ring knife method; soil pH was determined by pH meter (HQ40d); soil organic matter content was determined by oxidative external heating method with potassium dichromate; soil alkaline nitrogen was determined by alkaline diffusion method with 1.0 mol/L NaOH leaching; effective phosphorus was determined by molybdenum antimony antimony colorimetric method with 0.5 mol/L $NaHCO_3$ leaching; and quick potassium was determined by flame retardant method with 1.0 mol/L ammonium acetate leaching. Ammonium acetate extraction, flame photometric method; cation exchange capacity (CEC) refers to the ammonium acetate method[4].

2.6. Data analysis

The experimental data were processed using Excel 2019 software, and the data were analysed for the significance of differences using SPSS26.0 statistical analysis software, and the one-way ANOVA test was used for the comparison between multiple groups of data, with the significance level set at 0.05, and plotted using Origin2019.

3. Results and Discussion

3.1 Effect of different pyrolysis temperatures and times on the properties of the product carbon

3.1.1 Effect of pyrolysis temperature on char yield, pH and specific surface area.

As can be seen from figure 1, the char yield gradually decreased from 28.3 % to 25.4 % with the increase of temperature because the higher the pyrolysis temperature, the more intense the reaction is, the more volatiles are analysed, and thus the solid product decreases. The yield decreased by 2.2% from 400 to 500°C and only 0.7% from 500 to 600°C, which was because the char product decreased slowly after the final temperature of pyrolysis was nearly complete at 600°C. pH increased with the temperature, by 2.3% from 400 to 500°C and 0.8% from 500 to 600°C, which indicated that the pyrolysis temperature had a certain effect on the pH of the corncob carbon, and the temperature increased from 400 to 600°C, which had a certain effect on the pH of the corncob carbon, and the temperature increased from 400 to 600°C, and the temperature increased from 500 to 600°C. The specific surface area of the char at each temperature was 7.1, 11.6 and 13.5 g/cm^3, respectively, which means that the specific surface area of corn cob char increased with the increase of temperature.

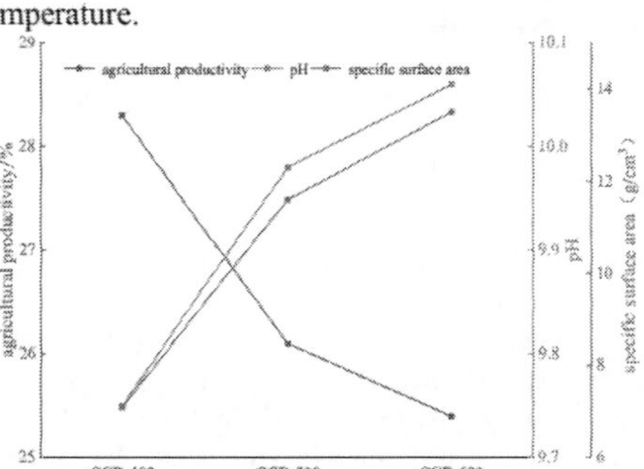

Figure 1. Effect of pyrolysis temperature on carbon yield、pH and specific surface area

3.1.2 Effect of pyrolysis time on carbon yield, pH and specific surface area.

As can be seen from figure 2, the corncob carbon yield showed a certain decreasing trend under different pyrolysis times, but all of them fluctuated above and below 26 % with little change; the pH showed an increasing trend, but the range of change was <0.1. This indicated that there was almost no effect of different retention times on the corncob yield and pH under 500 °C. The specific surface areas of the char at each pyrolysis time were 9.3, 11.6, and 12.1 g/cm3, respectively, indicating that the specific surface area of the corncob carbon increased with increasing time.

The study shows that alkaline biochar can increase soil pH and soil aggregation, thus improving crop yield; the larger specific surface area is because the charcoal pore structure is more abundant, which is of positive

https://doi.org/10.1051/e3sconf/202456002003

significance to improve soil pore density and enhance soil water retention. Considering the energy consumption, char yield, pH and specific surface area, the working conditions for soil improvement in this experiment were pyrolysis temperature of 500°C and pyrolysis time of 30 min.

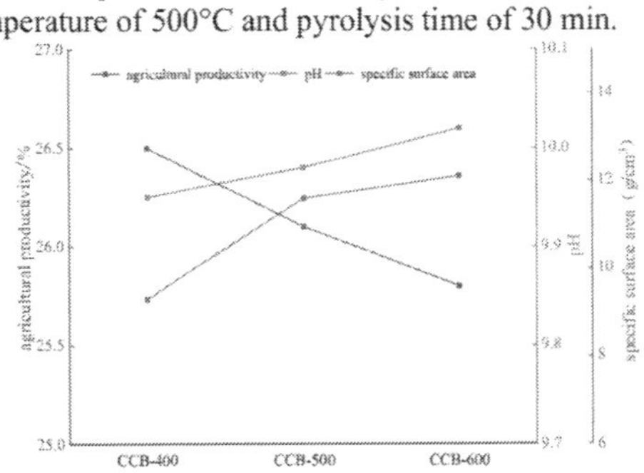

Figure 2. Effect of pyrolysis time on carbon yield、pH and specific surface area

3.2. Physicochemical properties of corncob carbon

3.2.1. Microstructure.

The microstructure of corncobs before and after charring is shown in figure 3. Before the charring of corncobs, the whole was in the form of large lumps with more folds on the surface, but almost no pores. After carbonisation, the original bulk shape of corncob gradually formed a clear grain bundle, forming a porous carbon framework structure with charred lignin as the main body. It can be seen that the charring of corn cobs has a rich pore structure and an increased specific surface area, which enhances its adsorption capacity and plays an important role in agricultural activities such as improving the soil environment, reducing nutrient loss, and enhancing water-holding capacity[5].

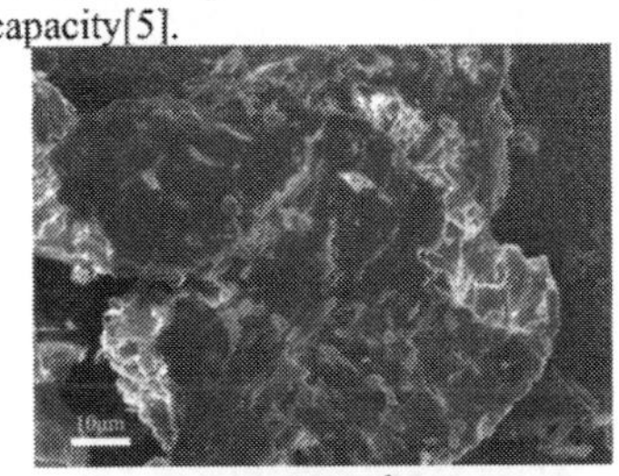

a. corncob

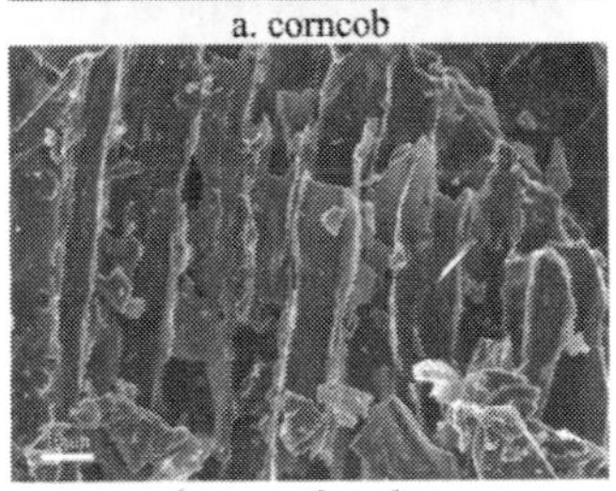

b. corncob carbon

Figure 3. Microstructure of corncobs before and after charring

3.2.2. Industrial analysis, elemental analysis and calorific value.

The industrial analysis and calorific value (dry basis) of corncobs before and after charring are shown in figure 4. The industrial analysis of corncobs before and after carbonisation had obvious changes, the volatile fraction content decreased by 92.22%, which was related to the susceptibility of corncobs to thermal decomposition and large loss of volatile components during pyrolytic carbonisation; the fixed carbon content increased by 10.25 times, which was conducive to the sequestration and emission reduction, and the improvement of the input of soil carbon sinks.

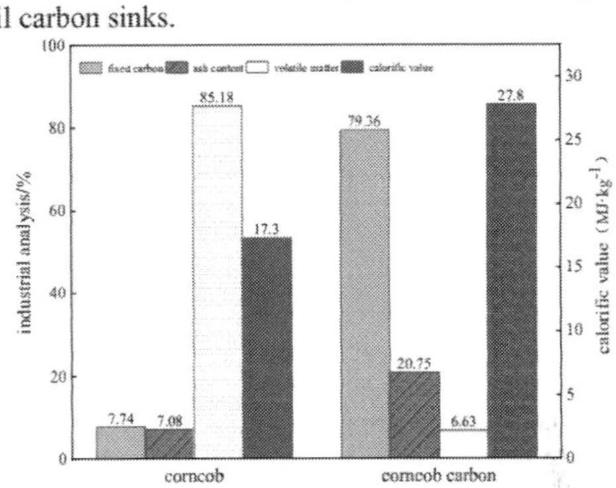

Figure 4 Industrial analysis and calorific value of corncobs before and after charring

The elemental analysis of corncobs before and after charring is shown in table 1. The C, S and N contents of corncobs increased by 1.74, 1.69 and 1.4 times respectively, and the H and O contents decreased by 3.62 and 2.38 times after carbonisation.The increase of C content and the decrease of H and O contents indicate that the organisation of the organic components of corncobs changed during pyrolysis and carbonisation, and accompanied by the breakage of the long-chain and the formation of thick ring, the resulting corncobs carbon is more stable than that of corncobs.H/C can reflect the aromaticity of the charcoal material, the H/C of corncobs is about 6.3 times lower than that of corncobs carbon, indicating that the unsaturated C of corncobs which is easy to be decomposed such as sugars and carbohydrates is transformed into the C with high aromaticity and saturation, which is relatively stable.(N+O)/C is a parameter for evaluating the polarity, and the decrease of about 4 times in the table indicates that the polarity of corncobs carbon is reduced and its water repellency is enhanced, and therefore it is not easy to be used by microorganisms, and it is not easy to be used by microorganisms[6]. The decrease of (N+O)/C indicates that the oxygen-containing functional groups such as hydroxyl, carboxyl and carbonyl groups in corncobs are burned away in large quantities, and the reduction of these highly reactive functional groups is conducive to the improvement of the stability of the biochar[7].

Table 1 Elemental analysis of corncob before and after carbonization (dry basis)

	corncob	corncob carbon
C	47.42	82.61
H	5.36	1.48
S	0.16	0.27

N	0.45	0.63
O	39.52	16.60
O/C	0.833	0.201
H/C	0.113	0.018
(N+O)/C	0.843	0.209

3.2.3. Other properties of corncob carbon.

Other properties of corncob carbon are shown in Table 2. It can be seen from Table 2 that the corncob carbon is alkaline, has a large specific surface area and CEC, and has a high content of N, P and K, which is suitable for soil improvement.

Table 2. Other properties of corncob carbon

	corncob carbon
pH	9.98
CEC/cmol/kg	21.49
specific surface area/m^2/g	11.6
organic matter/mg/kg	40.58
N g/kg	5.8
P g/kg	8.3
K/g/kg	49.5

In summary, corncob carbon's rich pore structure, large specific surface area, alkalinity, high stability, high fixed carbon content, rich in N, P, K nutrients and other characteristics are suitable for soil improvement to achieve the effect of yield increase and carbon sequestration and emission reduction.

3.3. Effect of corncob carbon soil amendment

3.3.1. Effect of corncob carbon on soil capacity weight.

The lightweight and porous structure of biochar makes it can directly affect the soil capacity after it is applied to the soil, which is important for improving the soil structure, tillability and soil microbiological environment. The effect of corncob carbon on soil capacity of peanut is shown in figure 5. The soil capacity of peanut at each carbon application was 1.281 g·cm^{-3}, 1.279 g·cm^{-3}, 1.239 g·cm^{-3} and 1.168 g·cm^{-3}, respectively, and the soil capacity of peanut treated with corncob carbon was lower than that of CK-P, and it gradually decreased with the increase of the amount of carbon applied, of which the differences between CCB5-P and CCB7-P treatments and CK-P were significant, and there was no significant change between the CCB3-P treatment did not vary significantly with CK-P. It indicated that the application of corncob carbon could reduce the soil capacity weight of potted peanut, and the larger the application amount, the better the reduction effect. The significant reduction of soil capacity weight improved the physical structure of peanut soil and made the peanut soil looser, which was conducive to the development of peanut root system and promoted the maturity of peanut fruits.

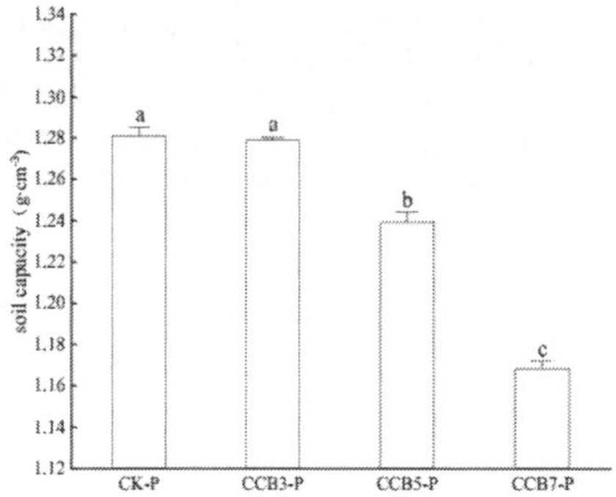

Figure 5. Effect of maize kernel charcoal on soil bulk density
Note: Different lower case letters in the graphs indicate significant differences between different treatment groups at the 0.05 level. Same as below

3.3.2. Effect of corncob carbon on plant height.

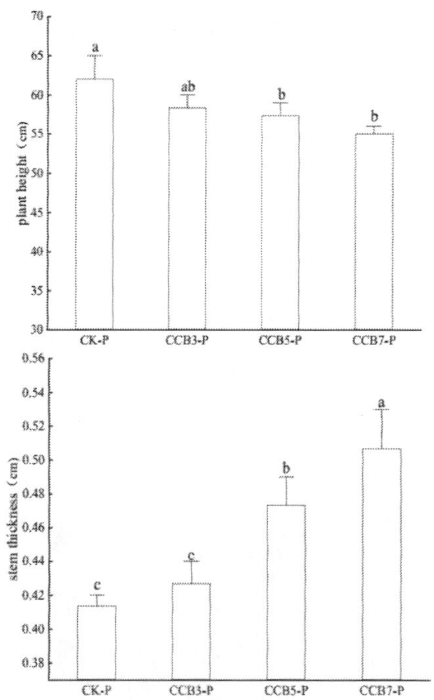

Figure 6. Effect of corncob carbon on peanut plant height and stem thickness

The effect of corncob carbon on plant height and stem thicknessis shown in figure 6. The height of peanut plants in each treatment group was 62、58.3、57.3 and 55 cm, respectively, and the corncob carbon treatments were all significantly lower than that of CK-P, indicating that corncob carbon had a certain inhibitory effect on the height of peanut plants, and the inhibitory effect was greater with the increase in the amount of carbon applied. Stem thickness reflects the degree of crop strength, and peanut stem thickness was significantly higher than that of the control group after applying corncob carbon, and the larger the amount of carbon applied, the more obvious the effect of stem thickness enhancement. Compared with the control group, peanut stem thickness increased by 3.1%, 12.7% and 18.4% in the corncob carbon treatments,

 https://doi.org/10.1051/e3sconf/202456002003

indicating that corncob carbon has a greater effect on crop stem thickness, and the best effect of promoting peanut stem thickness was achieved when the amount of carbon applied was 7%, which resulted in a fuller nutrient uptake for the crop.

3.3.3. Effect of corncob carbon on soil pH and CEC.

The effect of corncob carbon on soil pH and CEC is shown in figure 7. The pH values of peanut soil were 8.02, 7.96, 8.05, and 8.27 for each carbon application, respectively; compared with the control group, the pH decreased at low carbon application, but there was no significant difference, and the pH increased at high carbon application, and the higher the pH increased the more obvious with the increase of carbon application, which indicated that corn-core charcoal could increase the pH of the alkaline soil at the high carbon application. The CEC of peanut soil was 13.13, 13.95, 15.39 and 16.45 cmol·kg^{-1} for each charcoal application, respectively; compared with the control group, the peanut soil treated with corncob carbon was significantly higher than that of the control group, and the tendency of different carbon applications to increase the CEC was CK-P<CCB3-P<CCB5-P<CCB7-P, which indicated that the higher the amount of corncob carbon applied, the CEC of the soil was the greater the enhancement. It may be related to the surface functional groups and alkaline nature of corncob carbon. The application of corncob carbon changed the physicochemical properties and structure of soil, improved the living environment of soil microorganisms, and made the cation exchange activities more active, thus increasing the total amount of soil cation exchange. This has a positive effect on improving soil fertility and crop growth.

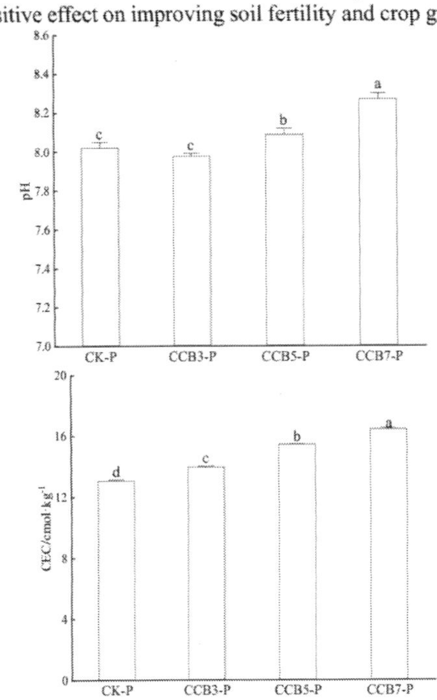

Figure 7. Effect of corncob charcoal on soil pH and CEC

3.3.4 Effect of corncob carbon on soil organic matter and effective nutrients

Figure 8 and Table 3 show the changes in soil organic matter and effective nutrients in soils with different amounts of corncob carbon applied. Organic matter is the sum of humus, plant and animal residues and microbial bodies formed by microbial action, and its content is a balance between the input of organic matter such as biological residues entering the soil and the loss of organic matter dominated by soil microbial decomposition. The soil organic matter under each charcoal application in figure 8 increased significantly and showed a gradual increase in the content of organic matter with the increase in charcoal application, which increased by 22.16, 44.34 and 52.03%, respectively, and the increase in organic matter increased soil fertility and had a promotional effect on the growth and development of crops. In table 3, soil effective phosphorus, quick-acting potassium and alkaline dissolved nitrogen increased with the application of corn cob charcoal, and the highest increase was observed when the charcoal application rate was 7%, which was 22.80、162.6 and 214.8 mg·kg^{-1}, respectively.These soil nutrient elements can be directly absorbed by the plants, which promotes the growth of the peanut crop and the increase in the yield of peanut fruits. The increase of organic matter and nutrients in the soil is due to the fact that corncob carbon itself is rich in organic matter and N, P and K elements. In addition, the special properties of corncob carbon improve the soil environment and increase the activity of soil microorganisms, resulting in an increase in soil organic matter and effective nutrients, which are absorbed and utilised by the crop.

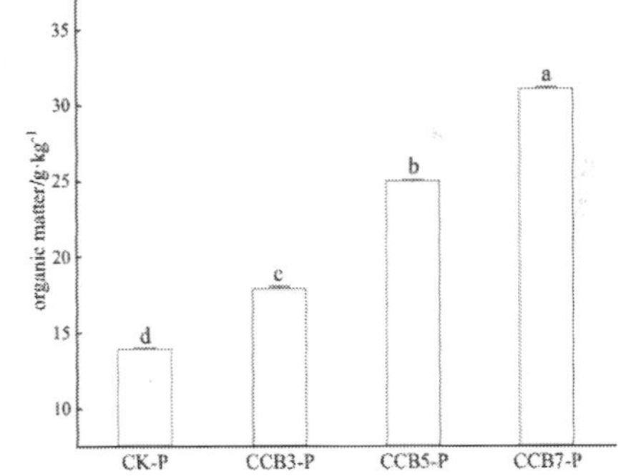

Figure 8 Effect of corncob carbon on soil organic matter

Table 3 Effect of maize corn cobcarbon on soil effective nutrients

	effective phosphorus	quick-acting potassium	alkaline nitrogen decomposition
0	12.92±0.137a	78.8±0.294a	92.5±0.125a
3%	16.60±0.207b	89.3±0.419b	141.4±0.648b
5%	18.58±0.145c	130.4±0.787c	165.7±0.531c
7%	22.80±0.083d	162.6±0.554d	214.8±0.531d

3.3.5. Effect of corncob charcoal on yield.

The peanut weights of the treatments were 126.9、136.2、181.8 and 214.3 g. Peanut yield increased by 6.8、30.2 and

40.8%, respectively, compared to the control, with a significant increase in yield at a carbon application of 5%, and a maximum increase in yield at a carbon application of 7%. The increase in peanut yield after carbon application was associated with a decrease in soil bulk density, increase in pH and CEC, which promoted rootstock development and increased peanut yield.

3.4 Discussion

The increase in peanut yield is related to the application of corncob carbon. Firstly, the unique microporous structure and large specific surface area of corncob carbon improved the pore density of the soil, reduced the soil bulk density, improved the soil structure, effectively alleviated the problem of soil compaction, promoted the development of peanut roots and stems, and made it easier for peanuts to absorb nutrients, thus increasing the peanut yield; at the same time, it could enhance the water retention of the soil, and reduce the loss of nutrients from the soil. Secondly, the high pH of corncob carbon improves the pH buffer capacity of the soil and enhances the soil's ability to resist acidity. The mechanism of biochar to improve soil pH is mainly related to its own chemical properties and the reaction process in the soil. On the one hand, biochar contains a large number of alkaline groups, which are able to combine with acidic substances in the soil and release OH- ions. This reaction process helps to neutralise the acids in the soil, thus increasing the pH of the soil. On the other hand, the surface of biochar is rich in oxygen-containing functional groups, and the anions of these weakly acidic functional groups are able to react with H+ to form neutral molecules. At the same time, this process releases previously adsorbed exchangeable salt-based cations into solution, thereby increasing the pH buffering capacity of the soil.

In addition, the application of biochar improves the soil carbon pool, increases soil salt base saturation, enhances soil fast-acting phosphorus, potassium, and effective nitrogen content, and enhances soil fertiliser retention capacity, thereby improving the plant growth environment. These positive effects further enhance the value of biochar application in agricultural production and soil improvement.

4. Conclusions

1) Pyrolysis temperature and time have an effect on the pH and specific surface area of corncob carbon; taking into account the charcoal yield, energy consumption and charcoal properties, the corncob carbon working condition suitable for soil improvement is 500°C and 30 min.

2) Corncob carbon is rich in pore structure, with a specific surface area of 11.6 $m^2 \cdot g^{-1}$; pH=9.98, alkaline, which can regulate soil pH; C content is as high as 82.61% with an H/C of 0.018, which is highly stable, with a fixed carbon content of 79.36%, which can be applied to the soil to sequester carbon and reduce emissions, and to improve soil organic matter; the contents of N, P, and K are 5.8 $g \cdot kg^{-1}$, 8.3 $g \cdot kg^{-1}$, and 49.5 $g \cdot kg^{-1}$, which are ideal agricultural biochar materials.

3) Corncob carbon can reduce soil capacity, improve soil structure, and promote peanut growth and soil pH and CEC, of which, the best yield increase was achieved when the carbon application rate was 7%. At this carbon application rate, peanut soil bulk weight decreased by 8.8%; plant height decreased by 5.5%, stem thickness increased by 18.4%; soil organic matter increased by 52.03%. pH increased by 3%, CEC increased by 20.2%, and finally peanut yield increased by 40.8%.

References

1. Statistical bulletin of the people's republic of china on national economic and social development for 2022. J. China Statistics, 2023(03):12-29.
2. Mašek O, Brownsort P, Cross A. (2013) Influence of production conditions on the yield and environmental stability of biochar. J. Fuel, 103: 151-155.
3. Zhang W M, Guan X C, Huang Y W. (2015) Biological effects of corncob biochar on soybean. J. Journal of Agricultural Environmental Science, 34(02):391-400.
4. Bao S. (2000) Soil agrochemical analysis . China Agricultural Science and Technology Press, Beijing.
5. Kang M W, Yibeltal M, Kim Y H. (2022) Enhancement of soil physical properties and soil water retention with biochar-based soil amendments. J. Science of the Total Environment, 836.
6. Singh B, Singh B P, Cowie A L. (2010) Characterisation and evaluation of biochars for their application as a soil amendment. J. Soil Research, 48(7): 516-525.
7. Özçimen D, Karaosmanoğlu F. (2004) Production and characterization of bio-oil and biochar from rapeseed cake. J. Renewable Energy, 29(5): 779-787.

https://doi.org/10.1051/e3sconf/202456002004

Study of seepage patterns and calculation of infiltration coefficients of loose stacked layers on slopes under rainfall action

Chengcheng He*

College of Geodesy, Chang'an University1, Xi'an, Shaanxi, China

Abstract. Rainfall is one of the most important factors inducing landslide occurrence, and previous researchers have carefully studied the stages of rainfall infiltration as well as the factors that affect infiltration (e.g., rainfall intensity, rainfall duration, initial water content and dry density). However, previous studies using homogeneous soils were overly idealised and ignored the effects of water content and dry density on infiltration, which vary with depth. In this paper, based on indoor rainfall infiltration tests under the above initial conditions and influencing factors, the following patterns were found: (1) dry density and water content of natural soils tend to increase along the direction of gravity under long-term gravity and repeated rainfall evapotranspiration; (2) stepped dry density increases infiltration resistance, which increases the total response time; stepped initial water content decreases infiltration water volume, which decreases the total response time; stepped dry density and water content have the shortest total response time;(3) The increase in dry density delayed the sudden change point in the permeability coefficient, and the initial moisture content had little effect on the sudden change point.

1 Introduction

Rainfall is an important cause of landslide disasters in the Qinba Mountains. Rainfall leads to spatial and temporal changes in slope soil moisture, causing soil deformation and leading to reduced stability. Therefore, the study of the infiltration law of the soil body in the accumulation layer is an important basis for revealing the mechanism of rainfall-type landslide disaster. Therefore, it is extremely important to study the unsaturated-saturated infiltration law of typical accumulation layer in Qinba Mountain area.

In terms of theory, Green and Ampt[1] proposed the infiltration theory in 1911, which simplified the water content depth profile into a wet zone with fully saturated water content and an initial water content zone below the infiltration front. Accordingly, the Green.Ampt infiltration model with some physical basis was established.In 1931, Richards[2] introduced Darcy's law to unsaturated soil water flow, and established the differential equation for unsaturated flow.Translated with DeepL.com (free version)Translated with DeepL.com (free version)The influencing factors of rainfall infiltration are divided into endogenous and exogenous factors, Zhu Wei (2006)[3] found that the infiltration process, i.e., the conversion of the flow boundary and the head boundary, and the control of the initial water content found that those who have a large initial water content, the point of ponding is more in front of the unsaturated infiltration stage infiltration rate decreases more quickly. Jian Wenbin (2020)[4] conducted a one-dimensional infiltration test, by controlling the rainfall duration or total rainfall, it was concluded that the greater the rainfall intensity and the stronger the duration, the greater the rainfall infiltration and the deeper the depth of infiltration. Mansida A (2021)[5] conducted an experiment on the effect of vegetation on surface runoff and infiltration in a watershed area using a rainfall simulator to analyse the extent to which vegetation contributes to rainfall intensity in relation to the effect of surface runoff and infiltration on rainfall intensity. The results showed that changes in land cover increased surface runoff and decreased infiltration. Yizhou Zhuang (2021)[6] conducted a physical simulation experiment on rainfall-induced landslides using Weber's criterion as the rainfall similarity criterion. The results showed that the fully weathered granite slopes responded rapidly to rainfall infiltration. Further, the water content increased sharply and the matrix suction dissipated rapidly. After dissipation, the matrix suction force was transformed into pore water pressure, which accelerated the deformation of the slope. The wet peak has a large infiltration depth within the slope, and the acceleration in the deep part is lower than that in the shallow part. Wang Qing (2021)[7] found that gravel interlayer accelerates infiltration and clay interlayer hinders infiltration through indoor modelling tests, and the closer the gravel interlayer is to the surface, the earlier the higher pore water pressure and hydraulic gradient are formed, which affects the infiltration rate.

* Corresponding author: 2271996578@qq.com

 https://doi.org/10.1051/e3sconf/202456002004

Xu Jiancong (2006)[8] found that gravel content and fine-grained soil content had the greatest positive and negative effects on infiltration through several sets of infiltration tests on different graded gravel soils. Cheng Guang (2023)[9] found a positive correlation between stone content and pore ratio and infiltration coefficient through experimental study. In order to study the response of soil moisture (θ) to rainfall, Xu Tianxiang (2023)[10] used continuous high-frequency θ monitoring data at depths of 5, 10, 20, 40 and 70 cm to reveal the θ dynamics and its response characteristics to rainfall in dryland (UF), paddy field (PF) and forest land (FL). Furthermore, the cumulative increment of soil water storage in FL (190.1 mm) was larger than that in UF (160.6 mm) and PF (143.8 mm).(#br)Conclusions(#br)The land use conversion from FL to UF and PF reduced the rainfall infiltration capacity and may increase runoff potential.

The above studies have ignored the fact that the physical properties of natural soils change with depth, this paper is based on an indoor one-dimensional rainfall infiltration test using non-homogeneous gravelly soils to investigate its effect on infiltration.

2 Experimental Design

2.1. Sampling and Preparation

Figure 1 Panoramic view of the Sai Yip Leung landslide

The soil samples taken in this test were gravel soil from the landslide monitoring site of Xiaohekou Village, Bamiao Town, Zhenba County, Hanzhong City, Shaanxi Province (Fig. 1 Saijianliang landslide), with a depth of 1.5 m. The soil had a natural water content of 23.2%, a density of 1.89 g/cm^3, and a pore ratio of 0.79.

When making samples, the design soil samples were first dried in an oven for 8h, and then hammered and dispersed through a 20mm sieve to remove particles with a particle size larger than 20mm (in order to eliminate the size effect). Then let the rest of the soil samples pass through 10, 5, 2 sieve in turn, the corresponding particle group separate storage, test according to the design of the proportion. A suitable amount of soil samples with particle size <2 mm were configured as 10%, 15% or 20% target mass moisture content soil samples and smothered for 24h and set aside. Finally, gravel with 27% stone content was added and mixed well. The soil samples were prepared as shown in Figure 2.

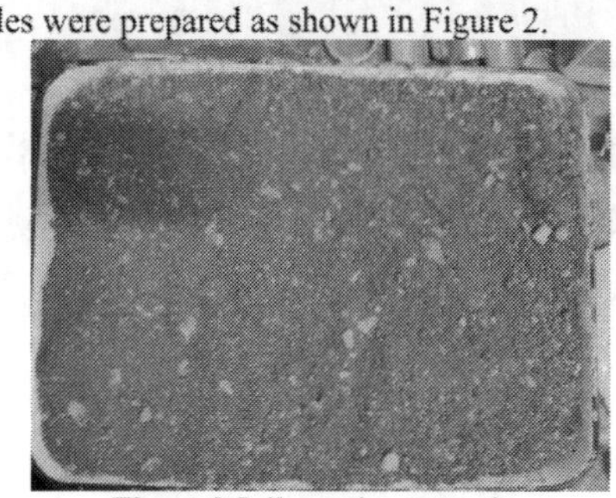

Figure 2 Soil sample preparation

2.2 Test equipment

As shown in Figure 3,This test apparatus is divided into three parts: the soil column mould is made of three plexiglass tubes with an inner diameter of 15cm and a length of 12cm spliced together. One side of the glass tube is 3cm and 7cm from the top of the reserved holes for the installation of moisture sensors and substrate suction sensors; soil water content sensor and soil water potential sensor model CS-655 and 253-L, respectively, the data acquisition instrument model CR1000; the water supply system for the series-connected Marschall's bottle and rainfall sprinkler, drainage system for the upper plane of the soil column moulds and the lower part of the drainage holes.

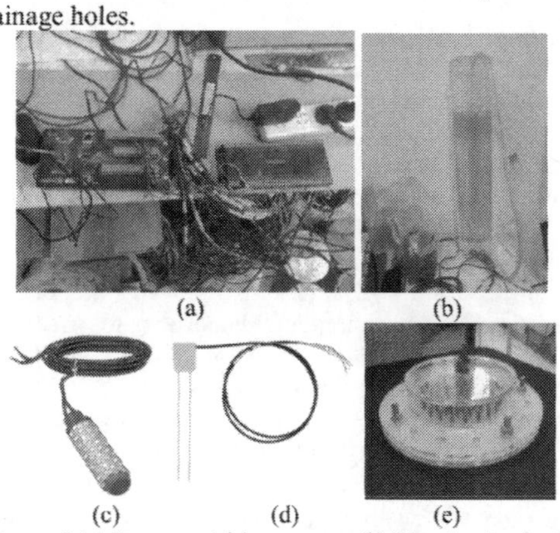

Figure 3 (a) Data acquisition system, (b) Mars bottle, the (c) 253-L soil water potential sensor, (d) CS-655 soil water content sensor, (e) rainfall nozzle

2.3 Test content

The aim of this test was to investigate the effect of non-homogeneous gravelly soils on the infiltration pattern, but it was too difficult to replicate soils with continuously varying water content and dry density in the field, so the soil columns were divided into three segments, with the physical properties of each segment remaining consistent. As a result, infiltration tests were carried out for the following conditions, with a uniform design rainfall intensity of 22 mm/h, rainfall until water flows out of the lower drainage holes or up to 8 h of rainfall, and a monitoring calendar of typically 12 h. The specific conditions of the soil column are shown in Table 1:

Table 1 Pilot programme table

programme	depths/cm	dry density g/cm3	moisture content/%
1	0-32	1.35	10
2	0-10	1.2	10
	10-22	1.35	10
	22-32	1.6	10
3	0-10	1.35	10
	10-22	1.35	15
	22-32	1.35	20
4	0-10	1.2	10
	10-22	1.35	15
	22-32	1.6	20

Scheme 1 is the control group; Scheme 2 keep 10% initial mass moisture content unchanged, let the dry density from bottom to top in order to reduce, specific for 1.6, 1.35 and 1.2 g/cm^3, respectively, through the control of different depths of the degree of compaction obtained; Scheme 3 keep 1.35 g/cm3 dry density unchanged, let the initial mass moisture content from bottom to top in order to reduce, specific for 0.1, 0.15 and 0.2, obtained by controlling the soil-water ratio during sample preparation; Scheme 4 changed the water content and dry density at the same time, while all the schemes left a 2 cm waterlogging space.

Rainfall conditions: Scenarios 1 and 2 end with 8h of continuous rainfall, Scenario 3 ends with 7h of continuous rainfall, Scenario 4 ends with 6h of continuous rainfall, and Scenario 5 has intermittent rainfall, with a 1h stoppage for every 2h of rainfall, and ends with 8h of net rainfall.

3 Test Results And Analyses

3.1 Basic Laws of Seepage

(1) The time-varying curves of water content and matrix suction can be divided into three stages, and the time of each stage coincides almost exactly, taking the sensor data at the depth of 14cm in Scheme 1 as an example: the soil body was in the first stage, i.e. the initial unchanged section, before 156min. At this time, the wet front did not arrive, the water content is almost unchanged, the matrix suction is unchanged or slowly increasing; the second stage is the rapid growth of the section, the wet front arrives, the water content increases rapidly to nearly the maximum, the matrix suction increases rapidly close to the maximum value; the third stage is the final stability of the section, the wet front continues to move downward, the water content grows slowly to the saturation or to reach the maximum value, if there is no subsequent rainfall replenishment, the water content will decline slowly and the matrix suction slowly climbs. If there is no subsequent rainfall recharge, the moisture content will slowly decrease, and the matrix suction slowly climbs to the maximum and then stabilises.

(2) According to the time-varying curve of water content and mechanism suction (Figure 4), take the root of its obvious change as the response time of the corresponding sensor. Programme 1, A1-A6 response time were 9, 50, 156, 207, 303 and 365min, A1A2 sensor response time difference of 41min less than the A3A4 sensor response time difference of 51min less than the A5A6 sensor response time difference of 62min, which and in the rainfall at the beginning of the soil body is drier seepage faster and with the infiltration of slower laws want to match. B1-B6 law is similar to the follow-up will not be described.

(3) The volumetric water content of the soil body in Scheme 1 is around 0.45, which is less than the saturated water content of 0.5. The reason for this is that the specimens of the soil-rock mixture are too large to be in the airtight condition, the seepage flow is always kept from high to low during the test and the bottom of the specimen is a permeable environment, so the test can't make the specimen completely saturated.The starting value of the measurement of the sensor of B1-B6 is around -175 kpa, which is influenced by the measurement range and accuracy of the sensor. The maximum substrate suction is around -15kpa due to the measurement range and accuracy of the sensor, and will not reach 0 or show a positive number even if rain continues to fall.

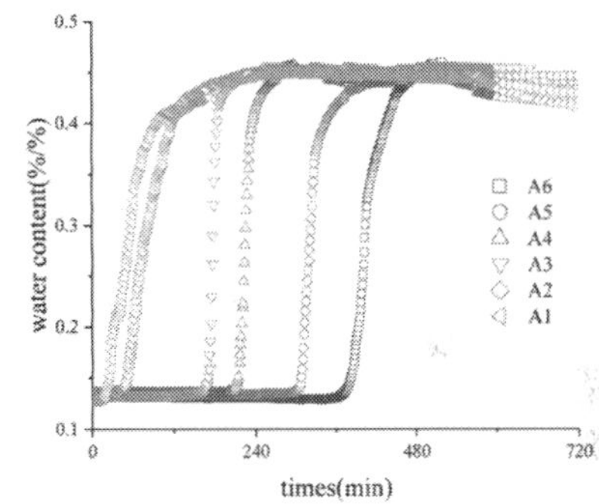

(a)

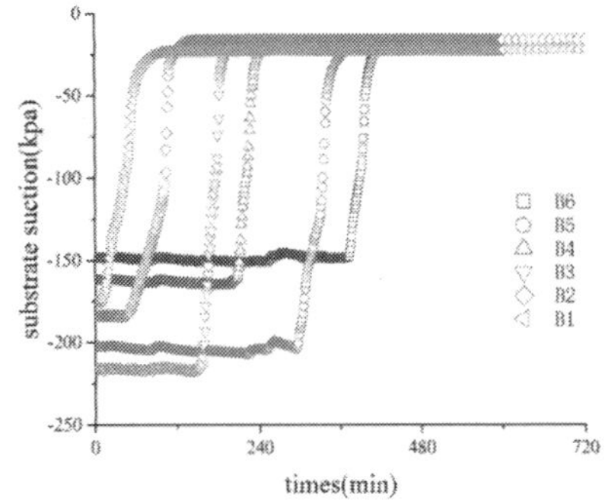

(b)

Figure 4 Scheme 1 (a) time-varying plot of volumetric water content, (b) time-varying plot of substrate suction

3.2 Seepage characteristics under stepped dry density conditions

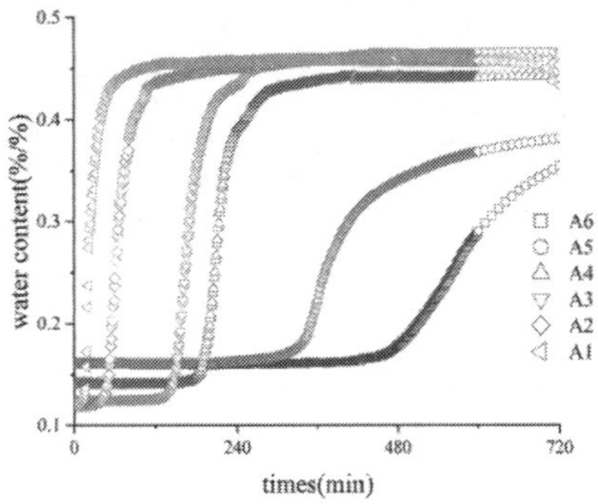

(a)

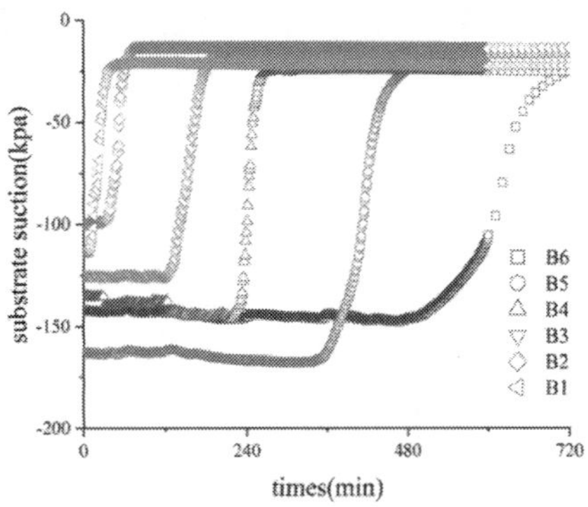

(b)

Figure 5 Scheme 2 (a) time-varying plot of volumetric water content, (b) time-varying plot of substrate suction

(1) As shown in Figure 5,A1-A6 response time were 9, 41, 131, 182, 329 and 466min, respectively, in the depth of 18cm before the programme 2 response time is earlier than the programme 1, because the specimen of the upper 10cm programme 2 dry density is less than programme 1, the middle 12cm programme 2 dry density is equal to the programme 1, the upper 22cm of the overall programme 2 dry density is smaller and therefore faster seepage. The total response time of scheme 2 is significantly larger than that of scheme 1 because the dry density of the lower 10cm scheme 2 is larger than that of scheme 1. The increase in dry density reduces the infiltration channel and narrows the cross section, which restricts the flow of water and greatly slows down the seepage rate.The response time difference of 32min for A1A2 is smaller than that of 51min for A3A4, and is much smaller than the response time difference of 137min for A5A6. This is another strong evidence that the increase of dry density will reduce the seepage velocity, and also shows that the increase of dry density and the slowing down of seepage velocity is not a linear relationship, which is similar to the exponential relationship.

(2) A1A2 maximum volumetric water content of about 0.46, much smaller than the saturated volumetric water content of 0.556, considering the upper layer of soil wetting brought about by seepage, the dry density of its deformation is about 1.35, the maximum volumetric water content of 0.5, the upper layer of the soil is close to saturated; A3A4 maximum volumetric water content of 0.465 and 0.445, respectively, the middle layer of the soil is close to saturated; A5A6 first 12h The maximum volumetric water contents were 0.38 and 0.355 respectively, but both were nearly saturated with a volumetric water content of 0.41 at 16h.

3.3 Seepage characteristics under stepped water content conditions

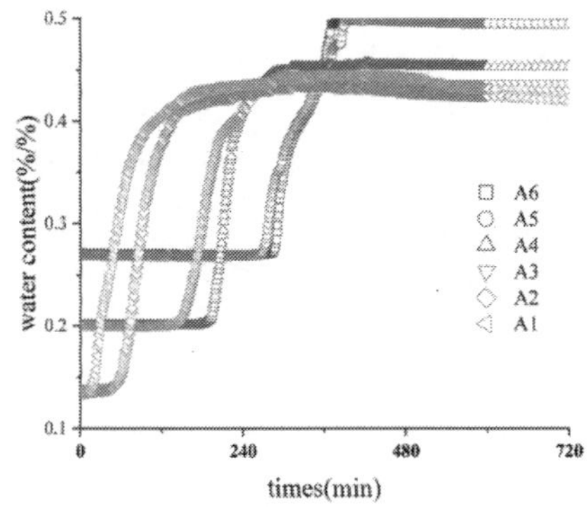

(a)

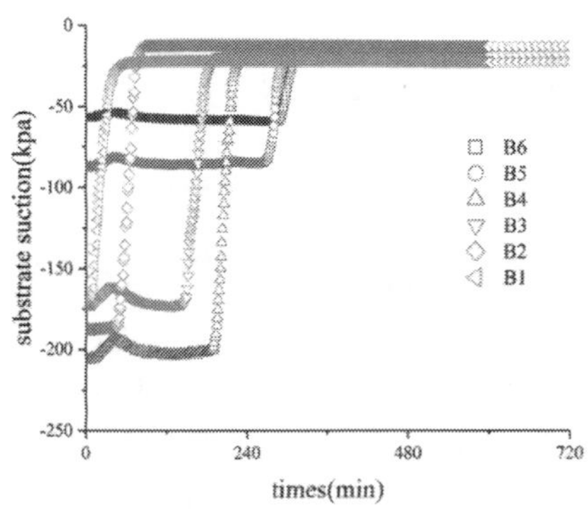

(b)

Figure 6 Scheme 3 (a) time-varying plot of volumetric water content, (b) time-varying plot of substrate suction

(1) As shown in Figure 6,The response times of A1-A6 are 10, 51, 148, 185, 267 and 284 min, respectively. Influenced by the initial state of the soil column, the response time of scheme 3A1A2 is approximate to that of scheme 1, whereas the response time of A3A4 is earlier than that of scheme 1, and the response time of A5A5 is even earlier than that of scheme 1, because the increase of the initial water content reduces the amount of water needed to reach the maximum, and the flow of water can be transported more quickly to the The response time difference of A1A2 is slightly larger than that of A3A4 by 41 min and 37 min, which is the result of the superposition of the decrease of infiltration rate

with the saturation of topsoil and the increase of seepage rate by the increase of initial water content, and the response time difference of A3A4 is larger than that of A5A6 by 17 min, which indicates that the latter's effect is larger than that of the former at this time.

(2) A1A2 from 0.135, A3A4 from 0.2, the maximum value of 0.44 or so, the upper and middle layers are always unsaturated seepage. A5A6 from 0.27 to the maximum value of 0.5, the soil body is saturated, presumably because of the mould of the lower part of the drainage holes of small caliber, and may be blocked by the migration of fine-grained soil, so that the rainfall below the column of the soil to pool, so the A5 in 0.45 and 0.475 Two stops make A6 reach the maximum before A5.

(3) The starting value of the B1-B4 sensors is smaller at around -175 kpa than that of the B5-B6 sensors at around -75 kpa, the latter approximating the substrate suction of the B3-B4 sensors at a moisture content of 0.27, with the seepage partially completed earlier.

3.4 Seepage characteristics at stepped water content and dry density

(1) As shown in Figure 7,The response times of A1-A6 are 8, 38, 98, 114, 155 and 163 min, respectively, and the response times are getting faster and faster. the response time difference of 30 min for A1A2 is almost twice as much as the response time difference of 16 min for A3A4, while the response time difference of A3A4 is equal to two times as much as the response time difference of 8 min for A5A6, and the response speed of water content rate is nonlinearly increasing.

(2) It can be seen that A1A2 from 0.12, the maximum value are around 0.48, observed that the seepage makes the upper layer of wetting about 1.1cm, the maximum volumetric water content after wetting 0.5, the upper layer of the soil is nearly saturated; A3A4 from 0.2, the maximum of 0.44 and 0.42, respectively, the middle layer of the soil is always in the non-saturated; A5A6 from 0.32 to the maximum value of 0.41, saturated state is reached.

(3) B1-B6 initial value is increasing, B5B6 initial value is only -35kpa less than half of the initial value of B5B6 of Scheme 3, because its initial volumetric water content has reached 0.32, and saturation is 0.762 greater than that of Scheme 4 saturation 0.52, which is in line with the results of the research of the previous soil-water characteristic curve, namely, the substrate suction and the effective saturation is a non-linear relationship.

(4) The total response time of Scheme 4 is smaller than that of Scheme 1 and Scheme 3, on the one hand, the increase of initial water content shortens the response time, on the other hand, although the increase of dry density increases the response time, but the dry density of the upper layer of soil in Scheme 4 is smaller, and according to the results of the previous research, the closer the top layer of soil, the greater the influence on the overall seepage rate, and the combined result is that the response time of Scheme 4 is the shortest.

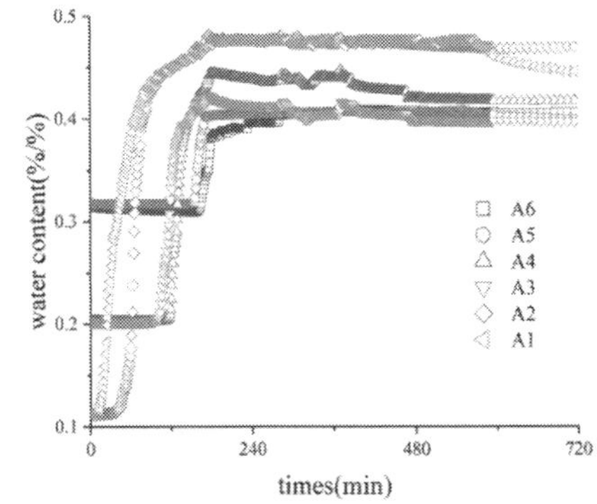

(a)

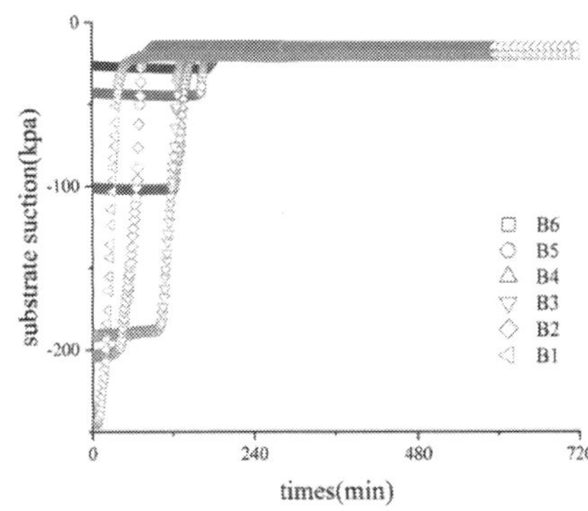

(b)

Figure 7 Scheme 4 (a) time-varying plot of volumetric water content, (b) time-varying plot of substrate suction

4 Calculation of Permeability Coefficient

4.1 calculation method

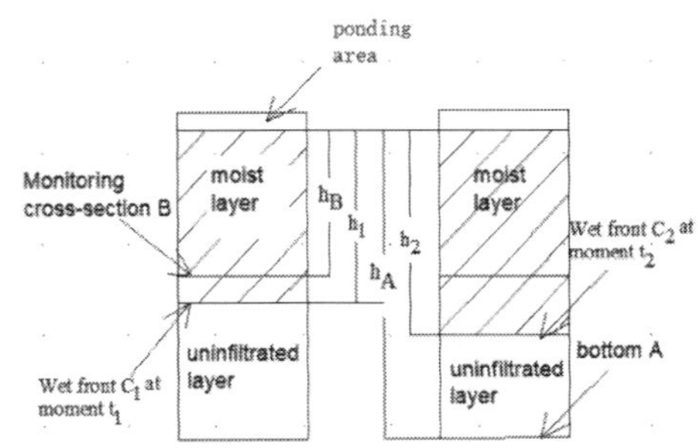

Figure 8 Wet front movement diagram

In this paper, the traditional wetting front advance method is extended to solve the unsaturated infiltration coefficient under waterlogged conditions. Fig. 8 shows the movement of the wetting front during rainfall infiltration, with B as the monitoring cross-section and A as the bottom surface, and the wetting front is located at C_1 and C_2, respectively, at moments t_1 to t_2. When the wetting front did not cross the bottom surface of the soil column, the flow rate through the monitoring cross-section B during the moments t_1 to t_2 was:

$$Q_B = \int_{h_B}^{h_A} \theta(h,t_2)Adh - \int_{h_N}^{h_A} \theta(h,t_1)Adh \quad (1)$$

Where: θ (h, t) is the volumetric water content distribution function of the vertical soil column, A is the cross-section area of the soil column, and h_B and h_A are the vertical distances of cross-section B and cross-section A from the infiltration surface, respectively.

It is assumed that the infiltrating front moves a distance of Δh , and that the volumetric water content distribution function θ(h, t) and the matrix suction distribution function ψ(h, t) are smooth for a smaller time $\Delta t = t_2 - t_1 = 10$ min. Eq. (1) can be deformed as:

$$Q_B = \frac{\left[\theta(h_B,t_2)+\theta(h_B,t_1)\right]A\Delta h - 2\theta_0 A\Delta h}{2} \quad (2)$$

Where: θ_0 is the initial volumetric water content of the vertical soil column.

Assuming that the permeability coefficient k of the unsaturated soil does not change during the time period Δt, according to Darcy's law

$$Q_B = \mathrm{k}iA\Delta t \quad (3)$$

Where: k is the coefficient of permeability of unsaturated soils; i is the hydraulic gradient and there are

$$i = \frac{dh + \frac{d\Psi}{r_w} + dH}{dh} = \frac{d\Psi}{r_w dh} + 1 + \frac{dHdt}{dtdh} = \quad (4)$$
$$\frac{\Psi(h_B,t_1) - \Psi(h_B,t_2)}{r_w \Delta h} + 1 + \frac{v_j}{v_s}$$

Where: r_w is the heaviness of water, v_j is the velocity of ponding, and v_s is the speed of advance of the wetting front. At the beginning of the rainfall, there is no accumulation of water and at the end of the rainfall, when the depth of accumulation of water is fixed, v_j is 0. In order to simplify the operation of the middle of the rainfall, v_j/v_s is set as a constant a. Substitute the flow rate of Q_B, the hydraulic gradient i, and $v_s = \Delta h / \Delta t$ into the Darcy's Law Expression (3), and simplify the coefficient of infiltration of unsaturated soils to be:

$$k = \frac{Q_B}{iA\Delta t} = \quad (5)$$
$$\frac{\left[\theta(h_B,t_1)+\theta(h_B,t_2)-2\theta_0\right]r_w v_s^2 \Delta t}{2\left[\Psi(h_B,t_1)-\Psi(h_B,t_2)+(1+a)r_w v_s \Delta t\right]}$$

4.2 Permeability coefficients under the influence of different factors

In Eq. (5), θ(h, t) and ψ(h, t) are obtained experimentally, v_s and v_j are converted from the recorded ponding moments and response times, θ_0 is obtained by converting the dry density to the water content, and S_e is obtained by converting the water content. r_w = 10kN/m3 and Δt = 10 minutes.

In scenario 2, v_s = 0.0644 cm/min and v_j = 0.025 cm/min, which are substituted into Equation 5 to obtain a plot of permeability coefficient versus saturation S_e for scenario 2 (Figure 9):

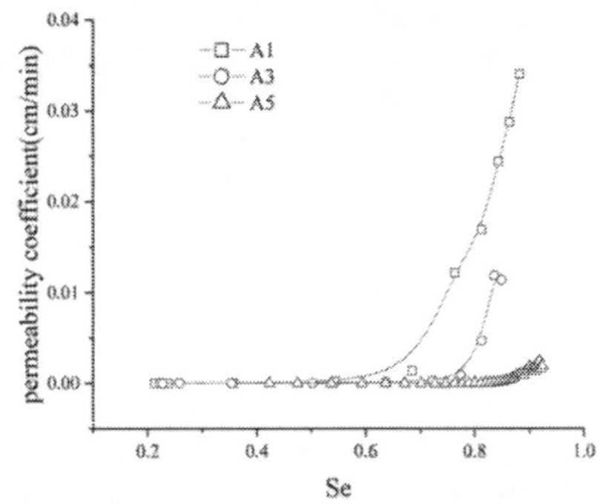

Figure 9 Scenario 2 permeability coefficient versus saturation plot

(1)The overall law of permeability coefficient and saturation degree is: when the saturation degree is lower than a certain value, the water flow is preferred to penetrate through the small pore space, and the small pore space is strong to water adsorption, resulting in water infiltration difficulties, the permeability coefficient is small and fluctuates up and down in a certain value. When the saturation degree is higher than a certain value, the small pore space is almost saturated, a large number of water flow through the large pore space infiltration, and the large pore space on the water adsorption is weak, so the permeability coefficient increases until the maximum, the value is defined as the limit value.

(2) Taking the positions 1, 3 and 5 of the scenario 2 sensors as cross sections, the cut-off values for each cross section are 0.6, 0.7 and 0.8, respectively, and it can be seen that the cut-off values increase with the increase in dry density. And the maximum permeability coefficient decreases rapidly with the increase of dry density.

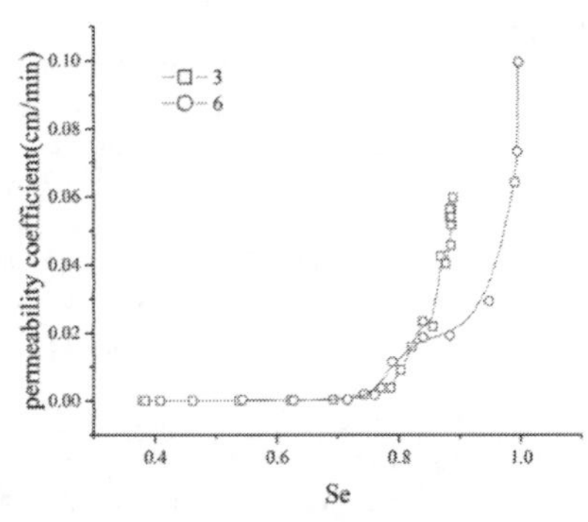

Figure 10 Scenario 3 permeability coefficient versus saturation plot

In scenario 3, v_s = 0.1056 cm/min and v_j = 0.032 cm/min, which are substituted into Equation 5 to obtain a plot of permeability coefficient versus saturation S_e for scenario 3 (Figure 10):

(3) Taking the cross sections at locations 3 and 6 of the sensor in Scheme 3, the cut-off values for both sections are 0.7, which shows that the initial moisture content does not have a significant effect on the cut-off values.

5 Conclusions

In this paper, a one-dimensional rainfall infiltration device, combined with a matrix suction sensor and a soil moisture sensor, was used to study the infiltration of gravelled soils in different initial states, and the following patterns were found:

(1) Natural soils tend to show an increase in dry density and water content along the direction of gravity under long-term gravity and repeated rainfall evaporation;

(2) The time-varying curves of water content and matrix suction can both be divided into three stages, and the time of each stage almost exactly coincides with the first stage, i.e., the initial invariant section. At this time, the wet front did not arrive, the water content is almost unchanged, the matrix suction is unchanged or slowly increasing; the second stage is the rapid growth of the section, the wet front arrives, the water content increases rapidly to nearly the maximum, the matrix suction increases rapidly close to the maximum value; the third stage is the final stability of the section, the wet front continues to move downward, the water content grows slowly to the saturation or to reach the maximum value, if there is no subsequent rainfall replenishment, the water content will decline slowly and the matrix suction slowly climbs. If there is no subsequent rainfall recharge, the moisture content will slowly decrease, and the matrix suction slowly climbs to the maximum and then stabilises.

(3) The increase in dry density reduces the infiltration channels and narrows the water cross section, limiting the flow of water and greatly slowing down the rate of seepage and lengthening the overall response time. The increase of initial water content reduces the amount of water needed to reach the maximum, and the water can be transported to the lower layer faster, which reduces the total response time. The response time is the shortest under the condition of stepped dry density and water content. On the one hand, the increase of initial water content shortens the response time, and on the other hand, although the increase of dry density increases the response time, the dry density of the upper soil layer of Scheme 4 is smaller, and according to the results of the previous research, the closer to the top of the soil layer, the more influence on the overall seepage velocity, and the combined result is that its response time is the shortest.

(4) The increase in dry density resulted in larger permeability coefficient cut-off values, and the initial moisture content had little effect on the cut-off values. At dry densities of 1.2, 1.35 and 1.6 g/cm^3, the soil cut-off values are 0.6, 0.7 and 0.8, respectively.

References

1. Green W H, and Ampt G A. Studies on soil physics: 1, Flow of air and water though soil[J]. J Agric Sci, 1911, 4(1): 1~24.
2. Richards L A. Capillary conduction of liquids in porous mediums [J]. Physics, 1931, 1: 318–333.
3. Zhu W, Cher X, Zhong X. Measurement and analysis of rainfall infiltration pattern [J]. Geotechnics, 2006, (11): 1873-1879.
4. Jian W, Huang C, Luo Y et al. Experimental study of wetting front transport in unsaturated slope residual soil under rainfall infiltration [J]. Geotechnics, 2020, 41 (04): 1123-1133.
5. A M ,F G ,A M Z , et al. Analysis of infiltration and surface runoff using rainfall simulator with variation of rain intensity and vegetation [J]. IOP Conference Series: Materials Science and Engineering, 2021, 1088 (1): 012108-.
6. Xin Q ,QingHua L ,YuYong J , et al. Experimental study on response law and failure process of slopes in fully weathered granites under precipitation infiltration [J]. Environmental Earth Sciences, 2021, 80 (20):
7. Wang Q. Influence of interlayers on rainfall infiltration in deep and thick piles[D]. Chengdu University of Technology, 2021.
8. Xu J, Shang Y. Influence of permeability characteristics of gravel soil on landslide stability [J]. Journal of Rock Mechanics and Engineering, 2006, (11): 2264-2271.
9. Cheng G. Correlation of soil-water characteristics and unsaturated permeability coefficient with microstructure of soil-stone mixtures in Qinba mountainous area[D]. Chang'an University, 2023.
10. Tianxiang X ,Hanjiang N ,Jun Y , et al. Characteristics of Soil Moisture Response to Rainfall under Different Land Use Patterns at Red Soil Region in Southern China [J]. Journal of Soil Science and Plant Nutrition, 2023, 23 (4): 6813-6826.

Quantifying the relative impacts of climate and human activities on vegetation change in Hunan Province between 2000 and 2015

Qingbin Zhang[1,*], Cuicui Qi[2], Hui Wang[1]

[1]Department of landscape architecture, College of Forestry, Xinyang Agriculture and Forestry University, Xinyang 464000, China
[2]Academic Affairs Office,College of Forestry, Xinyang Agriculture and Forestry University, Xinyang 464000, China

Abstract: Since1999, Hunan Province has embarked on a series of ecological projects to return farmland to forests across the province to explore the impacts of anthropogenic policies and actions on changes in vegetation cover between 2000 and 2015.It is important to understand the impacts of anthropogenic and climatic changes on regional vegetation cover change is an important guide for formulating reasonable environmental protection and restoration strategies in the region .this paper analysed the influence of climate and anthropogenic activities on the change of vegetation cover of Hunan Province during 2000-2015, which combined with an improved vegetation-climate residual regression model at the meta-scale. During 2000-2015, the contribution of anthropogenic activities to the increase of vegetation cover was 49%, 53%, 62% and 66% for unconverted forest land, grassland, unused land and water area respectively, while the contribution of climatic factors was 51%, 47%, 38% and 44% respectively. At the same time, all the land use types of contributions to vegetation cover change in Hunan Province is positive, and the land type without forest land change has the largest contribution to vegetation cover change in Hunan Province at 66.28%, while the land type where other land is converted to unused land has a lower contribution to vegetation cover change.

1. Introduction

Vegetation growth of changes are significantly influenced by climate (temperature, rainfall) and human activities (land use change, ecological restoration projects, etc.)[1]. At the same time, as the influence of anthropogenic activities on vegetation change increases worldwide, they are gradually becoming one of the important drivers of vegetation change. Therefore, exploring the impact of environmental change and anthropogenic activities on vegetation change , Which is from the view of anthropogenic factors is of importance to the environment. How to distinguish the effects of anthropogenic activities and climate change on vegetation change at a regional scale is now an important research direction.

In the past, Researchers have used datum from different websites to explore the interrelationships between climate, anthropogenic changes and vegetation change at regional scale[2]. Or different statistical analyses have been used to distinguish between climate and human activity effects on vegetation dynamics[3-4]. However, the above research methods fail to reveal the spatial scale of vegetation .With the rapid development of remote sensing technology, the product data from remote sensing satellites provide the possibility to solve the above problems; Normalized Difference Vegetation Index (NDVI) is one of the most important remote sensing product, can better respond to the vegetation growth condition at different scales[5].Many studies are now using the Normalized Difference Vegetation Index (NDVI) dataset to study the dynamic trends of vegetation at the world and regional scales and to conduct attribution analysis . Many researchers have used qualitative attribution methods to analyze the role of factors on vegetation change[6],However, such method does not allow for a quantitative assessment of the impact of each factor on vegetation.

* Correspondence: 2020290007@xyafu.edu.cn

Many researchers now use residual regression to quantify anthropogenic and climatic influences on vegetation change at a regional scale[7]; X. Tong et al. used the residual analysis method to develop a quantitative evaluation of ecological restoration projects in karst areas of southwest China[8]. The above studies used temporal trends in residuals to monitor trends in anthropogenic vegetation change, which is also commonly used in many vegetation driver change studie. Many studies now use residual analysis to develop quantitative analyses, mostly using regression analysis to quantitatively evaluate anthropogenic and climatic impacts on vegetation dynamics. However, regression analysis has shortcomings, as human activities can be very complex and diverse, and regression equations cannot take into account all anthropogenic factors and their complex interaction effects[7]. In addition, the regression analysis is only carried out at the administrative unit level due to limitations in the accessibility of human factor.

Combining the above factors, Rui Liu used the residual analysis method combined with LUCC theory to establish a new evaluation system[9]. The evaluation system completely eliminates the constraints of administrative boundaries, and at the same time, a wide range of human activities and their complex interactions do not generate uncertainty in this evaluation system. Since 1999, Hunan Province has responded to the call of the state to carry out forestry policies such as returning farmland to forests and forest nurseries throughout the province. In order to assess the changes in vegetation cover and the impacts of anthropogenic policies in Hunan Province after 1999. this study uses the theoretical framework improved by Rui Liu to conduct an attribution analysis of anthropogenic activities and climatic influences on vegetation cover change in Hunan Province.

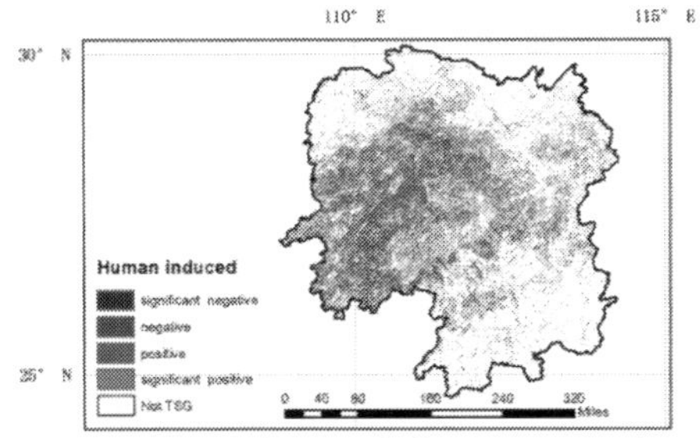

Figure 1. Impact of anthropogenic factors on vegetation cover change in Hunan Province, 2000-2015 *(P<0.05)*

2.Results

2.1. Influence of anthropogenic factors on NDVI changes

Figures 1 and 2 present the residual regression analysis for NDVI and climate factors and assess the influence of anthropogenic factors on NDVI changes at the meta-scale. Firstly, the value of grown season NDVI and climate regressions passed the significance test (95% level), and those with no significant relationship with climate were excluded from the TSG. partial squares regression analysis to obtain the distribution of residual slope values of GSN residuals versus time .

Figure 1 depicts the impact of human activities on the GSN (the white part of Figure 1 shows the images that did not pass the significance analysis. The area of the significantly positive trend grid is 130.2km^2 , accounting for 00.0615% of the study area, and the area of the significantly negative trend grid is 28.9km^2 , accounting for 00.014% of the study area. After constructing the climate-GSN model, it was found that the areas showing significant correlation between GSN and climate were mainly located in the southwest region of Hunan Province; through spatial overlay analysis in GIS, it was found that the spatial distribution of areas showing significant negative correlation with anthropogenic activities was basically consistent with the spatial distribution of areas showing significant decrease in GSN during 2000-2015.

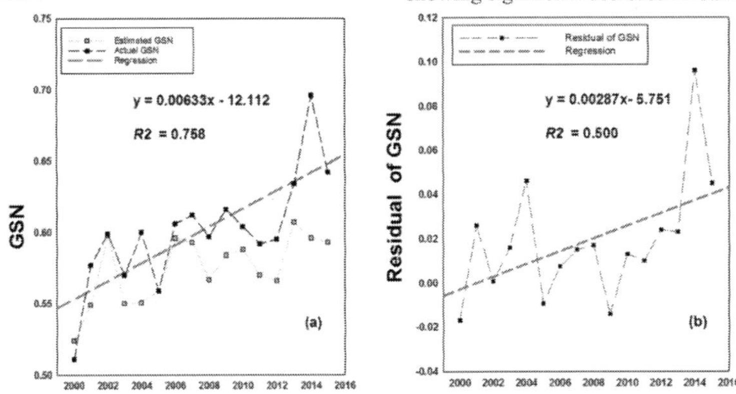

Figure 2. (a) Trend of annual means of GSN and GSN estimates tested for significance (b) Trend of annual means of actual value of grown season NDVI and estimated value of grown season NDVI residuals tested for significance

Figure 2 describes the trend of the mean actual value of grown season NDVI, estimated GSN and their residuals.The actual GSN and residuals increased linearly over time with coefficients of determination of 0.758 and 0.5, respectively. the actual GSN increased from 0.551 in 2000 to 0.642 in 2015, with a corresponding increase in residuals from -0.017 to 0.045. the estimated GSN increased from 0.524 in 2000 to 0.593 in 2015, but the estimated GSN , fluctuates less compared to the actual GSN, residual values between 2000 and 2015. The change in the line graph of the actual GSN, estimated GSN, and residual values reveals that the vegetation change in Hunan Province has increased significantly and that climate and anthropogenic activities have played a positive role in the change of vegetation cover in the Hunan region.

2.2 Relative anthropogenic and climatic contributions

According to the final analysis areas, and the calculation of the relative contribution of anthropogenic and climate factors was finalised by combining the areas where no land use change occurred from 2000-2015 to obtain Figure 3.

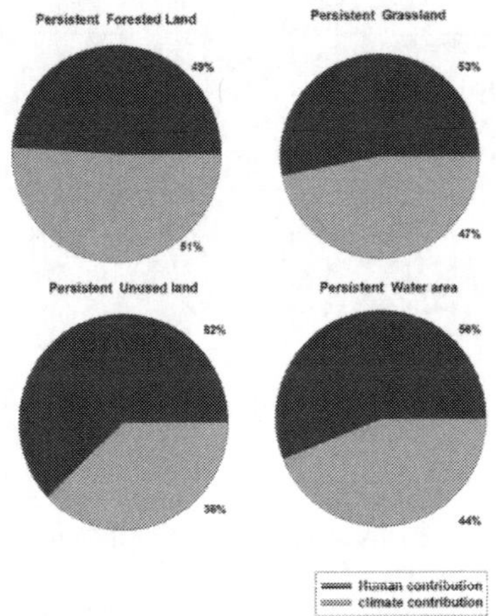

Figure 3. Quantitative analysis of anthropogenic and climatic effects on vegetation cover change

Figure 3 shows the contribution of anthropogenic and climatic factors to vegetation change in areas without land use change based on the construction principles of our framework and calculated by equations. As can be seen from the figure, the contribution of anthropogenic activities to vegetation cover change in unconverted forest land during 2000-2015 is 49%, and the contribution of climatic factors to vegetation cover change is 61%; the contribution of anthropogenic activities factors to vegetation cover change is greater than the role of climatic factors in the remaining three land types where no land use change has occurred. Thus, at the study area (TSG) level, the impact of human activities is stronger than the contribution of climate to vegetation change; the difference between the two values confirms the fact that the use of relevant human interventions to restore vegetation cover change in Hunan Province since 1999 has made a significant positive contribution.

3.Discussion.

3.1. The contribution of afforestation on vegetation cover

Many studies have been conducted to investigate land use and land cover change (LUCC) induced by related ecological restoration projects to explore the impact of land cover change on vegetation cover change at regional scales[10]. During 2000-2015, vegetation cover in Hunan Province showed a significant overall growth trend, and among the land use changes, the unchanged forest land made the largest contribution to vegetation cover change. The forested land that did not change contributed the most to the vegetation cover change rate in Hunan Province, while the various land use change types played a positive role in vegetation cover during 2000-2015, which also indicated that Changsha also took into account the need for green development in the process of urbanization. The GSN residual regression model, which strips out the anthropogenic activity factors, shows that the residual values are positive and the growth trend is consistent with the growth trend of the actual NDVI values during 2000-2015, indicating that the anthropogenic activity factors have a positive influence on the vegetation cover change in Hunan Province during 2000-2015. The largest contribution to the increase in vegetation cover was made by the sum of unchanged forest land and land use types where other land was converted to forest land, which may be due to forestry projects such as forest nurturing and returning farmland to forest in Hunan Province. This subsection calculates the correlation coefficients between the annual average GSN and the annual afforestation area, annual artificial afforestation area, annual cumulative afforestation area and annual cumulative artificial afforestation area data from 2000-2015 through Pearson correlation analysis, and the following results were obtained: the annual average GSN did not pass the significance analysis with the annual afforestation area and annual artificial afforestation area data, but showed a significant positive correlation with the annual cumulative afforestation area and annual cumulative artificial afforestation area data. The above indicates that due to the cyclical nature of tree growth, afforestation in the current year does not immediately have a significant effect on the vegetation cover in Hunan Province, while over time, trees become established and have a significant positive effect on the growth of vegetation cover in Hunan Province (Figures 4 (a),(b) indicate the annual afforestation area and annual cumulative afforestation area between 2000 and 2015, respectively).

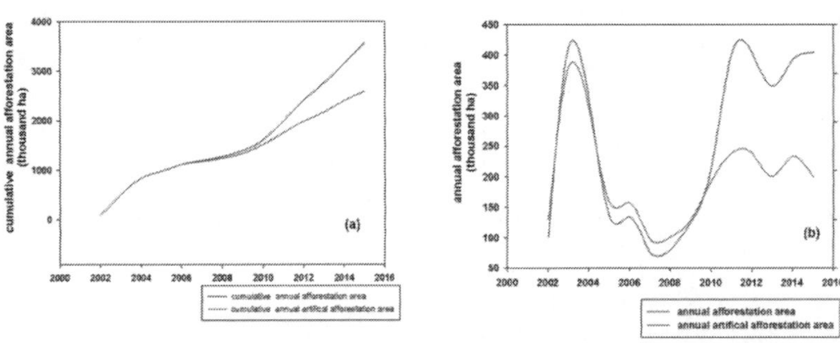

Figure 4. (a)Change trends of cumulative afforestation (b)Change trends of afforestation

By analyzing the correlation between vegetation cover in the growing season and the area of artificial afforestation and the cumulative area of artificial afforestation in Hunan Province during the period of 2000-

2015, it can be seen that the vegetation cover restoration and the cumulative area of artificial afforestation in Hunan Province are significantly correlated. Since 1999, Hunan Province has responded to the national call to carry out a series of forestry ecological restoration projects such as returning farmland to forests and planting trees in the province to solve the problems of rocky desertification and soil erosion[11], and the data from the statistical yearbook for the period of 2000-2015 show that the trend of growth of the cumulative afforestation area in Hunan Province is significant and plays a significant positive role in the change of Hunan Province's vegetation cover; land use and land use are significantly correlated with the cumulative area of afforestation.

3.2. Quantitative analysis of climate and anthropogenic impacts

There are two main factors affecting vegetation growth, namely climate change and human activities [12]. As an important indicator data that can reflect changes in forest vegetation, the normalised vegetation index (NDVI), a remotely sensed inversion index, has been frequently used in recent decades to reflect the monitoring of vegetation changes and to analyse the relationship with climate change[13-14]. In this thesis, the GSN-residuals regression model was constructed to calculate the residuals fractions of anthropogenic activities and the percentage contribution of anthropogenic activities and climatic factors to the NDVI changes of unchanged land types. After the above analysis, We have come to the following conclusions: the residual values of anthropogenic activities changed from negative to positive values during 2000-2015, indicating that anthropogenic factors played a positive role in the change of vegetation cover in Hunan during 2000-2015; meanwhile, by comparing the percentage contribution of anthropogenic and climatic factors to the change of NDVI for the four unchanged land types, the analysis reflects that anthropogenic factors have a stronger influence on the percentages of anthropogenic and climatic contributions to NDVI change in the four constant land types were also compared, reflecting that the influence of anthropogenic activities on vegetation cover was stronger than that of climatic factors. As vegetation cover change is influenced by both anthropogenic and climatic factors, in order to further clarify the influence of climatic and anthropogenic factors on vegetation cover change in Hunan Province, this subsection calculates the correlation coefficients between actual GSN values and rainfall, temperature and residual values to obtain Table 1. Table 1 shows that the role of anthropogenic factors on vegetation cover change in Hunan is stronger than that of climate.

Table 1. Correlation coefficient values for GSN

GS and residuals	GSN and rainfall	GSN and temperature
0.755**	0.253	0.571*

Note: ** means *P<0.01*,* means *P<0.05*.

4.Conclusions

This paper adopts the analytical framework proposed by Rui Liu [9] to conduct an attribution analysis of vegetation changes and their causes in Hunan Province during 2000-2015, which is different from the traditional residual regression analysis. The framework differs from the traditional residual regression analysis in that, on the one hand, it completely eliminates the restriction of administrative boundaries, and on the other hand, the traditional residual analysis lacks significant correlation between regional NDVI data and climate data, while the framework filters out significantly correlated pixels with meteorological data through significant regression analysis, which further improves the reliability of the later correlation analysis. Based on this framework for quantitative analysis between NDVI and climate in Hunan Province, We have come to the following conclusions: (1) Most regions in Hunan Province experienced significant vegetation growth between 2000 and 2015, and land types without forest land change had the largest contribution to vegetation change in Hunan Province, while the average annual NDVI values showed significant positive correlation with the cumulative afforestation area and cumulative annual artificial afforestation area data; the above indicates that The above indicates that the forestry projects adopted in Hunan Province, such as the afforestation programme, the closure of mountains for reforestation and the protection of permanent forest land, have contributed positively to the restoration of vegetation. (2) Through the evaluation of anthropogenic contribution at the meta-scale and the Pearson correlation analysis of annual average NDVI with residuals, rainfall and temperature data, in general, anthropogenic activities have played an increasingly positive role in vegetation cover change in Hunan Province, and the positive effect of anthropogenic activities on vegetation cover change in Hunan is stronger than the effect of climate.

There are several points in this study on the attribution analysis of vegetation change in Hunan that deserve further analysis (1) The framework analysis proposed by Rui Liu,which does not consider the residuals generated by the significance test (95% level) RESTREND of the grid climate and time trends, in the construction of the climate-NDVI residual model In the construction of the climate-NDVI residual model, although the significance correlation test makes up for the shortcomings of the previous climate-NDVI residual model, The improved residual regression model still has its own limitations, which only considers linear regression factors and does not incorporate nonlinear regression into the theoretical framework, which results in biased simulation predictions; In further research, nonlinear regression algorithms will be incorporated into the model to improve the accuracy of data fitting. (2) An ecological performance index (EPI) [15]needs to be established to assess the ecological benefits of Hunan Province between 2000 and 2015, and to make specific recommendations on ecological measures for different areas of Hunan Province according to the level of ecological benefits.

References

1. Sun, J, Cheng, G., Li, W., Sha, Y., Yang, Y(2013). On the variation of NDVI with the principal climatic elements in the Tibetan Plateau. Remote Sens. 5. 1894-1911.
2. Panxing He, Zongjiu Sun,Zhiming Han(2021).Dynamic characteristics and driving factors of vegetation greenness under changing environments in Xinjiang, China.Environmental Science and Pollution Research28,42516–42532.
3. Chang, X.L.; Lu, C.X.; Gao, Y.B(2003). Impacts of human economic activities on wind and sand environment in Kerqin sandy land. Resour. Sci25, 78 -83. -(In Chinese)
4. Lawler, J.J(2009). Climate change adaptation strategies for resource managementand conservation planning. Ann. N. Y. Acad. Sci 1162, 79 -98. -98.
5. Huete, A., Didan, K., Miura, T., Rodriguez, E.P., Gao, X., Ferreira, L.G(2002). Overview of the radiometric and biophysical performance of the MODIS Remote Sens. Environ. 83, 195-213.
6. Wang, H.; Liu, G.; Li, Z.; Ye, X.; Fu, B.; Lü, Y(2017). Analysis of the Driving Forces in Vegetation Variation in the Grain for Green Program Region, China. Sustainability 9, 1853.
7. Evans, J.; Geerken, R(2004). Discrimination between climate and human-induced dryland degradation. j. Arid Environ57, 535-554.
8. X. Tong et al(2017). International Journal of Applied Earth Observation and Geoinformation 54,105-113.
9. Rui Liu, Linlin Xiao,Zhe Liu, Jicai Daia(2018).Quantifying the relative impacts of climate and human activities on vegetation changes at the regional scale.Ecological Indicators.93, 91-99.
10. Zhang, G.L., Dong, J.W., Xiao, X.L., Hu, Z.M., Sheldon, S(2012). Effectiveness of ecological restoration projects in Horqin Sandy Land, China based on SPOT-VGT NDVI data. Ecological Engineering38(1), 20-29.
11. Sui L Y, Ming C B.(2009). The study framework of land use/cover change based on sustainable development in China. Geographical Research21(3),324-330.
12. Wang,J.Wang,K.Zhang,M.Zhang,C(2015).Impacts of climate change and human activities on vegetation cover in hilly southern China. Ecol.Eng81,451 -461.
13. Detsch,F.Otte,I.Appelhans,T.Hemp,A.Nauss,T(2016) ,Seasonal and long-term vegetation dynamics from 1-km GIMMS-based NDVI time series at Mt. kilimanjaro,Tanzania.Remote Sens.Environ.178,70-83.
14. Liu,L.Yang,X.Zhou,H.et al(2018).Evaluating the utiliy of solar-induced chlorophyll fluorescence for drought monitoring by comparison with NDVI derived from wheat canopy. sci.Total Environ.625,1208.
15. Song, Y.I.; Ma, M.G. A statistical analysis of the relationship between climatic factors and the normalized difference vegetation index in China. Int. J. Remote Sens. 2011, 32, 3947–3965.

 https://doi.org/10.1051/e3sconf/202456002006

Programmable Droplet Manipulation on Superhydrophobic Surfaces with Charge Gradient

Juncheng Xue, Haiyang Zhan and Shile Feng*

State Key Laboratory of High-performance Precision Manufacturing, Dalian University of Technology, Dalian 116024, P. R. China.

Abstract. The controllable manipulation of droplets is pivotal for diverse applications spanning from microfluidics to biotechnology. In this paper, we present a novel droplet manipulation method leveraging a charge gradient on superhydrophobic surfaces. By shielding the charges on a superhydrophobic surface with uniform charges, a charge gradient is generated, which can drive a neutral droplet to move along the charge gradient. By controlling the motion trajectory of the metallic shield, a programmable droplet motion in a desired manner is achieved. This programmable droplet manipulation accompanied by non-contact, fast-responsive, ultrafast, and convenient performances is promising in many fluidic systems.

1.Introduction

The controlled manipulation of droplets exhibits critical applications in drug delivery[1, 2], fog harvesting[3, 4], anti-icing[5], and heat conduction[6, 7]. Researchers have employed various external stimuli such as sound[8], light[9], electricity[10], heat[11], and magnetic fields[12] to manipulate droplets. Dielectric wetting, among these methods, has attracted considerable attention owing to its fast response and straightforward controllability. Sun et al.[10] pre-patterned charges on superhydrophobic surfaces to form a charge gradient, which could drive droplets along fixed trajectories. Xu et al.[13] utilized frictional charges on hydrophobic surfaces for droplet manipulation in the desired direction. Jin et al.[14] employed an electrostatic induction on superhydrophobic surfaces to achieve non-contact droplet driving. Despite great processes, the manipulation of charge distribution is always complicated. Therefore, proposing a simple method for controlling charge distribution on the surface to manipulate droplets in the desired manner poses a significant challenge.

Here, we report a simple charge-controlling strategy on superhydrophobic surfaces by leveraging the metallic shielding effect on charge distribution. By placing a metal rod below a superhydrophobic surface with uniform charges, a charge gradient is generated, which forms an asymmetric force to drive a neutral droplet to move along the charge gradient. Moreover, by controlling the mode and the motion trajectory of the metallic shield, the direction of the charge gradient is programmable, which achieves a droplet motion in the desired manner. This simplified approach for charge gradient-based droplet manipulation also manifests a non-contact, fast-responsive, ultrafast, and convenient performance, which has potential applications in microfluidic applications.

2.Results and Discussion

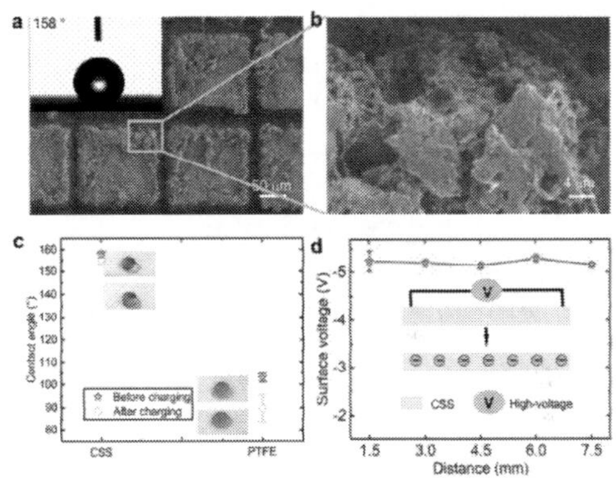

Fig. 1. The structure and function of CSS. (a, b) The hierarchical structure of CSS that consists of micro-scale pillars scattered with irregular nano-scale bumps and the water contact angle on CSS. The volume of the droplet is 4 μL. (c) Comparison of the wetting performances on PTFE plates with and without hierarchical structure before and after high-voltage charging. (d) The charge distribution of charged CSS and the schematic diagram of the high-voltage charging process.

Our droplet manipulation platform consists of two parts: the superhydrophobic surface with uniform charges (CSS) and the shielding metal (copper rod). The CSS was fabricated by laser-engraving microgrids on a PTFE plate followed by a high-voltage charging process. Figures 1a and 1b show that the designed surface manifests a hierarchical structure that consists of micro-scale pillars (~100 μm) scattered with irregular nano-scale bumps. This unique multi-scale structure endows a superhydrophobic performance with a water contact angle of 158° (inset image in Figure 1a). Comparing the wetting

* Corresponding author: fengshile@dlut.edu.cn

performances on PTFE plates with and without hierarchical structure before and after high-voltage charging, we found that the surface charges affect less to the superhydrophobic performance (with contact angles larger than 155°) of a PTFE plate with hierarchical structure, while slightly decreasing the contact angle on a PTFE plate without hierarchical structure (Figure 1c). This result indicates a stable superhydrophobic performance of charged PTFE plate with hierarchical structure (CSS). The schematic of the high-voltage charging process is shown in Figure 1d, which gives a uniform distribution of negative charges with a stable electrostatic voltage of ~5 kV. Note that the electrostatic charges are easy to remove by blowing with an electrostatic fan for 15 s.

We further examined the droplet motion process on our platform. First, a deionized water droplet (~20 μL) was placed on the CSS with a contact angle of ~ 158°. When controlling an uncharged copper rod (shielding metal) gradually approaching the bottom of the CSS reaching a distance of ~5 mm, the droplet swiftly departed from the copper rod, exhibiting an average velocity of 33.2 mm/s (Figure 2a). This fast and non-contact droplet-manipulated strategy is quite in contrast to previous studies. The relationship between moving distance and velocity to time is shown in Figures 2b and 2c. Furthermore, we investigated the impact of droplet volume on the velocity. With the volume of droplets increased from 10 to 50 μL, the maximum velocity decreased from 77.28 to 53.68 mm/s (Figure 2d).

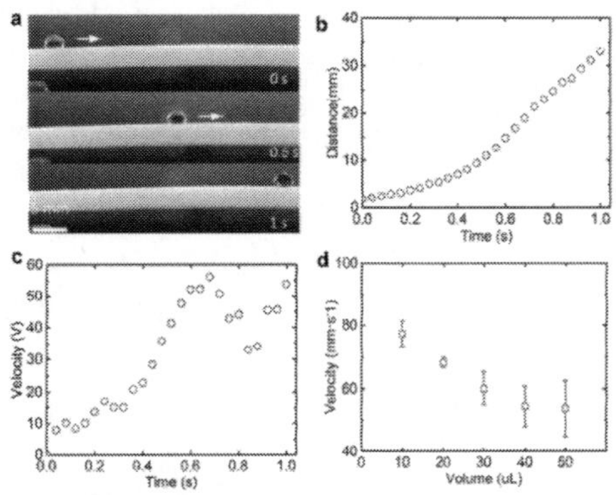

Fig. 2. Droplet manipulation on CSS by controlling an uncharged copper rod gradually approaching the bottom of the CSS, reaching a distance of ~5 mm. (a) Time-sequence images of a deionized water droplet with volume of 20 μL driven on CSS. (b) Distance-time relationship of water droplets moving on CSS. (c) Velocity-time relationship of water droplets moving on CSS. (d) The impact of droplet volume on driving velocity.

To examine the manipulating mechanism droplet on our designed platform, we measured the surface charge distribution when a copper rod gradually approached the bottom of the CSS with a distance of ~5 mm. The result manifests that the absolute value of the voltage increases gradually with the test point moving away from the copper rod (Figure 3a). This charge gradient is generated by the metal shielding effect[10]. For an uncharged droplet located on a surface with a charge gradient, it can be polarized with the positive charges moving toward the CSS surface and negative charges moving to the opposite side. After that, under the influence of electrostatic forces, the positively charged part of the droplet tends to move towards a higher negative charge region (Figure 3b).

We further conducted a force analysis on the droplet located on CSS with a charge gradient (Figure 3c). Due to the polarization, the positively charged side of the droplet tends to approach the CSS surface with negative charges. We divided the forces acting on the droplet into driving force F_d and resistance force F_r. The charge at the pillar nearest the front area of the droplet surface can be depicted as $\rho_d S_p$, which generates a driving force:

$$F_d = \frac{k_e \rho_d S_p Q_{wa}}{r^2} \tag{1}$$

Where r is the closest distance from the surface micropillar to the droplet, k_e is the Coulomb constant, ρ_d is the charge density, S_p is the area of the laser-printed grid block projection, and Q_{wa} is the droplet's polarization charge:

$$Q_{wa} = \nabla(\varepsilon_r^l - 1)\varepsilon_0 E_p \tag{2}$$

In the equation, ε_r^l is the dielectric constant of liquid, and E_p is the electric field generated by $\rho_d S_p$. Assuming that the charge on each microstructure grid block is a point charge, the electric field is:

$$E_p = \frac{k_e \rho_d S_p}{r^2} \tag{3}$$

Therefore, the driving force F_d acting on the liquid droplet is:

$$F_d = \nabla \frac{{k_e}^2 {\rho_d}^2 {S_p}^2 (\varepsilon_r^l - 1)\varepsilon_0 \cos\eta}{r^4} \tag{4}$$

Where η is the angle between the directions of droplet motion and the electric field force. As the liquid droplet moves forward, it also experiences a resistance F_r in the opposite direction. Similarly, we can define the charge resistance force acting on the liquid droplet as:

$$F_r = \nabla \frac{{k_e}^2 {\rho_r}^2 {S_p}^2 (\varepsilon_r^l - 1)\varepsilon_0 \cos\eta}{r^4} \tag{5}$$

Here $\rho_r S_p$ is the charge at the pillar nearest the rear area of the droplet surface. Note here that the friction between droplet and superhydrophobic surface can be neglected, for its minimal value compared to the charge-induced resistance force. Hence, the overall equation governing droplet motion can be written as:

$$ma = m\frac{dv}{dt} = F_d - F_r \tag{6}$$

Simplifies to:

$$dv = \nabla \frac{{k_e}^2 {S_p}^2 (\varepsilon_r^{l} - 1)\varepsilon_0 \cos\eta}{m{r_d}^4} (\rho_d + \rho_r) k D_c dt \tag{7}$$

From the above equation, it can be inferred that the droplet motion velocity v can be finely tuned by manipulating the charge gradient k on the CSS surface, as demonstrated in previous studies[10]. In equation 7, we found that the velocity v could be scaled as D_c/m for fixed droplet types and charge gradients. Considering that D_c and m could be depicted as $\pi r_w \sin\theta$ and $4\pi r^3_w/3$, we obtain that v could be scaled as $1/r^2_w$. Here r_w is the radius of droplets. We ignore the influence of r_w to the contact angle θ because that the θ changes little when r_w is not too large. It is clear to see that the velocity v is negative correlation with r_w as well as volume of droplets, which is consistent with the result in Figure 2d, verifying the validity of our theory.

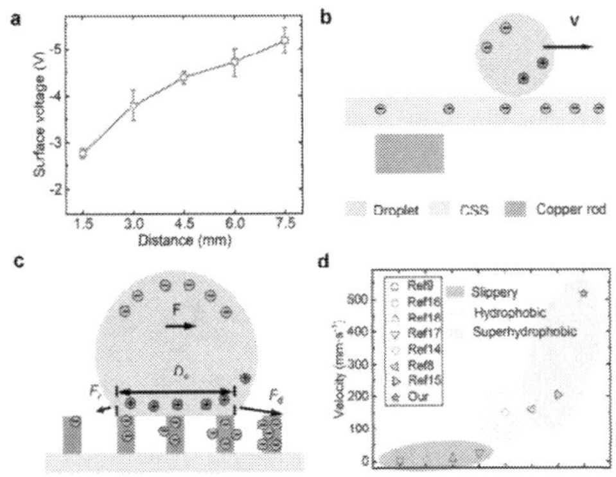

Fig. 3. The manipulating mechanism of droplet on our designed platform. (a) Charge gradient existing on CSS treated with a metal shielding effect. (b) Schematic diagram of droplet driven by charge gradient. (c) Force analysis of polarized droplets under the action of charge gradient. (d) Graph showing the moving velocities of droplets on our designed platform compared to other contrast surfaces.

To test the maximum velocity achievable in the experiment, we increased the surface charge gradient to increase the velocity, resulting in a final velocity of 516.7 mm/s. To demonstrate its superior driving performance, we compared the moving velocity on CSS with other contrast surfaces, including superhydrophobic surfaces (gray region), hydrophobic surfaces (yellow region), and slippery surfaces (red region)[15-17] (Figure 3d). The moving velocity on CSS surpasses that on slippery surfaces by ~90 times and exceeds that on superhydrophobic surfaces by a factor of 3, demonstrating ultra-fast droplet transport.

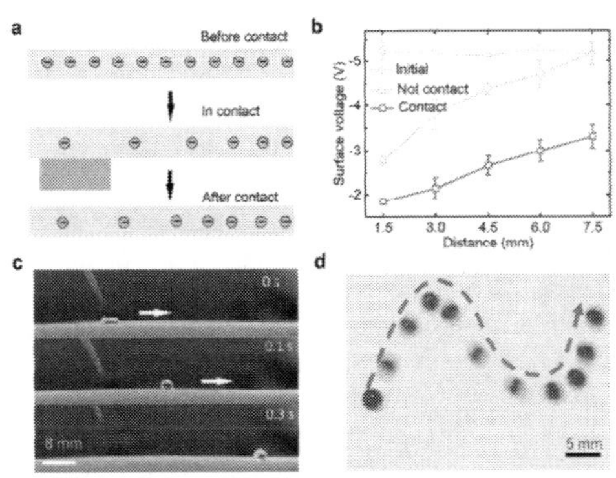

Fig. 4. Controlled droplet motion on the pre-treated trajectory of charge gradient. (a) Schematic diagram of charge distributions on the CSS, before, during, and after the copper rod touching it. (b)The voltage before, during, and after the uncharged copper rod contacting the CSS surface. The initial electrostatic voltage of CSS is -5kV. (c) Time-sequence images of 10 μL droplet movement on the CSS with a charge gradient. (d) Motion of droplet (10 μL) on a predetermined trajectory with s-type.

Furthermore, we observed that when the copper rod touched the bottom surface of the CSS, local electrostatic charges were shielded and transferred, forming a charge gradient even after removing the copper rod (Figures 4a and 4b). On this pre-treated trajectory of charge gradient, directional transport of water droplets was achieved with a velocity of about 71.16 mm/s (Figure 4c). We can also design a trajectory-based charge gradient with our desired shapes such as S-type trajectory. Also, the droplets exhibit a programmable transport along the predetermined paths (Figure 4d)

3.Conclusion

In this work, we have developed a method for designing programmed charge gradient trajectories on charged superhydrophobic surfaces using the metal shielding effect. Based on this, a programmable droplet motion in a desired manner is achieved. Compared to existing microfluidic techniques, this work demonstrates significant improvements for droplet manipulation in velocity and convenience, which manifests potential applications in microfluidic and microreactor devices.

4.Experiment

4.1 Materials

PTFE substrates measuring 10x60x2 mm and 60x40x2 mm were purchased from Jiayu Plastics (Shenzhen, China). Copper rods with a radius of 8 mm were obtained from a hardware store. Deionized water was fabricated by the Mili-Q water purification system (Summer-S2-20H, Delishi Technology Co., Ltd., Sichuan, China).

4.2 Preparation of CSS

The PTFE substrates measuring 10x60x2 mm and 60x40x2 mm were treated with laser-engraving methods to form micro- and nano-structures. This process resulted in superhydrophobic PTFE substrates. The substrates were then charged with high-voltage electrodes to create the superhydrophobic surface with uniform charges (CSS).

4.3 Characterization

The microstructure of fabricated samples was measured using field emission scanning electron microscopy (SEM, JSM-6700F, Japan). High-voltage charging was carried out by using a high-voltage power supply (GLOW 28720, Shenzhen, China). A contact angle measuring instrument (OCA25, Dataphysics GmbH, Filderstadt, Germany) was used to measure the wetting performances of designed surfaces. The surface electrostatic charge distribution of CSS was measured using an electrostatic tester (Shenzhen, China). The motion process of droplets was recorded by a single-lens reflex camera (EOS 5D MarkIV, Canon, Japan).

4.4 Data

All data in the figures and tables were processed and analyzed after being tested five times.

Acknowledgments

This work was supported by the National High-Level Talents Special Support Program, the Outstanding Youth Natural Science Foundation of Liaoning Province (2022-YQ-11), the Liao Ning Revitalization Talents Program (XLYC2203049), the National Natural Science Foundation of China (52005075), and Dalian Youth Science and Technology Star Project.

References

1. F. He, M.J. Zhang, W. Wang, Q.W. Cai, Y.Y. Su, Z. Liu, Y. Faraj, X.J. Ju, R. Xie, L.Y. Chu, Adv. Mater. Technol. **4**, 1800687 (2019).
2. H. Li, W. Fang, Z. Zhao, A. Li, Z. Li, M. Li, Q. Li, X. Feng, Y. Song, Angew. Chem. Int. Ed. **59**, 10535-10539 (2020).
3. J. Ju, H. Bai, Y. Zheng, T. Zhao, R. Fang, L. Jiang, Nat. Commun. **3**, 1247 (2012).
4. C. Liu, J. Ju, Y. Zheng, L. Jiang, ACS Nano **8**, 1321-1329 (2014).
5. C. Liu, Y. Li, C. Lu, Y. Liu, S. Feng, Y. Liu, ACS Appl. Mater. Inter. **12**, 25471-25477 (2020).
6. I. Banerjee, R.C. Pangule, R.S. Kane, Adv. Mater. **23**, 690-718 (2011).
7. M. Song, X. Liu, T. Wang, W. Xu, S. Zhao, S. Wang, Z. Wang, H. Zhao, Appl. Phys. Lett. **122**, 214102 (2023).
8. Z. Yuan, C. Lu, C. Liu, X. Bai, L. Zhao, S. Feng, Y. Liu, Sci. Adv. **9**, eadg2352 (2023).
9. H. Zhan, Z. Yuan, Y. Li, L. Zhang, H. Liang, Y. Zhao, Z. Wang, L. Zhao, S. Feng, Y. Liu, Nat. Commun. **14**, 6158 (2023).
10. Q. Sun, D. Wang, Y. Li, J. Zhang, S. Ye, J. Cui, L. Chen, Z. Wang, H. Butt, D. Vollmer, X. Deng, Nat. Mater. **18**, 936-941 (2019).
11. C. Liu, C. Lu, Z. Yuan, C. Lv, Y. Liu, Nat. Commun. **13**, 3141 (2022).
12. H. Kim, K. Lee, J.W. Oh, Y. Kim, J.E. Park, J. Jang, S.W. Lee, S. Lee, C.M. Koo, C. Park, Adv. Funct. Mater. **33**, 2210385 (2023).
13. W. Xu, Y. Jin, W. Li, Y. Song, S. Gao, B. Zhang, L. Wang, M. Cui, X. Yan, Z. Wang, Sci. Adv. **8**, eade2085 (2022).
14. Y. Jin, X. Liu, W. Xu, P. Sun, S. Huang, S. Yang, X. Yang, Q. Wang, R.H.W. Lam, R. Li, Z. Wang, ACS Nano **17**, 10713-10720 (2023).
15. N. Bjelobrk, H. Girard, S. Bengaluru Subramanyam, H. Kwon, D. Quéré, K.K. Varanasi, Phys. Rev. Fluids **1**, 063902 (2016).
16. H. Chen, P. Zhang, L. Zhang, H. Liu, Y. Jiang, D. Zhang, Z. Han, L. Jiang, Nature **532**, 85-89 (2016).
17. J. Lv, Y. Liu, J. Wei, E. Chen, L. Qin, Y. Yu, Nature **537**, 179-184 (2016).

Overview and Prospects of Sodium Battery Development in the Context of Accelerated Penetration of Electrification

Jiapeng Yang*

Polytechnic,Jinan University,Guangzhou, Guangdong Province,510632,China

Abstract. Electrification is an essential way to promote the green transformation of energy. Sodium power has attracted wide attention at home and abroad due to its abundant reserves, excellent low-temperature performance, and great safety. This paper firstly overviews the current development status of sodium batteries, analyzes the comparative advantages of sodium batteries over lithium batteries, and evaluates the future comprehensive positioning of sodium batteries. Secondly, the characteristics and progress of the current stage of technology routes of positive and negative electrode materials for sodium batteries are elaborated, and development suggestions are put forward. Finally, the application prospect and economy of sodium batteries are thoroughly analyzed, and the future market space of sodium batteries is preliminarily predicted.

1. Research Background and Status

In the context of global carbon neutrality, the search for safe and sustainable green energy has become an essential proposition of The Times for human society [1-3]. On the one hand, China's foreign dependence on crude oil has been elevated for numerous years, reaching 70.8% in 2021, which gravely affects national security and sustainable economic and social development. On the other hand, the combustion of traditional fossil energy will produce a large number of pollutants such as carbon dioxide, which will aggravate global climate change [4]. Driven by the twin goals of energy security and carbon neutrality, electrification has become an essential way to drive the green transformation of energy.

At present, lithium batteries are widely used in electric vehicles and large-scale energy storage. However, with the additional increase of downstream application requirements, the development of lithium batteries in the future will also face the bottleneck problem of lack of lithium resources. Therefore, it is necessary to develop a different battery technology with a higher balanced comprehensive performance [5]. Sodium batteries have received extensive attention due to their excellent physical and chemical properties and abundant reserves [6-7]. In terms of performance, compared with lithium batteries, sodium batteries have the advantages of higher safety, better low-temperature performance, and better peak-rate charge and discharge performance [8]. From the perspective of total resources, the abundance of lithium in the crust is only 0.0065%, while the reserves of sodium are extremely abundant, accounting for about 2.64% of the crustal reserves, which is more conducive to large-scale application [9-11]. Therefore, the core logic of developing sodium batteries is to ensure the safety of resource supply, so as to promote the industry to reduce its dependence on lithium resources and alleviate the imbalance between the supply and demand of lithium resources in China.

To better grasp the development trend of sodium batteries, it is necessary to conduct an in-depth analysis of their future positioning, key technologies, and application prospects. This paper compares the advantages and disadvantages of sodium and lithium batteries, comprehensively analyzes the progress of key technologies of sodium batteries, and defines the future development direction of sodium batteries. At the same time, based on the interdisciplinary perspective of current energy and economics, it analyzes the application prospect of sodium batteries and preliminarily predicts the future market space of sodium batteries.

2. Integrated positioning of sodium batteries

Recently, sodium batteries have made breakthroughs in the field of transportation and energy storage. On December 28, 2023, Jiangling Group Fresh Energy equipped with Funeng Technology sodium battery sodium electric model Jiangling easy to EV3 (youth version) officially rolled off the assembly line. On January 5, 2024, JAC Yttrium officially delivered the world's first sodium battery mass-produced model to users, marking the milestone of the large-scale application of sodium ion power batteries.

The development of sodium batteries began with the sustained high price of lithium, the rise in its comparative advantage over lithium batteries, the inner core logic is related to the security of supply of national resources. First of all, China is one of the countries with the richest sodium

*Corresponding author's e-mail: 3011378802@qq.com

resources in the world, and sodium electricity can promote the overall industry to reduce its dependence on lithium resources. Second, compared with lithium batteries, sodium batteries also have at least three advantages: first, the comprehensive cost is low, sodium batteries, due to their abundant reserves and simple mining methods, considerably reduce the cost of raw materials; second, the low-temperature performance is excellent, the normal working range of sodium batteries is -40℃ to 80℃, and some products can have a capacity retention rate of over 92% even in the very cold and low-temperature environment of -20℃, which is significantly better than the capacity retention rate of about 60%-70% of lithium iron phosphate; third, the safety is strong, the sodium chemical properties are stable, and the internal resistance of the sodium battery is higher than that of the lithium battery, and the instant heat is less in the safety test such as short circuit. In addition, the sodium battery can be discharged to its voltage of 0V, so it is more secure during transportation. However, the technical maturity of sodium batteries is still not as excellent as that of lithium batteries, and the energy density of sodium batteries is about 20% lower than that of lithium-ion batteries, and the cycle life of sodium batteries is about 65% of that of lithium batteries (Table 1). In summary, sodium batteries have comparative advantages over lithium batteries, and the strategic significance of energy security is more prominent, and the future is expected to form a complementary industrial pattern with lithium batteries.

Table1. Performance comparison of sodium battery, lithium battery and lead-acid battery.

Performance index	Sodium battery	Lithium battery	Lead-acid battery
Preparation cost	0.29 yuan/Wh	0.43 yuan/Wh	0.40 yuan/Wh
-20°C Capacity retention rate	<92%	< 70%	< 60%
Quick charging SOC 20% -80 %	~15 min	~45 min	5~6 h
Over-discharge resistance	As low as 0 V	Poor	Poor
Energy density	100-150 Wh/kg	150-220 Wh/kg	30-50 Wh/kg
Cycle life	1200-2000 times	> 3000 times	300-500 times

3. Key technologies and progress

The structure and working principle of sodium battery and commercial lithium battery are similar. Among the five most significant components of the battery, the diaphragm and fluid collector of the sodium battery can continue to use the related parts of the lithium battery, and the electrolyte only converts lithium salt into sodium salt, and the main replaced parts are the positive and negative electrode materials that account for more than 40% of the battery cost (Figure 1).

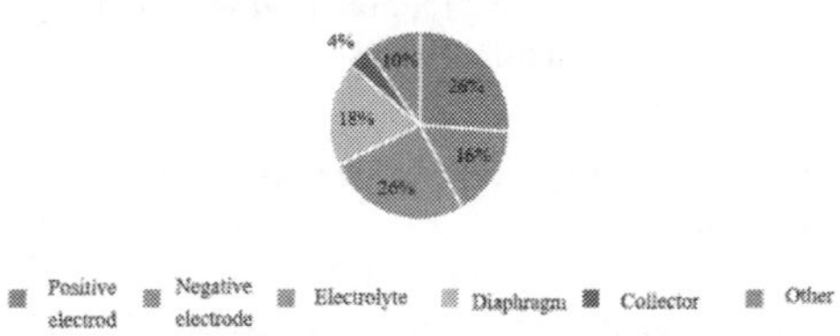

Figure1. Components and price ratio of sodium-ion batteries.

3.1 Positive electrode material technology

At present, there are three main types of cathode materials for sodium batteries: Prussian blue and its analogs, polyanionic compounds and layered transition metal oxides (Table 2).

Prussian blue and its analogs are sodium battery cathode materials with open three-dimensional stable structure and large ion tunneling, which are easy to synthesize and possess advantages such as high theoretical specific capacity. However, due to the structural defects and crystal water in the prepared Prussian blue and its analog crystals, the transmission of sodium ions will be hindered, therefore the ionic conductivity is poor. In addition, the reaction rate of Prussian blue materials during the preparation process is very rapid, resulting in a large number of Fe (CN) 64-vacancy defects in the formed Prussian blue materials, and the crystallization properties and controllability of the materials are considerably reduced.

Polyanionic compounds are connected by polyanionic polyhedra and transition metal ion polyhedra. They have a strong three-dimensional network structure, the highest thermal stability and electrochemical stability, and the best cycle life. In addition, sodium ions occupy the main channel position and have a peculiar 'induction effect', which makes the material have higher redox potential and higher energy density. However, the disadvantages of polyanionic compounds are excessive cost, low energy density and poor rate performance.

Layered transition metal oxides have two different crystal structures, single crystal and polycrystalline. The two crystal structures are closely related to the electrochemical properties of sodium batteries. Layered oxides are favored by numerous battery cathode companies because of their excellent physical and chemical properties and strong compatibility with lithium battery cathode processes and equipment. In addition, the theoretical gram capacity of layered materials is higher, there is a lot of room for improvement, the above characteristics make layered materials expected to become the mainstream development direction of sodium batteries, and will seize the first opportunity in industrialization.

Table2. Comparison of sodium battery cathode technology routes.

Positive electrode material	Advantages	Disadvantages
Layered transition metal oxides	Low cost, wide temperature range, great theoretical capacity, elevated energy density, etc.	Poor structural stability, excessive residual alkali, etc.
Polyanionic compounds	High cycle life, excellent thermal stability, etc.	Extreme cost, low energy density, poor rate performance, etc.
Prussian blue compounds	Low cost, high theoretical specific capacity, simple synthesis, elevated energy density	Short cycle life, tiny temperature range, poor conductivity, etc.

3.2 Anode material technology

According to the reaction mechanism of charge and discharge, the common anode materials are mainly divided into: embedded materials, alloying materials, conversion materials, and organic compounds.

Among them, the embedded materials are mainly carbon-based materials. Carbon-based materials have become the best anode materials for sodium batteries due to their wide sources and strong sodium storage capacity. Among carbon-based materials, tough carbon has additional advantages in sodium storage capacity and rate performance than graphite anode and soft carbon anode, so it has the most potential for industrialization in the field of sodium battery anode materials. In recent years, driven by the development of the energy storage industry, the demand for carbon-based anode materials for sodium batteries has also shown an increasing trend year by year (Fig.2). The remaining three types of anode materials have the problems of structural instability and volume expansion, which may lead to the risk of electrode fragmentation during charging and discharging, thus affecting the battery life. At present, industrialization progress is generally slow.

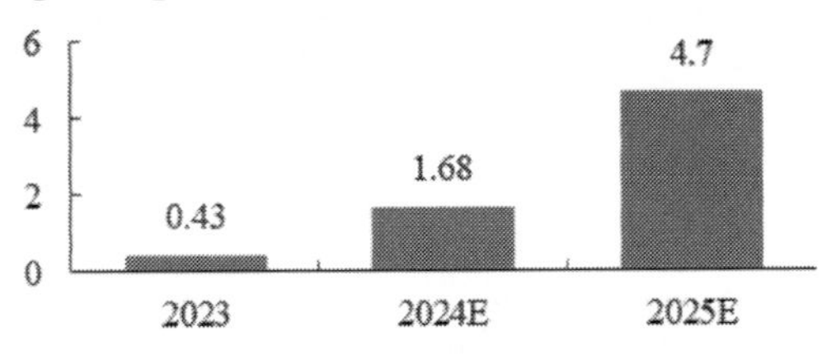

Figure2. Prediction of annual demand for carbon-based anode materials (unit: 10,000 tons)

4. Application prospect and economic analysis

4.1 Application prospect analysis

Sodium batteries, as a kind of 'emerging' battery that has entered the public view in less than two years, have developed more rapidly than lithium batteries. The reason is that sodium and lithium belong to the same family, and their physical and chemical properties are nearly the same. The existing related processes of a lithium battery can be applied to sodium battery only with a slight adjustment, so the production capacity of sodium battery can be rapidly increased in a short time.

In future application scenarios, the energy density of sodium batteries and lithium iron phosphate have a certain overlap, which is expected to replace some of the market share of lithium-ion batteries. Whether it is power battery or energy storage, the whole market scale is extremely large at the same time, there are also numerous market segments, and sodium electricity can be achieved in certain market segments, thus promoting the industrialization process. In the field of current energy vehicles, the analysis of the world's first two sodium battery vehicles, it can be found that the sodium battery capacity is below 50 kWh, and the energy density is below 200 Wh/kg. Therefore, in the short and medium term, sodium battery new energy vehicles are positioned as short-distance vehicles with low endurance. In the field of two-wheelers and three-wheelers, sodium batteries achieve a balance of performance and cost between lithium batteries and lead-acid batteries and are expected to replace lead-acid batteries in the short and medium term. In the field of energy storage, due to the excellent low-temperature performance and cycle performance of sodium batteries, its application scenarios with high safety requirements such as household energy storage, industrial and commercial energy storage are also expected to become an essential supplement to the lithium battery system (Figure 3).In addition, in recent years, a sodium-lithium hybrid power system has been proposed. On the one hand, it can alleviate the problem of insufficient energy density of sodium batteries. On the other hand, it can also make full use of the characteristics of high power and low-temperature resistance of sodium batteries, so as to further expand the application scenarios of sodium batteries.

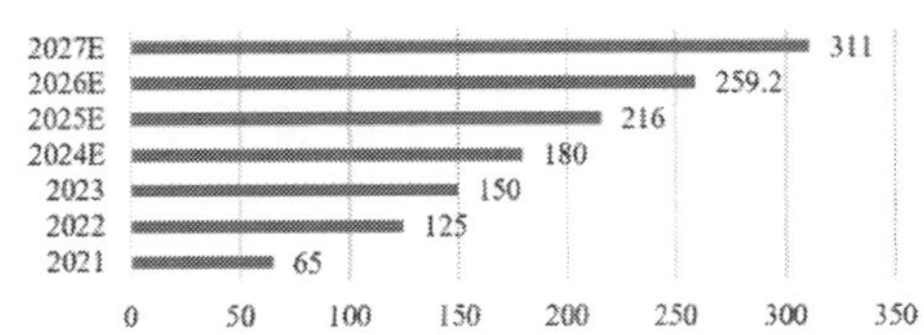

Figure3. Demand of energy storage system for sodium battery in 2021-2027 (unit: GWh)

4.2 Economic analysis

The price of sodium raw materials for sodium batteries is relatively low and stable, only 250 yuan/ton, while the price of lithium carbonate has soared to 400,000 yuan/ton in 2022. In addition to considering the cost of raw materials, the battery production cost also needs to take into account factors such as energy density, cycle life, and battery system composition. Although the energy density and cycle life of sodium batteries are not as excellent as lithium batteries, which narrows the cost gap between sodium batteries and lithium batteries to a certain extent, the cost of sodium battery systems is considerably lower than that of lithium batteries: First, the sodium battery current collector can use low-cost aluminum foil. Compared with the lithium battery using copper foil, the material cost for current collector in each 1 kWh battery is reduced by about 40 yuan; second, the cost of various cathode materials for sodium batteries is considerably lower than that of lithium battery cathode materials. According to the data of Nakai, the cost of cathode materials for sodium batteries is only about 40% of that of lithium batteries(Figure 4). Combining positive and negative electrode materials and fluid collection, the production cost of sodium batteries is about 370 yuan/kWh, which is still much lower than the cost of lithium batteries.

As of February 2024, the average price of lithium carbonate has fallen to 100,000 yuan per ton, and its plummeting price has in turn caused the cost of lithium battery cells to plummet. According to relevant statistics, the current production cost of lithium batteries still needs about 600 yuan/kWh, which is 62% higher than the production cost of sodium batteries. Therefore, despite the decline in lithium prices, the cost advantage of sodium batteries still exists. EV Tank expects that if material technologies such as positive and negative electrodes can be breakthroughs, the future cost of sodium electricity will be further reduced by 30%.

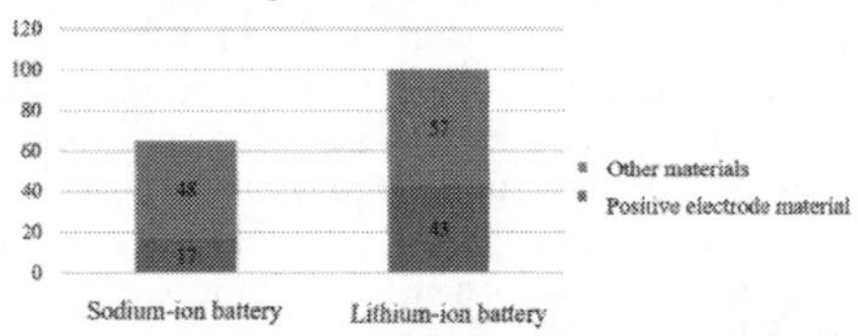

Figure 4. Cost comparison of cathode materials and additional materials for sodium-lithium batteries (lithium battery cost base is 100)

4.3 Market size forecast

With the rapid development of electric vehicles and the energy storage market, sodium batteries will mostly copy the high-speed development path of lithium batteries. Looking forward to 2024, the effective capacity of sodium batteries will reach 25 GWh under the condition of large-scale and continuous improvement of the manufacturing process. At the same time, the price of sodium batteries is expected to decline steadily, driving them to achieve large-scale application in additional fields. Based on the data of China Energy News and the White Paper on the Development of China's Sodium Ion Battery Industry (2023), it is predicted that by 2025, the shipment of sodium batteries will exceed 50 GWh; by 2026, the global demand for sodium batteries will reach 116 GWh; by 2030, global sodium battery demand will grow to 526 GWh (Figure 5). In recent years, the industrialization of sodium batteries has accelerated significantly. It is expected that the market size of sodium batteries will reach 156 billion yuan with an average annual growth rate of 11.6% in 2025 and continue to grow.

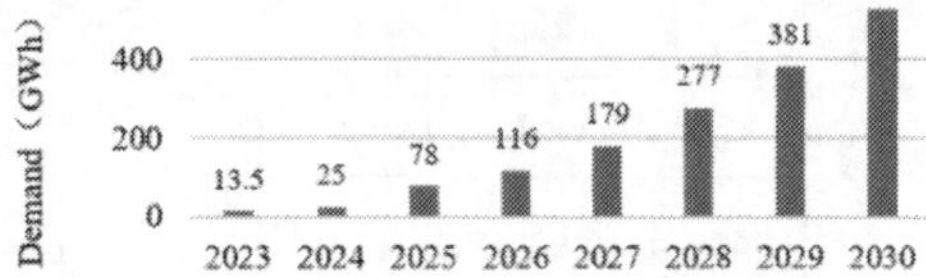

Figure5. Global sodium battery demand forecast.

5. Summarized and prospected

In this paper, the development status of sodium batteries is analyzed comprehensively from the perspective of new energy and economics. Research shows that the development of sodium batteries began with the continued high price of lithium, rising in its higher safety, low-temperature life better comparative advantages, the internal core logic is related to the supply of national resources security, the future sodium battery is expected to form a complementary industrial pattern with lithium. At the key technical level, the three major technical routes of the battery-positive electrode go hand in hand, and the layered oxide technology is the most promising to seize the lead in industrialization; Solid carbon anode material has additional performance advantages and is the mainstream technical route for sodium battery anode materials. In terms of application prospects and economy, sodium batteries are expected to be a cost-effective solution that will play an essential role in A00-class electric vehicles and energy storage. Global sodium battery demand is expected to grow to 526 GWh by 2030. Under the background of lithium price decline, the cost advantage of sodium batteries is still in place. With the advancement of industrial scale, sodium battery is expected to have 30% room for cost reduction.

References

1. Zhang Ping, Kang Libin, Wang Mingju, et al. Technology feasibility and economic analysis of Na-ion battery energy storage [J]. Energy Storage Science and Technology, 2022, 11(6): 1892-1901.
2. Liu Jinyu, Fan Fenglan, Jiao Liansheng, et al. Research Status and Prospect of Anode Materials for Sodium Ion Batteries[J]. Liaoning Chemical Industry, 2023, 52 (2): 238-245.
3. Zhao Yiwei, Zhang Fuhua, Yan Shun, et al. Research progress on the conductivity of positive electrode materials of Prussian Blue sodium-ion batteries [J]. Energy Storage Science and Technology, 2024,1-13.
4. Liu Yang, Wang Xiaodan. Research progress of sodium organic carbonyl ion battery materials [J]. Journal of Electric Sources, 2013,47(5):575-582.
5. Wu Jianbo, Zhou Tong, Li Wei, et al. Patent Review on Layered Cathode Materials for Sodium-Ion Batteries [J]. Mining and Metallurgical Engineering, 2023, 43 (4): 164-168 + 173.

6. Feng Renchao, Dong Yu, Zhu Xinyu, et al. Research progress on graphite oxide-based anode for sodium-ion batteries [J]. Energy Storage Science and Technology, 2024, 1-14.
7. Li Rui, He Xiangfei, Guo Meiqing, et al. Research progress of phosphorus-based polygenic cathode materials for sodium-ion batteries [J]. Journal of Changsha University of Science and Technology (Natural Science), 2023, 20 (03): 14-31.
8. Huang Huawen, Fan Shanshan. Exploration of Crystal Structure and Sodium Storage Performance of Transition Oxide Cathode Materials [J]. Guangdong Chemical Industry, 2023, 50(21): 1-4.
9. Li Hui, Wu Chuan, Wu Feng, et al. Sodium Ion Battery: A Promising Energy-storage Candidate for Supporting Renewable Electricity [J]. Acta Chimica Sinica, 2014,72 (1) : 21-29.
10. Hwang J Y, Myung S T, Sun Y K. Sodium-ion batteries: present and future[J]. Chemical Society Reviews, 2017, 46(12): 3529-3614.
11. Chayambuka K, Mulder G, Danilov D L, et al. Sodium-ion battery materials and electrochemical properties reviewed[J]. Advanced Energy Materials, 2018, 8(16): 1800079.

https://doi.org/10.1051/e3sconf/202456002008

Analysis of total nitrogen pollution load and identification of key pollution sources in Futuan River basin based on SWAT model

Yanjing Li [b]; Hui Peng[a*];

College of Environmental Science and Engineering, Ocean University of China, Qingdao 266100, China

Abstract. In order to identify the key source area of the Futuan River basin, the Futuan River basin is taken as the target area, establishes the Soil Water Assessment Tool (SWAT) model, simules the runoff and total nitrogen pollution process in the basin, and analyzes the spatiotemporal characteristics of total nitrogen pollution. The key source areas of total nitrogen pollution in the basin were identified by the cumulative load curve method. The results show that the total nitrogen pollution load in the Futuan River basin presents obvious seasonal distribution characteristics. The subbasins with high load intensity of total nitrogen into the river are mainly in the middle and upper reaches of the Futuan River.

1. Introduction

Nitrogen is a necessary nutrient for the growth of plants and animals, however, excess nitrogen in water can cause harm to the environment[1, 2].With the rapid development of distributed watershed models, some watershed hydrological models, for example, Soil Water Assessment Tool (SWAT), Hydrological Simulation Program Fortran (HSPF), and Storm Water Management (SWMM) Model[3-5], etc. These models have been used to simulate nitrogen pollution at catchment scale with good results. Among many models, SWAT model is an important tool for large-scale watershed hydrological simulation, many studies have demonstrated that the model is an effective tool for many hydrological and water quality processes such as pollutant transport and soil erosion, and can assess the impact of agricultural management measures and land use on pollution load[6, 7]. The SWAT model considers the complex migration and transformation process of non-point source pollution. Compared with the traditional empirical formula, the simulation results are more reliable.

The Futuan River originates from Rizhao City, Shandong Province. In recent years, the Futuan River basin is faced with serious pollution, so it is necessary to quantitatively analyze the pollution status of Futuan River basin and study how to effectively control the non-point source and point source pollution of Futuan River basin for the sustainable development of Rizhao's environment. This study takes the Futuan River basin as the research area, constructs SWAT model, simulates the total nitrogen pollution load, uses the pollutant accumulation load curve method to identify the key source areas of total nitrogen pollution in the basin, and provides data and technical support for the treatment of total nitrogen pollution in the Futuan River basin.

2. Materials and methods

2.1 Overview of the region

Futuan River is located in Rizhao City, Shandong Province, with a length of 58 km. It originates from Dama 'Anshan in Wulian County, Rizhao City, passes Rizhao Reservoir, and enters the sea in Donggang District, Rizhao City. It is the largest river flowing into the sea alone in Rizhao City. The total area of the Futuan River basin is 1319 km^2, and the geographical coordinates are 35°4'~35°58'N and 118°51' ~119°39'E, the location overview of the research area is shown in Figure 1. In terms of point source pollution in Futuan River basin, the main sources are sewage discharge from industrial enterprises and urban domestic sewage discharge. Among them, there are more than 50 industrial enterprises. In terms of non-point source pollution, the total cultivated land area of the Futuan River basin is 76023ha, accounting for 71.9%. The main crops are soybean, wheat and corn. The larger cultivated land area leads to more agricultural fertilization in the Futuan River basin. There are a lot of livestock and poultry farmers in the rural area of the Futuan River basin, including 506 large-scale farms. The pollution generated by livestock and poultry farming is mainly the discharge of livestock and poultry manure. In addition, there are 1.34 million permanent residents in the rural area of the Futuan River basin. Due to the imperfect sewage treatment facilities, the discharged domestic sewage may cause pollution to the Futuan River basin.

*Corresponding author: [a]pengh@ouc.edu.cn; [b]807546168@qq.com

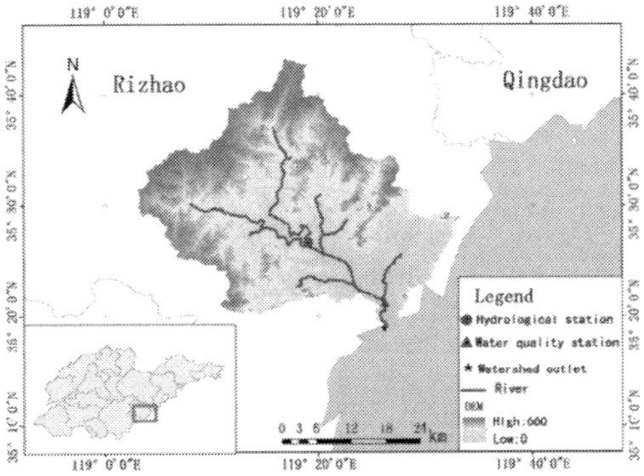

Fig.1 Study area location profile map

2.2 Research methods and data sources

2.2.1 Distributed watershed model

The distributed watershed model of Futuan River basin is constructed by SWAT model. According to the different combination of soil properties, land use types and slope of each subbasin, SWAT model further divides subbasin into many Hydrologic Research units (HRU). The hydrologic response unit can reflect the spatial heterogeneity of the basin, which is conducive to accurately identifying the key source areas of water environmental pollution in the basin.

2.2.2 Input data description

The simulation process of SWAT hydrological model is complicated and requires a lot of data information, the data types and sources are shown in Table 1.

Table 1 SWAT modeling data source

NO.	Data name	Data sources
1	DEM	http://www.gscloud.cn/
2	Landuse	http://data.ess.tsinghua.edu.cn/
3	Soil	FAO SOILS PORTAL
4	Meteorological	CMADS[8]
5	Pollution sources	Statistical yearbook of Rizhao City
6	Runoff	Rizhao City hydrographic bureau
7	Total nitrogen	Shandong ecological environment monitoring Center

2.2.3 Calibration and verification

In this study, SUFI-2 algorithm in SWAT-CUP software was used for automatic parameter calibration and co-calibration of 85 parameters related to runoff and total nitrogen flux. The evaluation parameters for selection and validation were R^2, NSE and PBIAS[9, 10], the calculation formula is shown as 1-3:

$$R^2 = \frac{\left[\sum_{i=1}^{n}(O_i-\bar{O})(S_i-\bar{S})\right]^2}{\sum_{i=1}^{n}(O_i-\bar{O})^2\sum_{i=1}^{n}(S_i-\bar{S})^2} \tag{1}$$

$$NSE = 1-\frac{\sum_{i=1}^{n}(O_i-S_i)^2}{\sum_{i=1}^{n}(O_i-\bar{O})^2} \tag{2}$$

$$PBIAS = \frac{\sum_{i=1}^{n}(S_i-O_i)}{\sum_{i=1}^{n}O_i} \tag{3}$$

Where O_i and S_i are respectively measured values and simulated values, $\bar{O}$ and $\bar{S}$ are measured average values and simulated average values.

2.2.4 Key source area identification

According to the operation results of the model, the total nitrogen inflow load of each subbasin was ranked from small to large, and its percentage was accumulated, and the cumulative load curve was drawn. The relevant calculation formula is as follows 4-6:

$$N_i = N_{org,i} + N_{amm,i} + N_{no3,i} + N_{no2,i} \tag{4}$$

$$N_{Ti} = N_i \times A_i \tag{5}$$

$$N_T = \sum_{i=1}^{n} N_{Ti} \tag{6}$$

Where N_i is the total nitrogen intensity (kg·ha^{-1}) input into the reach of subbasin i; $N_{org,i}$ is the organic nitrogen intensity (kg·ha^{-1}) input into the reach of subbasin i; $N_{amm,i}$ is the ammonia nitrogen intensity (kg·ha^{-1}) input into the reach of subbasin i; $N_{no3,i}$ is the nitrate nitrogen intensity (kg·ha^{-1}) input into the reach of the subbasin; $N_{no2,i}$ is the nitrous nitrogen intensity (kg·ha^{-1}) input from subbasin i to reach; A_i is the area of the subbasin (ha); N_{Ti} is the total nitrogen inflow load (kg) of subbasin i. N_T is the cumulative total nitrogen inflow load (kg) of each subbasin.

2.2.5 Source analysis of total nitrogen pollution in key source areas

The difference subtraction method can be used to analyze the source of total nitrogen pollution in the key source area, that is, different simulation scenarios are set, the input data of a pollution source is changed to 0 in the SWAT model, the calibration parameter is kept unchanged, and the simulation is performed again. By subtracting the simulation results with input data from all pollution sources and the simulation results without one pollution source, the resulting difference is the contribution value of the pollution source pair.

3. Results and discussion

3.1 Model calibration and validation

The calibration and verification results of runoff and total nitrogen flux are shown in Table 2. The R^2 and NSE of runoff rate in periodic and verification periods were higher than 0.6, and the R^2 and NSE of total nitrogen flux rate in periodic and verification periods were higher than 0.85, and $|PBIAS|$ was lower than 15%.

Table 2 Calibration and verification results

Variable	Calibration			Validation		
	R^2	NSE	PBIAS	R^2	NSE	PBIAS
River discharge	0.80	0.78	-10.8%	0.65	0.64	-12.5%
Total nitrogen	0.89	0.87	14.8%	0.90	0.87	13.4%

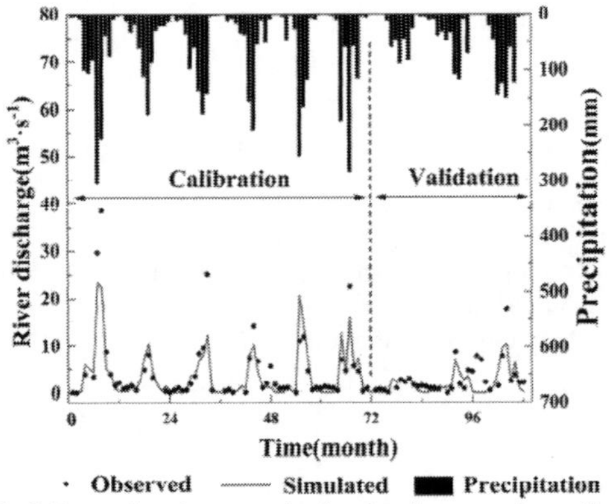

Fig.2 Comparison between simulated and measured runoff of Rizhao Reservoir Station

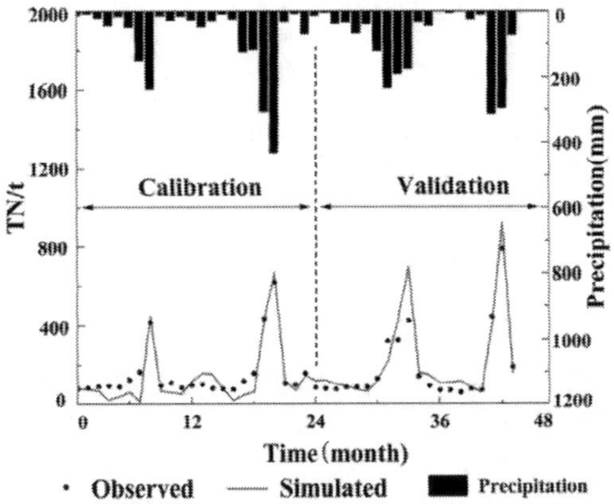

Fig.3 Comparison of simulated and measured values of total nitrogen in Dagu Station

As shown in Fig 2 and Fig 3, after SWAT-CUP parameter calibration, the simulated and measured values of runoff and total nitrogen are close to each other during the periodic and validation periods. In terms of runoff simulation, when the measured runoff value is medium and low, the model simulation results are better, but when the runoff is at its peak, the simulated value is slightly lower than the true value. In general, after the verification period, the model is well applied in the Futuan River basin.

3.2 Time distribution characteristics

During the simulation period, the monthly average total nitrogen flux of the Futuan River into the sea is shown in Figure 4. In the Futuan River basin, the total nitrogen flux into the sea in summer is 866.2 t, accounting for 46.7% of the annual total nitrogen flux, which is the highest in the whole year. In terms of rainfall, the summer rainfall is 381.9mm, accounting for 55.1% of the total annual rainfall, which is also the highest in the whole year. In winter, the total nitrogen flux into the sea was 272.4 t, accounting for 14.7% of the total annual nitrogen flux, and the rainfall was 48.5mm, accounting for 7.0% of the total annual rainfall, both of which were the lowest in the whole year. The fluxes of total nitrogen into the sea in spring and autumn accounted for 15.9% and 22.7%, and the rainfall accounted for 16.4% and 21.5%, respectively. The fluxes and rainfall of total nitrogen into the sea were between summer and winter.

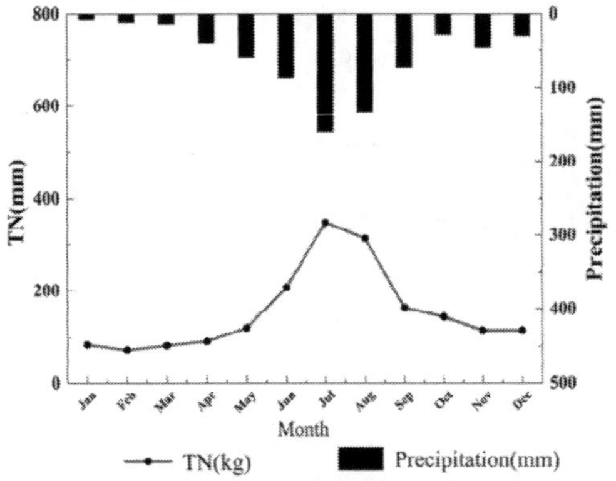

Fig.4 Time distribution of total nitrogen pollution in Futuan River basin

3.3 Spatial distribution characteristics

According to the results of the model, the annual total nitrogen inflow load intensity ($kg \cdot ha^{-1} \cdot a^{-1}$), the land use types and the average slope slope of each subbasin are shown in Figure 5. The land use type with high load intensity of total nitrogen into river is mainly cultivated land, and the average slope is larger. For example, subbasins 3 and 6 have the highest load intensity of total nitrogen inflow into the river, which is higher than 21.0 kg/ha. The analysis of land use type and slope shows that the cultivated land area of the two subbasins is about 70%, and the slope is large, with an average slope of about 18°. The load intensity of total nitrogen into the river in subbasins 2 and 9 is second only to that in subbasins 3 and 6, and the load intensity of total nitrogen into the river is about 20.0 kg/ha. The average slope of these two subbasins is lower than that in subbasins 3 and 6, but the proportion of cultivated land area is larger, reaching more than 80%. The average slope of subbasin with low total nitrogen inflow load intensity is below 10°, and the average slope of subbasin 14 with the lowest total nitrogen inflow load intensity is only 4°. The results show that the intensity of total nitrogen inflow load is related to the land use type and slope.

 https://doi.org/10.1051/e3sconf/202456002008

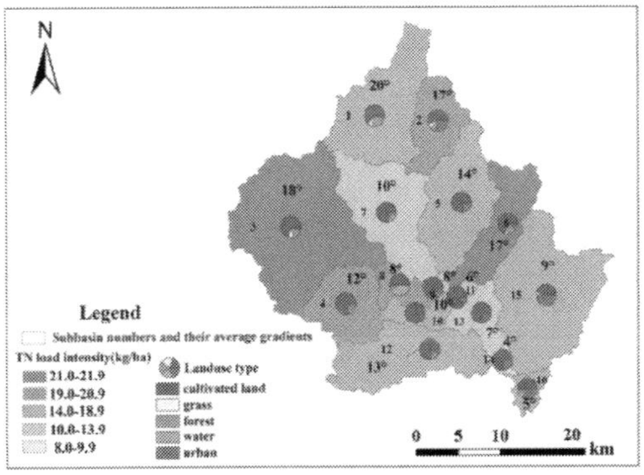

Fig.5 Spatial distribution of total nitrogen, proportion of land use and average slope in Futuan River basin

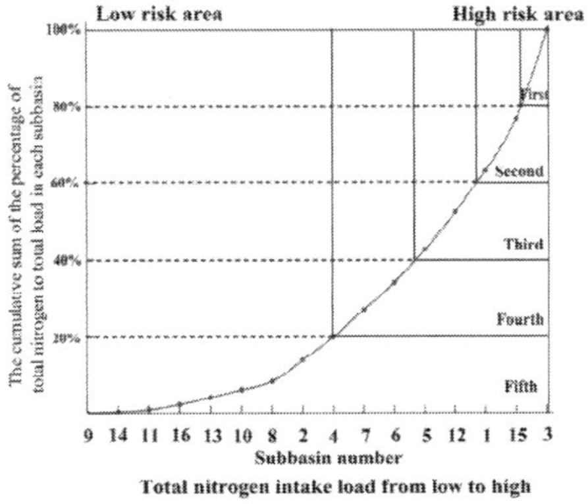

Fig.6 Percentage cumulative curve of total nitrogen loss load in subbasins

3.4 Key source area identification

In order to identify the key source areas of total nitrogen pollution and improve the treatment efficiency of total nitrogen pollution, the inflow load of each subbasin was ranked from small to large, and the cumulative curve of the percentage of total nitrogen inflow load in the subbasin was drawn, as shown in Figure 6. The subbasins with pollution levels of one and two are high-risk subbasins and the key source areas that need to be concentrated and focused. There are three key subbasins in the Futuan River basin, numbered 1, 15 and 3. The cumulative nitrogen load of these three subbasins is 901.4t, accounting for 47.6% of all subbasins. The reasons for the higher total nitrogen load of the three subbasins were further analyzed. According to the results of 3.3 total nitrogen inflow load intensity of each subbasin, the total nitrogen inflow load intensity of subbasin 3 was the highest, reaching 21.5 kg/ha, and the subbasin had a large area, so it was still the first level in the classification of the key source area. The load intensity of total nitrogen inflow into the river in the subbasin 15 is slightly smaller, but the area of the two subbasins is larger, especially the subbasin 15, whose area is second only to that of the subbasin 3 among the 16 subbasins in the Futuan River basin.

In order to verify the identification results of key source areas, samples were taken from the upper reaches of Liulang River and the lower reaches of Guhe River in subbasin 15 in August 2022, and the total nitrogen concentration was 4.9mg·L^{-1} in Liulang River and 7.3mg·L^{-1} in Guhe River. It can be seen that the total nitrogen concentration in the reach increased significantly after passing through subbasin 15, which can provide support for the identification results of key source areas in this study.

3.5 Sources of pollutants in key source areas and treatment measures

According to the difference subtraction analysis results, the proportion of total nitrogen pollution sources in each subbasin is output, as shown in Figure 7. The proportion of total nitrogen pollution sources in key source areas was discussed here. The main pollution source in subbasins 3 and 15 is fertilizer application. Especially in subbasin 3, fertilizer application accounted for 60%, mainly due to the high proportion of cultivated land in this subbasin and the large amount of fertilizer applied by crop planting. The main pollution source in subbasin 1 was livestock and poultry breeding, accounting for 53%. Subbasin 1 is located in Wulian County, Rizhao City. According to the statistical yearbook of Rizhao City, Wulian County has a large number of livestock and poultry breeding, which is much higher than other districts and counties. Subbasin 15 is located in Rizhao city, the point source pollution accounts for 35%, and the main point source is the sewage discharge from the urban sewage treatment plant. Lots of sewage treatment plants are located in subbasin 15, totaling 7. Its sewage treatment capacity is large, and the total nitrogen flux discharged into the river is also large.

According to the identification of key subbasins and their sources, the treatment measures of total nitrogen pollution in Futuan River basin are put forward. We should pay more attention to non-point source pollution in the Futuan River basin, which mainly includes the control of total nitrogen pollution caused by fertilizer application and livestock pollution. For the total nitrogen pollution caused by fertilizer application, the fertilization intensity in this area can be adjusted reasonably, and the scientific fertilization guidance can be strengthened. For the total nitrogen pollution caused by livestock pollution, livestock management can be strengthened in the region, such as strengthening the treatment of livestock and poultry breeding sewage in the region. In the treatment of point source, we should pay more attention to the subbasin 15, where point source pollution is relatively high, and the main pollution source is urban domestic sewage discharge. In this regard, the total nitrogen pollution of urban domestic sewage can be reduced by improving the treatment process of urban sewage.

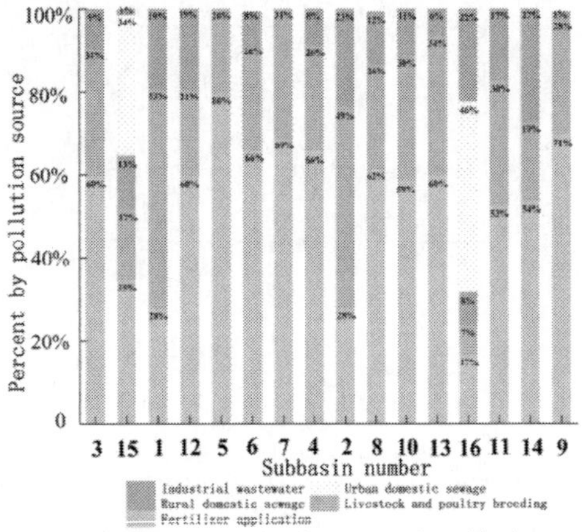

Fig.7 Proportion of pollution sources in subbasins

4. Conclusions

In this research, the SWAT model of the Futuan River Basin was constructed, and the simulation results were verified by the measured data. According to the running results of the model, we get the following conclusions:

(1) The total nitrogen pollution showed obvious time distribution characteristics, high in summer and low in winter, and the total nitrogen loss load was positively correlated with rainfall..

(2) The spatial distribution characteristics show that the load intensity is higher in the middle and upper subwatershed, and the load intensity is related to the land use type and slope.

(3) Three subbasins in the Futuan River basin are identified as key source areas, and the sources in each key source area are different. Targeted treatment of each key source area can improve treatment efficiency.

Future studies can develop more detailed treatment measures, and use the SWAT model to explore the treatment effect and cost of various treatment measures in the Futuan River basin, optimize the treatment measures according to the cost and benefit, and provide scientific basis for the pollution control of the Futuan River basin.

References

1. Ma, G.W., Wang, Y.Y., Bao, X. (2015) Nitrogen pollution characteristics and source analysis using the stable isotope tracing method in Ashi River, northeast China. J. Environmental Earth Sciences., 73(8): 4831-4839.
2. Qiu, B., Li, P.P., Zhong, C.Y. (2012) Characteristics and spatial distribution of the rural non-point source pollution in Haihe River Basin. J. China Environmental Science., 32(3): 564-570.
3. Zhang, Z.S., Montas, H., Shirmohammadi, A. (2023) Effectiveness of BMP plans in different land covers, with random, targeted, and optimized allocation. J. Science of the Total Environment., 892: 164428.
4. Tuoyo, J., Kumarasamy, Vellaisamy, M. (2018) Assessing non-point source pollution models: a review. J. Polish Journal of Environmental Studies., 27(5): 1913-1922.
5. Zhu, D.T., Cheng, X.J., Li, W.H. (2023) Estimating the impact of temperature and streamflow change on river nitrogen pollution using the land-river integrated modeling system. J. Journal of Hydrology., 626.
6. Tan, M.L., Gassman, P.W., Yang, X.P. (2020) A review of SWAT applications, performance and future needs for simulation of hydro-climatic extremes. J. Advances in Water Resources., 143: 103662.
7. Sattari, M.T., Joudi, A.R., Kusiak, A.A. (2016) Estimation of Water Quality Parameters With Data-Driven Models. J. American Water Works Association Journal., 108(4): 72-72.
8. Meng, X.Y., Wang, H. (2017) Significance of the China Meteorological Assimilation Driving Datasets for the SWAT Model (CMADS) of East Asia. J.WATER., 9(10).
9. Moriasi, D.N., Arnold, J.G., Van Liew, M.W. (2007) Model evaluation guidelines for systematic quantification of accuracy in watershed simulations. J. Transactions of the Asabe., 50(3): 885-900.
10. Abbaspour, K.C., Johnson, C.A., van Genuchten, M.T. (2004) Estimating uncertain flow and transport parameters using a sequential uncertainty fitting procedure. J. Vadose Zone Journal., 3(4): 1340-1352.

Research on Online Flocculation Purification Method for The Centralized liquid supply and chip removal system

Qiang Xu [1], Shihua Yuan [1], Ruihua Guo [1], Bin Dai [1], Weiwei Zhang [1], Liangqiang Dai [1], Yu Wang [2,*]

[1]Chengdu Aircraft Industry (Group) Co., Ltd, Chengdu, Sichuan, China
[2]Tianjin High-end Equipment Resereach Institute of Tsinghua University, Tianjin, China

Abstract. This work analysed the principle of flocculation purification method and the flocculation purification effect of fully synthesized water-based cutting fluid under experimental conditions, and the anti-corrosion performance of the purified cutting fluid was studied. Then, the engineering application of flocculation purification method in the centralized liquid supply system was carried out to form a flocculation purification method suitable for the Centralized liquid supply and chip removal system of fully synthesized water-based cutting fluid. This work can provide an effective evolutionary method for the fully synthetic water-based chip cutting fluid of the centralized liquid supply system.

1.Introduction

With the development of modern mechanical processing technology towards high-speed, precision, composite, intelligent, and green lighting, coupled with the wide application of modern water treatment methods, the cutting fluid supply system is also developing towards a centralized cutting liquid system characterized by automation, high-quality assurance, environmental protection, and energy conservation[1]. The centralized cutting fluid system is a circulation system that centrally processes and supplies cutting fluid from multiple processing equipment and one or more processing production lines. It is usually composed of cutting fluid centralized filtration system, reflux system, supply system, and auxiliary system. After used by machine tool , the cutting fluid is collected by the reflux system and pumped to a centralized filtration treatment station. Multiple levels of filtration treatment are carried out within the station as needed, and then the pump group sends filtered cutting fluid that meets the process requirements back to the machine tool through the supply pipe network. This cycle of operation continuously assists the processing equipment of the machine tool.

The centralized liquid supply system has the following advantages compared to the traditional single machine circulation mode of one pump and water tank in the machining workshop: 1) saving resources, reducing pollution emissions, and improving the production environment; 2) Centralized green treatment of cutting fluid effectively ensures the machining efficiency of machine tools and improves the machining quality of parts; 3) Centralized and automated management of waste and liquid waste to improve the overall efficiency of the workshop. Upgrade workshop management. Meanwhile, with the proposal of the dual carbon strategic plan, many CNC machining enterprises are more inclined to use fully synthetic cutting fluids with better environmental performance for centralized fluid supply systems, in order to truly meet the requirements of green manufacturing[2]. During the use of fully synthetic cutting fluid in the centralized liquid supply system, as the usage time increases, cutting particles and impurities will continue to accumulate inside the system. During the circulation process, the filtering device of the system will filter out large particle sizes, but small particle sizes will continue to increase over time, suspended in the liquid, and will bring significant harm when reaching a certain level. Firstly, a large amount of suspended particles can reduce the transparency of the liquid and make its appearance cloudy. Secondly, a large amount of suspended particles will adhere to the filters, sensors, and other parts of the machine tool after reaching the equipment end, causing frequent alarms on the machine tool. In addition, a large number of suspended particles may react with some components in the cutting fluid to form some precipitates, foam, etc. At the same time, the reduction of effective components in the fluid will cause the lubrication, extreme pressure, corrosion inhibition performance of the cutting fluid to decline, resulting in quality problems of processed products[3]. To remove these small particles from the liquid without stopping machine tools, an external filter can be used, but a high-precision filter element is required to achieve the purpose of filtration. In general, the particle size of suspended particles in liquid is mostly within the range of 40nm to 1000nm. Using a filter element with such precision level will bring about a large amount of consumables, high cost, and slow filtration speed, making it difficult to meet practical application needs.

*Corresponding author:Wang Yu,Engineer ,email:wangyu2@tsinghua-tj.org

 https://doi.org/10.1051/e3sconf/202456002009

2.Flocculation principle.

When a spherical particle undergoes natural sedimentation in a stationary liquid under the action of gravity, its sedimentation velocity is proportional to the square of the particle diameter. Generally, the diameter is less than 100μm, the natural sedimentation velocity of small particles is extremely slow. However, the solid suspended solids contained in industrial wastewater, domestic sewage, and other water suspensions belong to the colloidal dispersion phase or group dispersion phase type substances that are difficult to settle naturally. Therefore, using flocculants with flocculation functions for solid-liquid separation and removing suspended substances in the liquid phase is an extremely effective separation method. In order to accelerate the sedimentation of suspended particles, it is necessary to try to disrupt the stability of particles in the system, promote their collision, and achieve the goal of increasing their particle diameter. This is the basic principle of flocculation[4]. The types of flocculant bridge can generally be divided into ring type bridge erection, tail type bridge erection and train type bridge erection, as shown in the figure 1.In a stable colloidal dispersion system, a long chain macromolecule can adsorb two or more colloidal particles simultaneously, or a single colloidal particle can adsorb simultaneously two polymer chains. Therefore, the adhesive particles can be wrapped together and agglomerated and precipitated in the form of bridging[3]. It may also be due to the irregular adsorption of polar groups in the macromolecular chain on the surface of the colloidal particles, resulting in the aggregation of the colloid[5] , as shown in the figure 2.

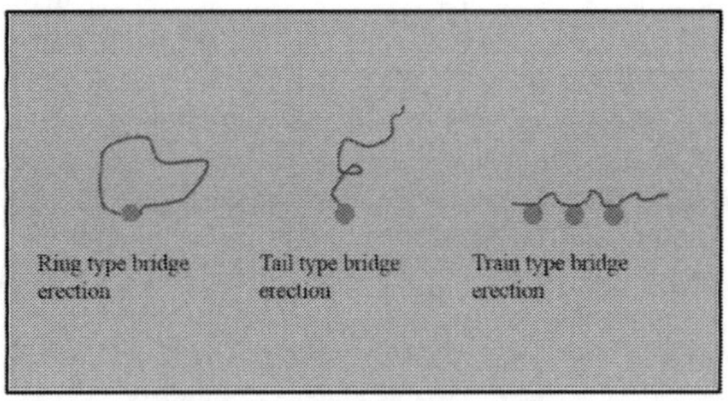

Figure 1. Bridging in flocculants.

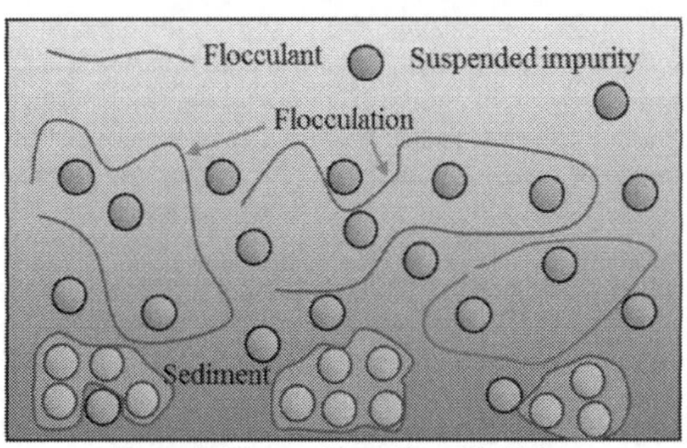

Figure 2: The scrolling effect of bridging.

3. Application of flocculants.

3.1 Type of flocculants

1) Anionic type

Anionic type flocculants suitable for inorganic salts (hydroxides with heavy metals, etc.), with a pH requirement of neutral or alkaline (under strong alkalis and high temperatures).The flocculation effect still does not decrease, for example, mineral processing, electroplating, machinery, coal washing, pulp wastewater, etc..

2) Cationic type

Cationic type flocculants suitable for organic matter, with strong acidic or neutral pH requirements, such as domestic sewage, food processing enterprise wastewater, oil wastewater, etc.

3) Non ionic type

Non-ionic type flocculants suitable for inorganic or inorganic organic mixed systems, with pH requirements of weak acidity, weak alkalinity, and neutrality (not easily affected by pH and metal ions, especially in acidic media), such as wastewater from ore processing.

3.2 Select and use of the anionic type flocculants

Anionic flocculants mainly include Polyacrylamide and sodium polystyrene sulfonate, etc.

Molecular formula of polyacrylamide is as follows.

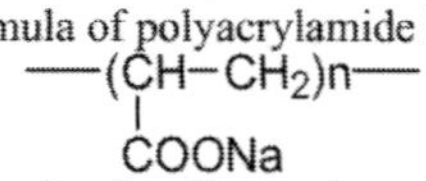

Molecular formula of sodium polystyrene sulfonate is as follows.

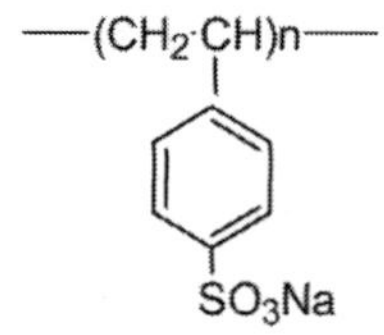

4. Cutting fluid flocculation experiment.

The pH value of fully synthesized cutting fluid is generally between 7-9, belonging to neutral or alkaline solutions, within the effective range of anionic flocculants In addition, impurities such as metal particles and metal ions in the cutting fluid usually contain a certain amount of positive charge, which can be removed by anionic flocculants. Taking into account factors such as viscosity, flocculation speed, economic cost, and chemical composition of anionic flocculants, a widely used anionic polyacrylamide was selected as the target flocculant, and its purification effect was verified under experimental conditions. Take three parts of the working solution of the 200ml centralized supply system, add 1mg, 10mg, and 100mg of anionic polyacrylamide respectively, and then use a magnetic stirrer to stir the solution at a speed of 100r/min for 15min. Observe the purification effect, and the experimental results are shown in the table1.

Table 1. Results of polyacrylamide flocculation test

Flocculant addition amount (mg)	Mixing time (min)	Test results
1	15	Remains basically unchanged
10		Significant increase in transparency
100	15	Increased transparency and flocculence

Perform corrosion testing after filtration of the flocculent precipitate in a liquid containing 10mg of polyacrylamide, and compare the test results with the corrosion results of the original working solution, as shown in the figure 3.

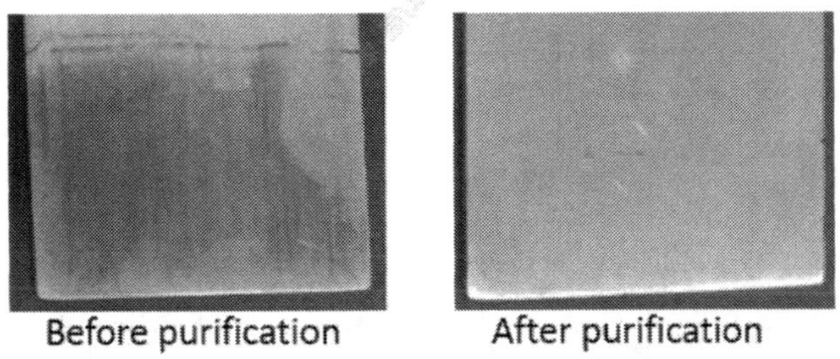

Figure 3 . Surface condition of parts before and after flocculation treatment.

5. Flocculation purification process for fully synthetic cutting fluid in the centralized liquid supply system.

The centralized liquid supply system provides cutting fluid supply for all CNC production equipment in the workshop. Once the system shuts down, it will affect the operation of the entire workshop equipment. Therefore, the use of flocculants for purification treatment of online working fluids requires the development of a more rigorous process. As shown in the Figure 4, the centralized liquid supply consists of a liquid supply system (including a liquid supply pump group and pipeline network), a return liquid and cutting chip system (including a return liquid tank, suction pump and pipeline network), a cutting fluid purification system, a chip processing system, a cutting fluid management system, an electrical control system, etc. It is mainly used to achieve centralized chip processing and the circulation and filtration of cutting fluid used in the processing process.

The cutting liquid supply system mainly consists of a liquid supply pipeline network, a liquid supply pump group, etc. The return system consists of a return tank, a return pump, and a return pipe network. The chips and cutting fluid processed by the machine tool are transported from the return tank to the centralized filtration system for treatment. The working status of the return system can be fed back to the centralized liquid supply system to achieve fault monitoring of the return system by the centralized liquid supply system. The cutting fluid purification system is used for filtering and purifying cutting fluid, mainly including pre-treatment filters, precision filtration systems, purification tanks, and floating oil removal devices. In order to enable the flocculant to better capture particles, it is necessary to choose a location with a faster liquid flow rate, and at the same time, the flocs should be easily filtered and discharged after agglomeration. Therefore, the area near the return port of the coarse filter box is chosen as the area for adding flocculant. On the one hand, the liquid after use flows back to the system from here, which gives the liquid a large kinetic energy. On the other hand, the bottom of the coarse filter box is generally equipped with a scraper chip conveyor. After the flocculent reaches a certain degree, it will sink into the water bottom due to its own gravity and be discharged into the system by the scraper chip conveyor. As shown in the figure 5, in order to improve the flocculation speed of the flocculant, the flocculant powder is fully dissolved in deionized water before being added to the centralized liquid supply system, forming a uniform and highly viscous solution, then slowly join the system. As shown in the figure 6, after adding an anionic polyacrylamide flocculant solution, the impurity particles in the fully synthesized cutting fluid quickly aggregate and become larger. As shown in the figure 7, as flocculation and sedimentation continue to occur and large particle impurities are continuously removed, the transparency of the cutting fluid in the Centralized liquid supply and chip removal system gradually increases. Based on the processing of parts on the production line, it can be seen that after purification treatment with anionic polyacrylamide flocculant, the corrosion phenomenon of parts is significantly reduced.

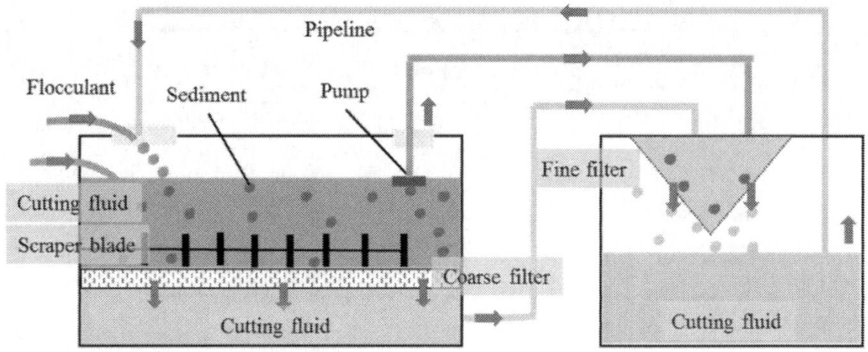

Figure 4 . Flocculation Purification in centralized liquid supply system

Figure 5. Flocculation dilution and stirring

Figure 6. Condensed material

Figure 7 . The filtered cut fluid

6. Conclusion

(1) Anionic polyacrylamide flocculant can effectively agglomerate suspended particles in the fully synthesized cutting fluid, separate the suspended solids in the cutting fluid with high turbidity, and achieve good purification effect. The appropriate addition amount is 10mg/liter.

(2) Anionic polyacrylamide flocculants are suitable for online purification treatment of fully synthetic cutting fluid in Centralized liquid supply and chip removal system. Adding flocculants at the return of the coarse filter box can effectively capture suspended particles and facilitate the discharge of flocs generated after flocculation.

(3) Before adding coagulant to the central supply and drainage system, it can be thoroughly stirred and dissolved to form a uniform coagulant solution to improve the flocculation speed of the flocculant.

References

1. Liu J.A., Xie S.S. (2004) Application and technical development of aluminum alloy materials. Metallurgical Industry Press, Beijing.
2. Jia Xiaoming, Zhang Xiuling. Corrosion of water-based cutting fluid on machine tool and maintenance measures[J]. Machine Tool & Hydraulics | Mach Tool Hydrau, 1998(4):76-76.
3. Liu Z.C. (2008) Cutting fluid technology. Mechanical Industry Press, Beijing.
4. Piotr Ma'cczak,Halina Kaczmarek*,Marta Ziegler-Borowska.Recent Achievements in Polymer Bio-Based Flocculants for Water Treatment[J]. Materials,2020,13(18).
5. Tao Wang,Xiaomin Tang,Shixin Zhang,et.Roles of functional microbial flocculant in dyeing wastewater treatment:Bridging and adsorption[J].Journal of Hazardous Materials,384(1),1 2020.7

Research Progress of the Application of UV Light-Emitting Diodes for Water Disinfection

Jinchi Wang[1a], Xiaolin Feng[2, b*]

[1]School of Computer and Communication Engineering, Northeastern University at Qinhuangdao, Qinhuangdao, 066004, The People's Republic of China
[2]Faculty of Engineering and Life Sciences, Universiti Selangor, Bestari jaya, Selangor 45600, Malaysia

Abstract: With the increasing global demand for safe drinking water, the search for effective and economical water treatment technologies has become an urgent need. Ultraviolet (UV) light sterilization is a proven method for eliminating bacteria in water, but traditional UV systems often rely on mercury lamps, which pose potential environmental and health risks. In recent years, the development of UV-LED technology has offered a new solution for water treatment. This article provides a detailed explanation of the working principles, technological advantages, and applications of UV-LEDs in water treatment. By comparing them with traditional UV systems, UV-LEDs not only demonstrate lower energy consumption and longer lifespan, but also avoid the use of toxic mercury. Despite current challenges in cost and technological maturity, the environmental friendliness and efficiency potential of UV-LED technology indicate a promising market outlook. This article aims to provide a comprehensive review of UV-LED technology for practitioners and researchers in the water treatment industry, and to explore its future development directions.

1. Introduction

Water is the source of life, and the presence of pathogens such as typhoid bacillus, parasites, and viruses in drinking water poses significant health risks[1]. Particularly in developing countries and rural areas, millions of people annually face the threat of waterborne diseases due to a lack of safe drinking water resources[2]. Although traditional water treatment technologies such as chlorination have been widely used for sterilization and purification, these methods often suffer from drawbacks like chemical residues. Ultraviolet disinfection technology offers significant advantages over traditional water disinfection methods; it utilizes specific wavelengths of UV light to directly disrupt the DNA of microorganisms, achieving instant sterilization and efficiently inactivating bacteria, viruses, and other pathogens without the need for chemical additives, thus demonstrating great potential to replace traditional water treatment methods[3]. However, traditional UV disinfection uses mercury lamps, which pose problems related to environmental pollution, high energy consumption, and difficult equipment maintenance[4]. In contrast, UV-LED technology exhibits greater potential for application due to its lower energy consumption, longer lifespan, and smaller equipment size, offering a safer and more environmentally friendly solution for water treatment. This article will explore in detail the application of UV-LEDs in water sterilization, summarize the effects and potential of UV-LED technology in practical water treatment, and discuss the challenges and future directions of this technology. Through this approach, this article aims to provide valuable information and insights for practitioners and researchers in the water treatment industry.

2. Advantages of UV-LEDs in Water Disinfection Applications

UV LEDs offer significant advantages over traditional mercury lamps in the field of water disinfection, making them a more environmentally friendly and efficient choice[5]. As illustrated in Table 1, UV LEDs do not contain mercury or other hazardous chemicals, reducing the risk of environmental pollution and avoiding the generation of toxic waste during disposal. Additionally, UV LEDs can be activated and deactivated instantly without any warm-up time, enabling faster system response and on-demand operation, which minimizes energy waste[6]. UV LEDs typically boast a longer lifespan than mercury lamps, which decreases the frequency of replacements and maintenance costs. They are also more energy-efficient, converting a greater proportion of electrical energy into effective ultraviolet light, thus lowering operational costs. Due to their smaller size, the devices built using UV LEDs can be more

[b*]Corresponding author: 18345148026@163.com
[a]Jinchi Wang, 2460811425@qq.com

compact and are easier to install in space-constrained areas. The miniaturization and flexibility of UV LEDs can be tailored to specific application needs. Additionally, UV LEDs generate less heat during operation, which is particularly beneficial for water treatment applications that require controlled operational temperatures.

Table 1. Comparison of advantages and disadvantages of UVLEDs and Mercury Lamps

Feature	UV LED Advantages	Mercury Lamp Disadvantages
Environmental Safety	Mercury-free and no chemical usage	Contains mercury, generates hazardous waste
Operational Efficiency	Instantaneous operation with no warm-up	Requires warm-up period for optimal operation
Service Life	Extended lifespan with minimal maintenance	Shorter operational life, frequent replacements needed
Energy Efficiency	Higher energy conversion efficiency	Lower energy efficiency
Physical Size	Compact design allows flexible installation options	Bulkier, rigid design constraints
Thermal Management	Emits minimal heat, suitable for sensitive environments	Generates significant heat, may require cooling
Long-Term Costs	Lower maintenance and operating costs	Higher long-term expenses

3. Current status of UV LEDs

The basic operating principle of UV LEDs is similar to that of standard LEDs, both of which are based on a semiconductor p-n junction. When current passes through the LED, electrons move from the n-type semiconductor to the p-type semiconductor[7]. During this process, energy is released when electrons combine with holes. If this released energy falls within the ultraviolet spectrum, ultraviolet light is emitted. Therefore, the emission wavelength of the LED is determined by the bandgap width of the material in the emission region. UV LEDs commonly utilize materials such as Gallium Nitride (GaN) or Aluminum Gallium Nitride (AlGaN) in the active region because these materials have wider bandgaps, allowing for the emission of shorter wavelengths of light[8]. By altering the proportion of components in the active region material—specifically the ratio of Aluminum in AlGaN—the emission wavelength of UV LEDs can be precisely controlled to range between 210 nm and 360 nm. UV LEDs are categorized based on their wavelengths into UVA (320-400 nm), UVB (280-320 nm), and UVC (100-280 nm). Among these, UVC LEDs, with their shorter wavelength and higher energy, are particularly suited for disinfection and sterilization applications.

Fabricating ultraviolet LED packaging devices is a process involving advanced technology and precision engineering, typically divided into the following steps: substrate preparation, epitaxial growth, chip fabrication, and testing and packaging.

a. Substrate preparation: The substrate serves as the foundation for LED manufacturing, commonly using materials like sapphire, silicon, or silicon carbide. Sapphire is the most frequently used substrate material due to its excellent thermal stability and transparency. The main function of the substrate is to provide a stable and clean surface for subsequent material growth. Before manufacturing begins, the substrate must be cleaned and pre-treated to ensure it is free from dust, grease, or other impurities.

b. Epitaxial growth: Epitaxial growth involves depositing one or more layers of semiconductor materials on the substrate using Metal Organic Chemical Vapor Deposition (MOCVD) technology[9]. For ultraviolet LEDs, typical semiconductor materials include GaN or AlGaN, which are effective at emitting ultraviolet light. During the epitaxial growth process, temperature, pressure, and gas flow must be precisely controlled to ensure the grown material layers have good crystalline quality and the required electronic properties. Figure 1 shows the common epitaxial structure of an ultraviolet LED, including the substrate, AlN intrinsic layer, n-AlGaN electron injection layer, quantum well active region, electron blocking layer, and p-AlGaN hole injection layer. Figure 2(a) displays an image of the epitaxial wafer of a UV LED after growth.

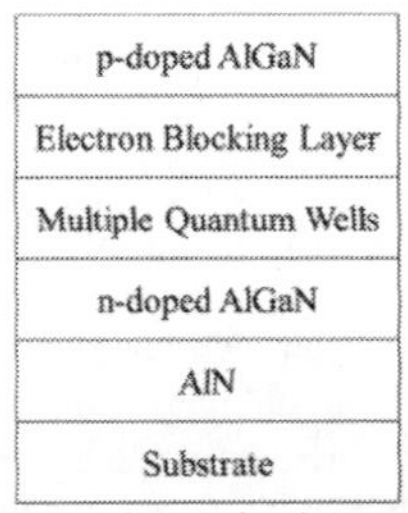

Fig. 1. Schematic diagram of the epitaxial structure of UV LEDs

c. Chip fabrication: After the epitaxial layer growth, the next step is to transform these layers into individual LED chips. This includes several steps: transferring circuit patterns to the epitaxial layer using photoresist materials and ultraviolet light, removing excess material to form the LED structure, and creating electrodes on the epitaxial structure so the chip can connect to a power source and emit light. Figure 2(b) shows an image of the epitaxial wafer after it has been processed into chips. A 2-inch epitaxial wafer can yield nearly 8,000 ultraviolet LED chips.

d. Packaging and testing: Packaging involves mounting the LED chip into a protective casing to prevent exposure to water and air during operation, affecting stability. Typically, the LED chip is fixed onto a heat-conductive base, such as copper or ceramic. Electrodes are connected to external circuits through soldering or other methods. The chip is covered with transparent epoxy resin or silicone to protect it from physical damage and environmental impacts while allowing light to pass through[10]. After packaging, the LED is subjected to final tests, including long-duration operation and stress tests, to ensure its stability and reliability in practical applications. Figure 2(c) shows an image of an ultraviolet LED chip packaged onto a ceramic substrate. The ceramic substrate has good thermal conductivity, which enhances the chip's heat dissipation capabilities.

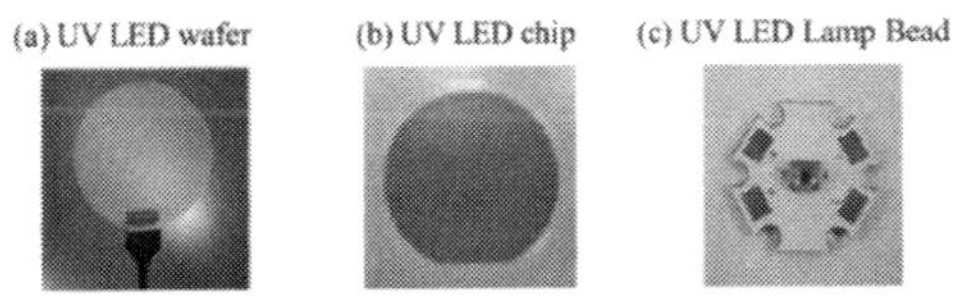

Fig. 2. Schematic diagram of the epitaxial structure of UV LEDs

The efficiency of current ultraviolet LEDs can generally be expressed by formula (1):

$$\eta_{EQE} = \eta_{RAD} \times \eta_{INJ} \times \eta_{LEE} = \eta_{IQE} \times \eta_{LEE} \quad (1)$$

where η_{EQE} represents the external quantum efficiency, which describes the overall efficiency of converting input electrons into output photons in an LED[11]. η_{RAD} stands for radiative recombination efficiency, indicating the efficiency of light emission when electrons and holes recombine in the semiconductor material. η_{INJ} is the injection efficiency, or the proportion of carriers (electrons and holes) effectively introduced into the active region. η_{LEE} is the light extraction efficiency, which is the percentage of generated photons that successfully escape from the LED structure. η_{RAD} is primarily affected by factors such as the dislocation density of the material and polarization effects. LEDs are generally heteroepitaxial, with significant lattice mismatch between the substrate and the epitaxial layer, leading to numerous dislocations within the LED material. These dislocations form non-radiative recombination centers, reducing the radiative recombination efficiency. $\eta_{INJ}$is related to the device's doping concentration and the design of its band structure. If the band structures of the quantum wells and the electron blocking layer are mismatched, electrons may cross the electron blocking layer and emit light in the P-type region, thereby reducing the carrier injection efficiency. η_{LEE} depends on factors such as the TE/TM light emission modes, absorption by the P-GaN layer, total internal reflection at interfaces, and losses. Typically, the light extraction efficiency of deep ultraviolet LEDs is less than 10%. Currently, the η_{EQE} of UV LEDs is generally around 8%. The primary reasons for this low efficiency include material and technological limitations. Firstly, ultraviolet LEDs typically were grown on sapphire substrates, where lattice mismatch and high dislocation density lead to increased non-radiative recombination, thus reducing the efficiency of radiative recombination.

In the quantum wells of UV LEDs, not all injected carriers emit photons; some undergo non-radiative recombination instead. There are primarily two types of non-radiative processes: the first is defect-assisted Shockley-Read-Hall (SRH) recombination, and the second is Auger recombination. The rate of radiative recombination is proportional to the product of the carrier concentrations. In contrast, SRH non-radiative recombination depends on the concentrations of either electrons or holes, as they can interact independently with trap levels. As the density of material defects increases, defect-assisted SRH recombination is enhanced, which is a primary reason for the impact of defect density on LED efficiency. However, non-radiative recombination can still occur in the absence of defects, with Auger recombination serving as an example. In Auger recombination, the energy released by electron-hole recombination is absorbed by another electron or hole, which then transitions to a higher energy state and subsequently relaxes back to the original energy level by emitting phonons. This makes Auger recombination a three-particle process dependent on the product of three carrier concentrations. According to the ABC theory, assuming a current injection efficiency of 100% for deep ultraviolet LEDs, the internal quantum efficiency (IQE) can be expressed as follows:

$$\eta_{IQE} = \frac{Bn^2}{An+Bn^2+Cn^3} \quad (2)$$

Here, n represents the carrier density, while A, B, C represent the coefficients for SRH non-radiative, radiative, and Auger recombination, respectively. The right side of the equation indicates the proportion of radiative processes to total recombination processes, which defines the internal quantum efficiency. This formula can be transformed to:

$$\eta_{IQE} = \frac{B}{\frac{A}{n}+B+Cn} \quad (3)$$

At $n = \sqrt{\frac{A}{C}}$, formula (3) achieves its maximum value, where the maximum internal quantum efficiency exists:

$$\eta_{max} = \frac{B}{B+2\sqrt{AC}} \quad (4)$$

From equation (4), it is evident that the LED's external quantum efficiency is influenced by the coefficients A, B, C. The internal quantum efficiency of UV LEDs decreases as the non-radiative recombination coefficient A and the Auger recombination coefficient C increase. Conversely, when the radiative recombination coefficient B is higher, the internal quantum efficiency of ultraviolet LEDs increases. Due to the relatively large values of and in ultraviolet LEDs, their quantum efficiency tends to be lower. Moreover, the light extraction efficiency of UV LEDs is generally low, because in the ultraviolet spectrum, photons are more readily absorbed by the semiconductor material, and the device structure exhibits strong total internal reflection. Key factors for enhancing the efficiency of UV LEDs include precise control over the thickness and composition of the epitaxial layer, optimized design of quantum well structures, and improvements in electrode and packaging technologies[12].

4. Application of UV LEDs in water sterilization

UV dosage plays a crucial role in the water sterilization process. The UV dose can be defined as the product of UV intensity and exposure time, essentially the cumulative effect of UV irradiance over time, represented by the equation:

$$Q = E \times t \tag{5}$$

where Q is the UV sterilization dose, E is the UV irradiance, and t is the exposure time. With a sufficient UV dose, deactivated viruses and bacteria will not reactivate. However, if the dose is insufficient, many of the microbes deactivated by UV exposure can repair their damaged structures with the aid of light. For the same UV dose, sterilization can be achieved by employing either a high-intensity, short-duration exposure or a low-intensity, long-duration exposure. Although the external quantum efficiency of UV LEDs is not high, there has already been research conducted globally on the application of UV LEDs in water disinfection. Chatterley and Linden demonstrate that UV-LEDs, emitting at 265 nm, are as effective as traditional UV lamps in inactivating Escherichia coli in water samples. This finding is significant as it supports the feasibility of UV-LEDs for microbial control in decentralized water treatment applications. The efficacy of UV-LEDs in water disinfection is explored by Bowker et al through the development of a UV-LED collimated beam system[13]. The study focused on assessing the UV fluence-response of three non-pathogenic microorganisms: Escherichia coli, MS2 bacteriophage, and bacteriophage T7. By comparing the effectiveness of 255 nm and 275 nm UV-LEDs against a conventional low-pressure mercury lamp at 254 nm, the researchers aimed to understand how different UV wavelengths influenced microbial inactivation. Aoyagi et al. delves into the utilization of UV-LEDs at wavelengths of 280 nm and 255 nm for disinfecting water contaminated with bacterial viruses such as MS2, $Q\beta$, and $\varphi X174$. The research highlights that 280-nm LEDs are more suitable for water disinfection applications due to their significantly higher external quantum efficiency compared to 255-nm LEDs, despite the slightly lower disinfection efficiency[14]. Luo et al. explore the challenges of removing 2-Methylisoborneol (2-MIB), a common odor-causing compound in water treatment facilities, which is notoriously resistant to conventional treatment methods due to its chemical stability[15]. This research contributes to understanding how LED technology can be optimized for microbial disinfection, particularly using the UV-C spectrum (100–280 nm), known for its germicidal properties. In 2023, Oguma investigated the application of UV-LED for water disinfection at small, decentralized water facilities, focusing on practical implications and effectiveness in remote locations in Japan[16]. The study, conducted over a year, assessed a flow-through UV-LED disinfection module's performance in three distinct environments targeting indigenous Escherichia coli, total coliforms, standard plate count, and heterotrophic plate count (HPC) bacteria. The study conducted by Wang et al. (2021) addresses the efficiency improvement of a flow-through water disinfection reactor using UV-C LEDs. The research demonstrated that using 24 UV-C LEDs optimized the irradiance distribution within the reactor, leading to the highest Escherichia coli (E. coli) inactivation efficiency. This was attributed to a more uniform light exposure across the water flow, essential for effective microbial inactivation. Further research is recommended to explore the scalability of this technology for larger water treatment applications and to evaluate the long-term operational stability and cost-effectiveness of UV-C LED-based disinfection systems. Sala et al. (2021) focuses on addressing the critical global issue of clean water accessibility due to environmental pollution, by detailing the design and optimization of a portable UV-LED water disinfection system powered by piezoelectric energy. This innovative system employs piezoelectricity to harvest energy through mechanical actions, such as shaking, to power UV-LEDs that disinfect water, making it a sustainable and portable solution ideal for areas lacking in infrastructure[17]. Mohaghegh Montazeri and Taghipour (2022) have significantly enhanced UV radiation collimation through the development of an optical module specifically tailored for large-scale reactors, a critical advancement for improving the efficiency of water treatment processes[18]. The essence of their research lies in the use of multi-parabolic aluminum reflectors, which generate collimated UV light beams. These reflectors enhance both the uniformity and intensity of radiation across greater distances within the reactor. Lohan et al. provides an in-depth examination of innovative approaches to enhancing clean and sustainable water access in Africa, emphasizing smart UV/LED disinfection, solar energy utilization, and wireless positioning support. The study offers a multidisciplinary perspective that integrates various technological innovations to address the persistent challenge of water scarcity in remote and underserved regions[19].

Current research on UV-LEDs for water disinfection is still limited, and more studies are needed to better understand the application of UV-LEDs in water disinfection. For instance, to accurately evaluate the effectiveness of UV-LEDs, it is crucial to conduct detailed characterization and establish standard microbial inactivation protocols. Particularly, devising precise methods for measuring UV doses from UV-LEDs is essential for generating reliable dose-response data. Additionally, the wavelength diversity and pulsing capabilities of UV-LEDs significantly impact microbial inactivation; these phenomena require further investigation to determine the most effective usage conditions. Finally, studies on the design of reactors for UV-LED water disinfection systems could advance the practical application of this technology. Compared to traditional UV mercury lamps, UV-LEDs offer characteristics such as smaller size, greater portability and intensity, diversity in wavelength selection, and pulsing capabilities, all of which open up extensive possibilities for designing innovative reactors and also provide opportunities for new applications of UV-LED reactors.

5. Conclusion

UV LEDs, as an emerging water treatment technology, are particularly significant due to their high efficiency, long lifespan, and environmental friendliness. We have presented the current state of this technology, including case studies of its application in actual water treatment scenarios, demonstrating significant improvements over traditional ultraviolet lamp technology. Nevertheless, the adoption of UV LED technology in the field of water disinfection still faces several challenges, including high initial investment costs, varying levels of technological maturity, and stability issues under different water quality conditions. These challenges limit its broader application potential. Future research should focus on optimizing the photoelectric efficiency of LEDs, reducing production costs, and further validating the performance of UV LED systems in various real-world water quality settings. Additionally, interdisciplinary collaborative research, integrating material science, optical design, and microbiology, will be crucial in advancing this technology.

References

1. M. Qv, D. Dai, D. Liu, Q. Wu, C. Tang, S. Li, and L. Zhu, Bioresource Technology **370**, 128574 (2023)
2. H. Khan, J. Alzabut, A. Shah, Z.-Y. He, S. Etemad, S. Rezapour, and A. Zada, Fractals **31**, 2340055 (2023)
3. M. A. Khan, N. Maeda, Y. Itokazu, M. Jo, K. Iimura, and H. Hirayama, Physica Status Solidi (a) **220**, 2200621 (2023)
4. B. Ma, S. Burke-Bevis, L. Tiefel, J. Rosen, B. Feeney, and K. G. Linden, Science of The Total Environment **869**, 161848 (2023)
5. P. O. Nyangaresi, T. Rathnayake, and S. E. Beck, Science of The Total Environment **859**, 160256 (2023)
6. M. Martín-Sómer, C. Pablos, C. Adán, R. van Grieken, and J. Marugán, Science of The Total Environment **885**, 163963 (2023)
7. J. Lang, F. Xu, J. Wang, L. Zhang, X. Fang, Z. Zhang, X. Guo, C. Ji, C. Ji, F. Tan, Y. Wu, X. Yang, X. Kang, Z. Qin, N. Tang, X. Wang, W. Ge, and B. Shen, Adv Elect Materials 2300840 (2024)
8. J. Wang, N. Xie, F. Xu, L. Zhang, J. Lang, X. Kang, Z. Qin, X. Yang, N. Tang, X. Wang, W. Ge, and B. Shen, Nat. Mater. **22**, 853 (2023)
9. H. Amano, R. Collazo, C. D. Santi, S. Einfeldt, M. Funato, J. Glaab, S. Hagedorn, A. Hirano, H. Hirayama, R. Ishii, Y. Kashima, Y. Kawakami, R. Kirste, M. Kneissl, R. Martin, F. Mehnke, M. Meneghini, A. Ougazzaden, P. J. Parbrook, S. Rajan, P. Reddy, F. Römer, J. Ruschel, B. Sarkar, F. Scholz, L. J. Schowalter, P. Shields, Z. Sitar, L. Sulmoni, T. Wang, T. Wernicke, M. Weyers, B. Witzigmann, Y.-R. Wu, T. Wunderer, and Y. Zhang, J. Phys. D: Appl. Phys. **53**, 503001 (2020)
10. W. Liu, M. Peng, Y. Yu, Z. Zheng, P. Jian, Y. Zhao, Y. Zeng, D. Xu, M. Chen, Y. Luo, C. Chen, J. Dai, and F. Wu, Advanced Optical Materials 2302746 (2024)
11. M. Kneissl, T.Y. Seong, J. Han, and H. Amano, Nature Photonics **13**, 233 (2019)
12. S. Zhou, Z. Liao, K. Sun, Z. Zhang, Y. Qian, P. Liu, P. Du, J. Jiang, Z. Lv, and S. Qi, Laser & Photonics Reviews **18**, 2300464 (2024)
13. C. Bowker, A. Sain, M. Shatalov, and J. Ducoste, Water Research **45**, 2011 (2011)
14. Y. Aoyagi, M. Takeuchi, K. Yoshida, M. Kurouchi, N. Yasui, N. Kamiko, T. Araki, and Y. Nanishi, J. Environ. Eng. **137**, 1215 (2011)
15. Z.N. Luo, T.Y. Zhang, M.Y. Xu, R.F. Fang, X.-X. Wang, Z.-X. Zheng, H.-X. Zhao, Z.-C. Li, Y.-L. Tang, and B. Xu, Environmental Technology **45**, 2132 (2024)
16. K. Oguma, Journal of Water and Health **21**, 1369 (2023)
17. D. E. Şala, Y. Dalveren, A. Kara, and M. Derawi, Applied Sciences **11**, 3007 (2021)
18. M. Mohaghegh Montazeri and F. Taghipour, Journal of Photochemistry and Photobiology A: Chemistry **433**, 114184 (2022)
19. E. S. Lohan, K. Bierwirth, T. Kodom, M. Ganciu, H. Lebik, R. Elhadi, O. Cramariuc, and I. Mocanu, IEEE Access **11**, 81882 (2023)

Temporal-spatio Changes of Vegetation Cover and Its Relationship with Terrain Factors in Jinzhai County

Chuanbin Dong[1,2,a], Chris Rey M. Lituanas[1,b*]

[1]Department of Biology, Central Mindanao University, Musuan, Bukidnon 8710, Philippines
[2]School of Environment and Surveying Engineering, Suzhou University, Suzhou 234000, China

Abstract: By monitoring and studying changes in vegetation coverage, we can assess the stability and biodiversity of ecosystems, and promptly identify changes and issues in the ecological environment. This study was based on Landsat data and selects data from 2003 to 2023, including 2003, 2007, 2011, 120, 152, 2019, and 2023.The pixel binary model was used to calculate the vegetation coverage of Jinzhai County for the past 20 years and 6 periods. Combining trend analysis to study the temporal-spatio changes characteristics of vegetation cover in Jinzhai County. Combining DEM data and distribution index methods to explore the relationship between vegetation coverage and terrain factors. Results showed that: (1) The vegetation cover in Jinzhai County showed an overall increasing trend, mainly with MH-FVC and H-FVC. (2) The vegetation coverage was closely related to terrain factors. With the continuous increase of elevation and slope, the vegetation coverage in Jinzhai County continues to increase.

1. Introduction

Vegetation cover refers to the ratio of the vertical projection area of leaves, stems, and branches of vegetation to the ground in the total eigenvalue of the area. It is a comprehensive measure of the vegetation cover on the ground. It has a good indication of the surface environment and is widely used in ecological environment remote sensing monitoring[1]. Vegetation cover is an important part of monitoring and evaluating the ecological environment such as soil erosion[2]. Timely and accurate acquisition of surface vegetation cover and its change information, revealing the temporal and spatial distribution of surface vegetation cover, has important practical significance for the evaluation of regional ecological environment[3,4].

2. Materials and methods

2.1 Study area

Jinzhai County (115°22′E-116°11′E, 31°06′N-31°48′N) is located in the west of Anhui Province, in the hinterland of the Dabie Mountains in China (Figure 1). The geologic structure of Jinzhai County has formed the landform structure of middle mountains, low mountains, hills, basins and valley plains. The mountains, hills, elongated basins, and river valleys are alternately distributed. It belongs to the north subtropical humid monsoon climate in Jinzhai County. Affected by terrain and climate, there are ecological problems such as soil erosion within the territory. There are problems such as soil erosion, natural disasters, and decreased ecosystem service value in some areas. This study has important practical significance for the protection and management of vegetation in Jinzhai County.

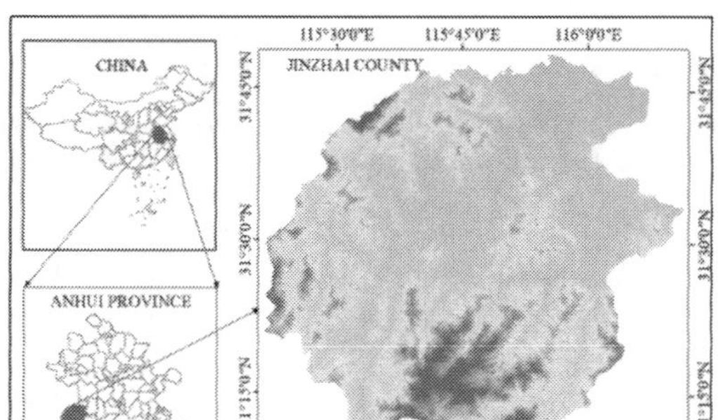

Figure 1 The location of the study site.

2.2 Calculation of vegetation cover

We used the pixel binary model to calculate vegetation cover[5]. The model assumes that each mixed pixel can be decomposed into two parts, pure vegetation, and bare soil, and the spectral information obtained is a linear combination weighted by the area ratio of the two pure components. The percentage of area occupied by pure

*Corresponding author: [a]email: ledinghamcarroll@gmail.com, [b]email: f.chrisrey.lituanas@cmu.edu.ph

vegetation is the fractional vegetation cover (FVC). The equation can be represented as:

$$FVC = \frac{NDVI - NDVI_{soil}}{NDVI_{veg} - NDVI_{soil}} \quad (1)$$

To facilitate the analysis and description of FVC changes, uniform interval classification method[6] was adopted in this study, dividing the FVC into five levels: low FVC (L-FVC) (0-0.2), medium-low FVC (ML-FVC) (0.2-0.4), medium FVC (M-FVC) (0.4-0.6), medium-high FVC (MH-FVC) (0.6-0.8), and high FVC (H-FVC) (0.8-1.0).

2.3 Trend analysis

The unary linear regression method would be used to analyze the trend of vegetation cover over time[7]. Its advantage was that it could simulate the vegetation cover change trend of each pixel. The calculation formula was:

$$K = \frac{n \times \sum_{i=1}^{n}(i \times V_i) - \sum_{i=1}^{n} i \sum_{i=1}^{n} V_i}{n \times \sum_{i=1}^{n} i^2 - (\sum_{i=1}^{n} i)^2} \quad (2)$$

In formula (2), K represented the change trend (slope); n was the research time period (year), n=21 in this paper; i ranged from 1 to 21, such as i=1 in 2003; V_i was vegetation cover in year i. A positive slope of K meant an increase, and vice versa meant a decrease, the larger the |K|, the greater the change, and vice versa meant a smaller change. According to trend analysis, the change in vegetation cover trends in this paper was divided into significant better (increase) (0.05—0.25); general better (increase) (0.01—0.05); unchanged (-0.01—0.01); general degradation (decrease) (-0.05 — -0.01); Significant degradation (decrease) (-0.25 — -0.05) category 5[8].

2.4 Distribution index

Vegetation changes exhibit different characteristics in different regions. However, when evaluating the characteristics of vegetation changes, there is often a situation where a certain type of change has a small area in a specific region but a relatively large proportion compared to the overall study area, which affects the validity of the evaluation[9]. To eliminate the influence of regional and proportional differences, a distribution index is used in this study to describe the distribution characteristics of vegetation changes in different regions[10]. The calculation formula is as follows:

$$K = \frac{S_{ie} / S_e}{S_i / S} \quad (3)$$

Where K is the distribution index of the i-th change type in region e. S_{ie} is the area of the i-th change type in region e. S_i is the total area of the i-th change type in the study area. S_e is the total area of region e in the study area. S is the total area of the entire study area. The distribution index K is a standardized, dimensionless index, ranging from 0 to S. If K > 1, it indicates that the distribution of the i-th change type in region e is dominant, and the larger the value of K, the more dominant the distribution. If K < 1, it indicates a non-dominant distribution of the i-th change type in region e. Using Arcgis10.8, the elevation, slope, and slope orientation distribution of FVC in Jinzhai County from 2003 to 2023 were statistically analyzed.

2.5 Data source

The Landsat data that were mainly used in this study were Landsat 5 TM and Landsat 8 OLT data, obtained from the geospatial data cloud platform (http://www.gscloud.cn/). In this study, the vegetation cover were selected from images from May to September, and the spectra were relatively stable (Table 1) .

Table 1 List of remote sensing image data.

Serial Num-ber	Get tim-e	Strip Num-ber	Line Num-ber	Satellite / sensor	Cloud Cover (%)
1	May 26, 2003	122	38	Landsat 5 TM	0
2	May 5, 2007	122	38	Landsat 5 TM	0
3	Sep 21, 2011	122	38	Landsat 5 TM	3
4	Oct 2, 2015	122	38	Landsat 8 OLT_TIRS	0.01
5	May 22, 2019	122	38	Landsat 8 OLT_TIRS	0.38
6	May 9, 2023	122	38	Landsat 9 OLT_TIRS	0.68

The Digital Elevation Model (DEM) data were derived from the geospatial data cloud platform (http://www.gscloud.cn). The format was Tiff, the projection was longitude and latitude, and the resolution was 30 m.

The downloaded ASTER GDEM V3 data was spliced, projectively transformed, resampled and cropped , and the resulting elevation and slope maps were graded[10]. When grading elevations, below 200 m and above 1,500 m were each regarded as one grade, and between 200 m and 1,500 m were graded at intervals of 100 m, with a total of 15 grades. According to the topographic slope classification criteria, the slope was divided into five levels as 0-5°, 5-15°, 15-25°, 25-35° and >35°[9].

3. Results and discussion

3.1 Temporal and spatial changes in FVC

3.1.1 Temporal changes in FVC

Based on Table 2, Figure 2, it can be observed that from 2003 to 2023, regarding the trend of FVC changes in

Jinzhai County, the proportions of L-FVC, ML-FVC, and M-FVC remained relatively stable with a fluctuating declining trend. However, significant changes were observed in the MH-FVC and H-FVC levels. The MH-FVC showed a fluctuating declining trend, while the H-FVC exhibited a fluctuating increasing trend. Specifically, the L-FVC slightly decreased from 2003 to 2011, remained relatively stable from 2011 to 2015, experienced a subsequent increase from 2015 to 2019, and then continued to decline. The ML-FVC showed a continuous declining trend from 2003 to 2011, followed by a slight increase from 2011 to 2019, and remained relatively stable thereafter. The M-FVC declined from 2003 and showed minor fluctuations with a slight increase after 2015. The MH-FVC exhibited an overall declining trend, with a significant decrease from 2003 to 2007, followed by a period of stabilization from 2011 to 2015. A noticeable decline occurred from 2015 to 2019, followed by a subsequent increase. The H-FVC demonstrated a fluctuating increasing trend from 2003 to 2023. Specifically, there was an upward phase from 2003 to 2011, a slight decline from 2011 to 2015, a significant increase from 2015 to 2019, and a declining phase from 2019 to 2023.

Table 2 Area and Percentage of FVC in different stages of Jinzhai County.

Lev-el	Distribu-tion	2003	2007	2011	2015	2019	2023
L-FVC	Area /km ^{2}	322.19	323.82	268.28	270.30	307.19	286.66
	Percentage/%	8.22	8.27	6.85	6.90	7.84	7.32
ML-FVC	Area /km^2	240.80	216.25	103.42	133.97	166.55	165.37
	Percentage/%	6.15	5.52	2.64	3.42	4.25	4.22
M-FVC	Area /km^2	426.37	348.96	207.25	257.33	233.03	265.66
	Percentage/%	10.88	8.91	5.29	6.57	5.95	6.78
MH-FVC	Area /km^2	839.89	605.33	759.28	744.37	418.04	656.41
	Percentage/%	21.44	15.45	19.38	19.00	10.67	16.76
H-FVC	Area /km^2	2088.17	2422.75	2579.20	2511.44	2792.61	2543.33
	Percentage/%	53.30	61.85	65.84	64.11	71.29	64.92

In summary, from 2003 to 2023, Jinzhai County exhibited a slight decline in the L-FVC, ML-FVC, and M-FVC categories, while the MH-FVC experienced a significant decrease, and the H-FVC showed a noticeable increase. This indicates an overall improvement in FVC in Jinzhai County.

Table 3 Classification criteria, area, and proportion of FVC change trends in Jinzhai County.

Number	Slope range	Area (km^2)	Percentage (%)
1	-0.25 — -0.05	108.33	2.77
2	-0.05 — -0.01	523.53	13.37
3	-0.01 — 0.01	2052.23	52.39
4	0.01 — 0.05	951.05	24.28
5	0.05 — 0.25	281.96	7.20

3.1.2 Spatial changes in FVC

From Figure 2, it can be observed that from 2003 to 2023, the areas with L-FVC and ML-FVC were mainly distributed in the central, northeastern, and southern parts of Jinzhai County. The central part primarily consisted of water areas such as valleys, Meishan Reservoir, and Xianghongdian Reservoir. The northeastern part corresponded to the region where the county town is located. The southern part exhibited a narrow strip-like distribution in the valley areas. The areas with M-FVC were generally distributed around the periphery of the L-FVC and ML-FVC regions. The regions with MH-FVC and H-FVC were widely distributed throughout the entire area, mainly concentrated in mountainous and hilly areas with significant topographic variations.

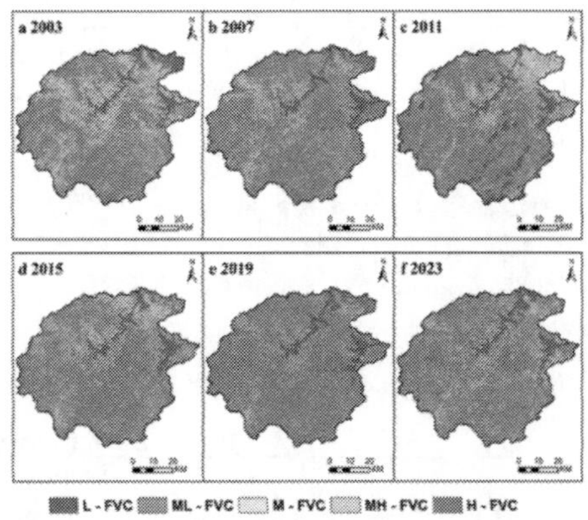

Figure 2 Spatial distribution of FVC level in each interval in Jinzhai County.

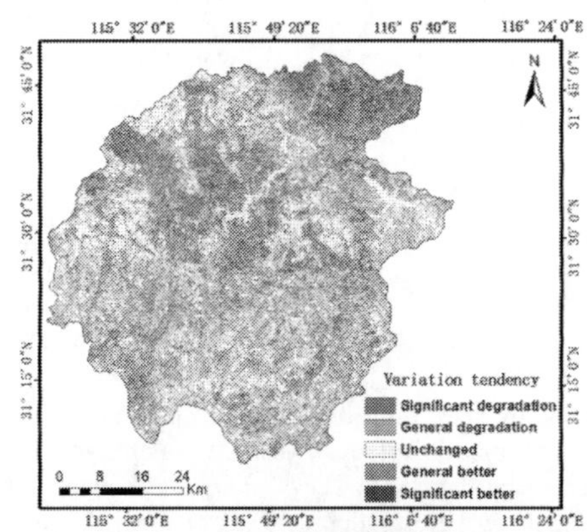

Figure 3 Spatial distribution of FVC change trend in Jinzhai County.

3.1.3 Trend Analysis

From Table 3, Figures 3, it can be observed that the FVC in Jinzhai County from 2003 to 2023 mainly exhibited an increasing trend, accounting for 31.48% of the total area studied. The proportion of degraded vegetation cover areas was relatively small at 16.14%. The largest proportion was observed in the unchanged areas, reaching 52.39%, which exceeds half of the total area.

Regarding the spatial distribution of FVC changes, there is no apparent concentration of different change categories in specific regions. Areas with improved vegetation cover are generally distributed in the northeast and western parts of Jinzhai County. Degraded areas are primarily located in the county town area, with scattered occurrences in some valleys and low-lying zones. The southern part of Jinzhai County is predominantly characterized by areas with little to no change in FVC.

3.2 The terrain effect of FVC

3.2.1 Elevation distribution

With increasing elevation, the distribution of L-FVC, ML-FVC, and M-FVC gradually decreases, while the distribution of MH-FVC initially increases and then gradually decreases, and the distribution of H-FVC gradually increases. Different FVC types have dominant distribution areas at different elevations. L-FVC is mainly distributed in areas below 300 meters above sea level. ML-FVC coverage is distributed in areas below 400 meters, while M-FVC and MH-FVC are found in areas below 500 meters. The dominant distribution area for H-FVC is located above 400 meters above sea level. Specifically, below 300 meters, L-FVC has a significant dominance. In the 300-500 meter range, MH-FVC is dominant. Above 400 meters, H-FVC exhibits a clear dominance. The dominance of ML-FVC and M-FVC coverage is lower than other levels at different elevations(Figure 4). In summary, FVC gradually increases with elevation. Below 300 meters, areas heavily influenced by human activities such as urban construction, agricultural land, and water bodies are characterized by L-FVC and ML-FVC. In the 300-500 meter range, there is a higher proportion of agricultural land, shrubs, and grasslands, leading to M-FVC and MH-FVC dominance. Above 500 meters, forests and grasslands are prevalent, resulting in H-FVC.

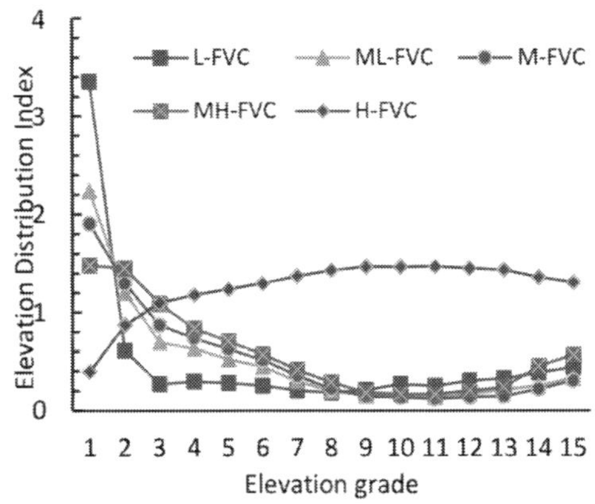

Figure 4 Elevation Distribution Index of FVC in Jinzhai County.

3.2.2 Slope distribution

From Figures 5, it can be seen that the distribution of L-FVC rapidly decreases as the slope increases, while the decrease in ML-FVC is relatively slow. M-FVC shows a trend of initially increasing and then decreasing. MH-FVC exhibits an initial increase followed by a slower decrease, while H-FVC gradually increases. Different FVC types have dominant distribution areas across different slope ranges. L-FVC is mainly distributed in areas with slopes less than 5° . ML-FVC is distributed in areas with slopes less than 15°. The distribution of M-FVC does not vary significantly below 25°, and its occurrence is relatively low in areas above 25°. There is no significant variation in the distribution of MH-FVC across different slope ranges. The dominant distribution area for H-FVC is found in areas with slopes above 25°.

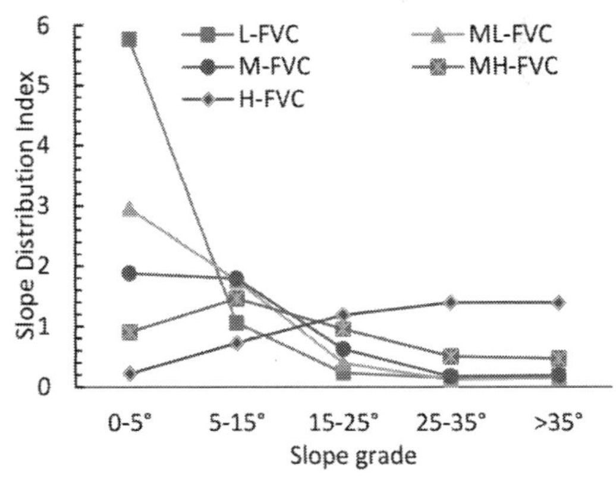

Figure 5 Slope Distribution Index of FVC in Jinzhai County.

In summary, FVC gradually increases with slope. In areas below 5°, which are characterized by relatively flat terrain, there is a significant presence of human activities such as urban construction, agricultural land, and water bodies, resulting in L-FVC, ML-FVC, and M-FVC dominance. The 5-15° range represents gentle slopes with relatively small surface undulations, suitable for various land uses, and heavily influenced by human activities, with little variation in FVC levels. In areas above 15°, characterized by steeper slopes, forests and grasslands dominate, and FVC rapidly increases, with H-FVC being the primary type and a small portion representing MH-FVC.

4. Conclusion

(1) The vegetation cover in Jinzhai County showed an increasing trend in interannual variations, while in terms of spatial distribution, mountainous areas exhibit higher vegetation cover compared to lowland regions.

(2) The overall trend of vegetation cover in Jinzhai County was increasing, with the improved area accounting for 31.48% of the total area of the study area. The degraded vegetation cover area accounted for 16.14%.

(3) In terms of the impact of terrain factors on vegetation coverage, with the continuous increase of elevation, the vegetation coverage in Jinzhai County continues to increase. Below an altitude of 400m, the vegetation coverage was mainly L, ML, and M-FVC, while above 400m, the vegetation coverage was mainly MH and H-FVC. As the slope increases, the vegetation coverage in Jinzhai County continues to increase. Below a slope of 15°, the vegetation coverage was mainly L, ML, and M-FVC, while above a slope of 15°, the vegetation coverage was mainly MH and H-FVC.

References

1. Leprieur, C., Ker, Y.H., Mastorchio, S., & Meunier, J.C. (2000). Monitoring vegetation cover across semi-arid regions: Comparison of remote observations from various scales. International Journal of Remote Sensing, 21(2), 281-300.
2. Wang, Y., Hao, L.N., Xu, Q., Li, J.Q., & Chang, H. (2023). Spatio-temporal variations of vegetation coverage and its geographical factors analysis on the Loess Plateau from 2001 to 2019. Acta Ecologica Sinica, 43(06), 2397-2407.
3. Wang, B., Xu, G.C., Li, P., Li, Z.B., Zhang, Y.X., Cheng, Y.T., Jia, L., & Zhang, J.X. (2020). Vegetation dynamics and their relationships with climatic factors in the Qinling Mountains of China. Ecological Indicators, 108, 105719.
4. Zhao, Z.J., Fan, B.L., Wang, Y.T., & Zhou, Q.B. (2021). Analysis on the characteristics of spatial-temporal changes and influencing factors of vegetation coverage in the Xiliao River basin from 2000 to 2018. Chinese Journal of Agricultural Resources and Regional Planning, 42(12), 75-88.
5. Adams, J.B., Smith, M.O., & Johnson, P.E. (1986). Spectral mixture modeling: A new analysis of rock and soil types at the Viking Lander 1 Site. Journal of Geophysical Research Solid Earth, 91(B8), 8098-8112.
6. Pang, J.T., Duan, J.L., Zhang, R., & Lu, J.T. (2021). Spatial-temporal evolution characteristics and climate response of vegetation cover in the Weihe River Basin from 2000 to 2019. Soil and Water Conservation Research, 28(5), 230-237.
7. Wang, Y., Hao, L.N., Zhao, M.L., Xu, X.L., & Li, C.H. (2021). Variation of vegetation NDVI and its response to climatic factors and human activities in Chongqing from 2001-2018. Research of Soil and water Conservation, 28(5), 222-229.
8. Han, J.C., Yu, S.L., Yang, Q.L., Shao, H.Y., & Liu, E.Q. (2019). vegetation cover Characteristics and Its Response to Climate and Topography in the Upper Yangtze River Basin from 1999 to 2015. Journal of Yangtze River Academy of Sciences, 36(9), 51-57.
9. Tong, X.W., Wang, K.L., Yue, Y.M., Liao, C.J., Xu, Y.F., & Zhu, H.T. (2014) Vegetation change trend and its response to climate and topography in the karst region of Northwest Guangxi. Journal of Ecology, 34(12), 3425-3434.
10. Zhang, H.X., Li, Y.M., & Zhang, N. (2022). Relationship between temporal and spatial characteristics of vegetation coverage and topographic factors in Fenhe River basin. Bulletin of Soil and Water Conservation, 42(1), 353-359.

Utilization of mango seed waste and lemongrass leaf extract as biomaterials for edible coating product

Hamidah Harahap*, Angelia P Galatea, Ricky R E Gultom, Halimatuddahliana, Ayuni Yustira, M Hendra S Ginting

Department of Chemical Engineering, Faculty of Engineering, Universitas Sumatera Utara, Medan, North Sumatera 20155 Indonesia

Abstract. Mango seed waste is waste from mango fruit that is not utilized and is a concern because it causes environmental problems. Utilizing mango seed waste as a raw material for making biomaterials for edible coating products is the right step. To support the use of mango seed waste as a good edible coating biomaterial, it is necessary to add lemongrass leaf extract as an antibacterial so that the resulting edible coating biomaterial has good mechanical characteristics and can extend the shelf life of food. This research used 100 mL of coating solution consisting of mango seed waste starch with a weight percent variation of 5; 10; 15%, lemongrass leaf extract with a weight percent variation of 0; 3; 6; 9%, glycerol 2%, CMC 1%, and distilled water. The characteristics of the film produced, the best tensile strength and Young's modulus were obtained from variations of 5% mango seed waste starch and 0% lemongrass extract with a tensile strength value of 0.000486 MPa and a Young's modulus of elasticity value of 0.0103 MPa. Meanwhile, for the elongation at break value, the film containing 5% mango seed waste starch and 9% lemongrass extract had the best value, namely 2.24%. The results of SEM (Scanning Electron Microscopy) analysis also show that the morphology of the film with the addition of lemongrass leaf extract has a smoother and flatter surface so that the film with the addition of lemongrass leaf extract is better than without lemongrass leaf extract

1 Introduction

Mango (Mangifera indica) is one of the most popular tropical fruits for cultivation and consumption in the world [1]. The demand for mangoes is quite high, namely 2.3 million tonnes in 2021. The increase in demand and production of mangoes has resulted in an increase in mango seed residue which is a concern because it causes environmental problems. After processing mango fruit, mango seed waste will become lignocellulosic residue or the equivalent of 35-55% of the mass of the mango fruit [2]. Mango fruit contains active components in the form of tannins and natural polymers such as protein and starch [3]. Mango seed starch waste also contains minerals, fatty acids, vitamins and bioactive compounds [4]. The starch contained in mango seed waste is 65% [1]. The starch in mango seed waste contains amylose, where the higher the amylose content, the higher the cohesive and mechanical properties as well as the ability of the coating to act as a permeable membrane biomaterial that is selective for CO_2 and O_2 exchange [5].

Lemongrass (Cymbopogon citratus) contains chemical compounds such as polyphenols, flavonoids, terpenes and others [6]. Lemongrass leaf extract is effective and efficient in controlling phytogenic fungi [7]. Biologically, lemongrass leaf extract has antibacterial, antifungal, antioxidant and antiseptic properties [8-11]. Lemongrass leaf extract can inhibit the growth of Phytophthore species mycelium: P. Capsici, P. drechsleri, and P. melonis [12]. The addition of antimicrobials is carried out indirectly, namely the antimicrobial agent is spread from the coating solution to the surface of the food ingredient [13]. By adding lemongrass extract as an antibacterial, the resulting edible coating product biomaterial will be able to maintain the shelf life of food coated with the edible coating product biomaterial.

Edible coating biomaterial is a material in the form of a thin layer used to wrap food with the aim of increasing shelf life and preventing the movement of moisture, dissolved substances and oxygen with a thickness of around 0.25 mm which acts as an edible coating biomaterial which aims to preserve food in an environmentally friendly and environmentally friendly manner safe for consumers [14]. One method of application of edible coating biomaterials on food can use a coating technique, namely a method of applying a thin layer to the surface of the food which functions as a barrier to prevent the release of gas, water vapor and contact with oxygen, thereby preventing spoilage and extending the shelf life of food [15].

This research focuses on evaluating the use of starch from mango seed waste and lemongrass leaf extract as biomaterial raw materials for edible coating products, considering that there is still limited research that utilizes agroindustry residues for biomaterial purposes for edible coating products. Therefore, this research seeks to explain the potential of mango seed waste and the contribution of lemongrass leaf extract as an antibacterial in edible product coating biomaterials. Edible coating biomaterials will be analyzed for the mechanical characteristics of

* Corresponding author: hamidah.harahap@usu.ac.id

edible coating biomaterials including determining tensile strength, elongation at break, Young's modulus and SEM analysis.

2 Materials and methods

2.1. Materials

The material used in this research was Aquadest (H_2O) obtained from CV. Rudang Jaya, Arumanis Mango Seeds *(Mangifera indica L.)* were obtained from fruit juice sellers in Medan City, Strawberry and lemongrass leaves were obtained from Brastagi Supermarket, CMC *(Carboxymethyl Cellulose)* and Ethanol (C_2H_5OH) 96% was obtained from CV. Rudang Jaya.

2.2 Mango seed starch waste processing

Mango seeds waste obtained from fruit juice sellers in the city of Medan, washed with running water then boiled for 15 minutes to remove the sap from the seeds, then peeled and cut then put into a blender containing water in a ratio of 1:1 (weight of mango: weight of water). After mixing, the starch slurry is filtered to separate the dregs from the juice. The starch essence is put into a basin to settle for 24 hours. After that, the precipitated starch was taken, then dried in an oven at 40 °C for 20 hours. The dried starch is then sieved with a 100 mesh/inc sieve.

2.3 Lemongrass leaf extract processing

The lemongrass is cleaned with running water, then dried using an oven at 50 °C for 22 hours, then ground with a blender until it becomes powder. Lemongrass leaf powder was macerated in a bottle containing 96% ethanol with a powder: solvent ratio (1:10) for 3 days. After that, the solution is filtered to separate the filtrate from the dregs. The filtrate was put into a rotary evaporator at 50 °C for 3 hours to separate the solvent from the solution then re-concentrated using a water bath at 50 °C for 2 hours to produce lemongrass extract with a more concentrated concentration.

2.4 Product of edible coating biomaterials

Aquadest is heated to a temperature of 70 °C on a hot plate. After that, variations in weight percent of mango seed starch waste were added with treatment P1: 5%, P2: 10%, P3: 15% and variations in weight percent of lemongrass leaf extract with treatment E1: 0%, E2: 3%, E3: 6%, E4: 9% and stirred until homogeneous using a magnetic stirrer on a hotplate at 70 °C for 15 minutes. Once homogeneous, the solution was added with 1% CMC and 2% glycerol for each treatment variation while heating and stirring for 15 minutes until homogeneous. The solution will be used to make 100 mL of coating and film each. The strawberry were cleaned and weighed and the initial weight recorded. After that, dip the strawberry in the coating solution for 3 minutes until it covers all the surface of the strawberry skin, then drain and leave to dry in the air, then weigh again.

2.5. Film mechanical properties and morfology

In testing the mechanical properties of film, there are 3 tests carried out, namely tensile strength, elongation percentage and young's modulus. The tensile strength testing process can be used with the WEW-300A Universal Testing Machine (UTM). Tensile strength can be determined based on the maximum load at which the film breaks. Then, for the elongation percentage based on the elongation of the film at break, and for elasticity based on the comparison of tensile strength with elongation. The SEM *(Scanning Electron Microscope)* test aims to examine and analyze the surface and study the morphological properties of the sample in the form of a film. Samples were placed on double-sided cellophane adhesive glued to aluminum, magnification was carried out at 1000X.

2.6 Weight Loss

Weight loss in strawberry edible coating products can be measured using the gravimetric method, namely comparing the difference in weight before storage and after storage for 7 days. Weight loss during storage can be calculated using equation 1.

$$Weight\ loss = \frac{Wo - Wt}{Wo} \times 100\% \quad (1)$$

Where :
Wo : initial sample weight (g)
Wt : final sample weight (g)

3 Results and discussion

3.1. Tensile Strength

Tensile strength is the maximum pull that a film can withstand until the film breaks. The greater the tensile strength, the better the film is at resisting mechanical damage [16]. The effect of variations in mango seed starch waste and lemongrass extract on the tensile strength of the film can be seen in Figure 1

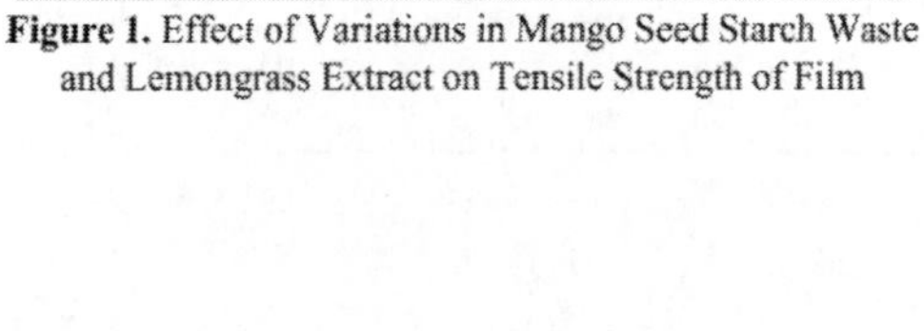

Figure 1. Effect of Variations in Mango Seed Starch Waste and Lemongrass Extract on Tensile Strength of Film

From Figure 1 it can be seen that the higher the starch percentage, the higher the tensile strength value of the film. The tensile strength values for each variation of 5%, 10% and 15% starch for 0% lemongrass extract are respectively 0.000486 MPa, 0.000836 MPa and 0.00228 MPa. This happens because the greater the concentration of starch added, the amylose content in the film solution also increases. This causes the number of polymers in the matrix formation to increase and the bonds between polymers to become stronger so that the strength provided to support external loads is also greater.

However, the tensile strength will decrease as more active ingredients are added to the film. From Figure 1 it can be seen that there was a decrease in tensile strength for each variation of starch added with lemongrass extract with variations of 0%, 3%, 6% and 9%. At 5% starch, the tensile strength values obtained for each variation of lemongrass extract were 0.000486 MPa, 0.000421 MPa, 0.000380 MPa, and 0.000132 MPa. At 10% starch it was 0.000836 MPa, 0.000550 MPa, 0.000393 MPa, and 0.000177 MPa. At 15% starch it was 0.000228MPa, 0.00110 MPa, 0.000515 MPa and 0.000338 MPa. The weakening of the tensile strength as the lemongrass extract increases is caused by the presence of more and more active ingredients so that the interaction between starch molecules weakens. A similar thing also happened in research conducted by Lim et al (2010) where the tensile strength of the film became smaller as the concentration of grape extract increased [17]. This is also supported by the results of the Scanning Electron Microscopy (SEM) test which can be seen in Figure 4. From the image, it can be seen that edible film with a starch concentration of 15%; 9% lemongrass extract has an even film surface.

So from the explanation above, it can be concluded that the increase in tensile strength is directly proportional to the increase in mango seed starch waste concentration, but inversely proportional to the increase in lemongrass extract concentration.

3.2. Elongation at break

Elongation is the elongation value of a material. The higher the elongation value, the more elastic the material. This value is obtained from a comparison between the additional length of the film and the initial length of the film [18]. The elongation value for each variation can be seen in Figure 2.

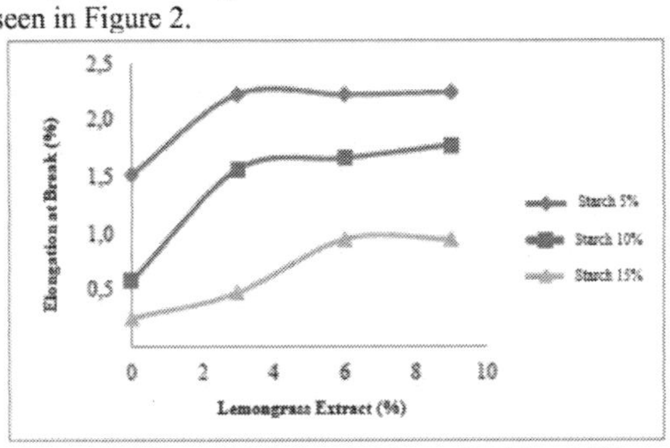

Figure 2. Effect of Variations in Mango Seed Starch Waste and Lemongrass Extract on Film Elongation

From Figure 2 it can be seen that the higher the percentage of mango seed starch waste, the lower the film elongation value. The elongation values for each variation of 5%, 10%, and 15% starch in 0% lemongrass extract were respectively 1.52%, 1.34%, and 0.25%. This happens because the greater the percentage of starch, the denser the film matrix will be, resulting in reduced film flexibility.

Meanwhile, the higher the percentage of lemongrass extract, the higher the film elongation value. From Figure 2 it can be seen that there was an increase in the elongation value for each variation of starch added with lemongrass extract with variations of 0%, 3%, 6% and 9%. At 5% starch, the elongation values obtained for each variation of lemongrass extract were 1.52%, 2.22%, 2.22% and 2.24%. At 10% starch it was 0.58%, 1.56%, 1.66% and 2.18%. At 15% starch it was 0.25%, 0.47%, 0.94% and 0.95%. This occurs because the addition of greater amounts of active ingredient causes stretching of the intermolecular space of the film matrix network and a decrease in the number of internal hydrogen bonds, thereby reducing film brittleness and increasing the percent elongation. A similar phenomenon also occurred in the research of Fama et al (2005), where the percent elongation of films with the addition of active ingredients was higher than without active ingredients [19]. This is also supported by the results of the Scanning Electron Microscopy (SEM) test which can be seen in Figure 4. From the image, it can be seen that edible film with a starch concentration of 15%; 9% lemongrass extract has an even film surface.

So from the explanation above, it can be concluded that the increase in elongation is inversely proportional to the increase in mango seed starch waste concentration, but is directly proportional to the increase in lemongrass extract concentration.

3.3. Modulus Young

The modulus of elasticity is a measure of the stiffness of a material. Modulus Young's compares the stress and strain in a material. Young's modulus can be seen from the comparison between tensile strength and elongation test values. The elastic modulus value for each variation can be seen in Figure 3.

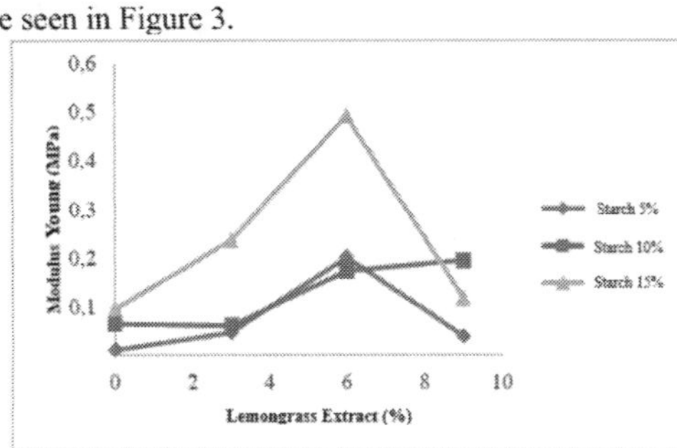

Figure 3. Effect of Variations in Mango Seed Starch Waste and Lemongrass Extract on Modulus young's of Film

From Figure 3 it can be seen that the graphs produced by each variation of lemongrass starch and extract fluctuate. Young's modulus values for 5% starch in each variation of 0, 3, 6 and 9% lemongrass extract respectively are 1.03x10-2 MPa, 4.62x10-2 MPa, 2.00x10-1 MPa and 3.84x10-2 MPa. Then, the Young's modulus values for 10% starch in each variation of 0, 3, 6 and 9% lemongrass extract respectively are 6.39x10-2 MPa, 5.98x10-2 MPa, 1.72x10-1 MPa and 1.94 x10 -1 MPa. Then, the young's modulus values for 15% starch in each variation of 0, 3, 6 and 9% lemongrass extract are respectively 9.44x10 2MPa, 2.36x10-1 MPa, 4.92x10-1 MPa and 1.16x10-1 MPa.

The young's modulus value is inversely proportional to the elongation value and directly proportional to the tensile strength. The young's modulus value is influenced by hydrogen bonds which form long chains so that the resulting film is elastic. The smaller the young's modulus value, the more elastic and flexible the edible film formed.

3.4 Scanning Electron Microscopy (SEM)

The Scanning Electron Microscopy (SEM) test was carried out to see the morphology of the edible film. SEM test results of mango seed starch concentration of 5% with variations in lemongrass extract concentration of 0% and 9% at 3000x magnification can be seen in Figure 4.

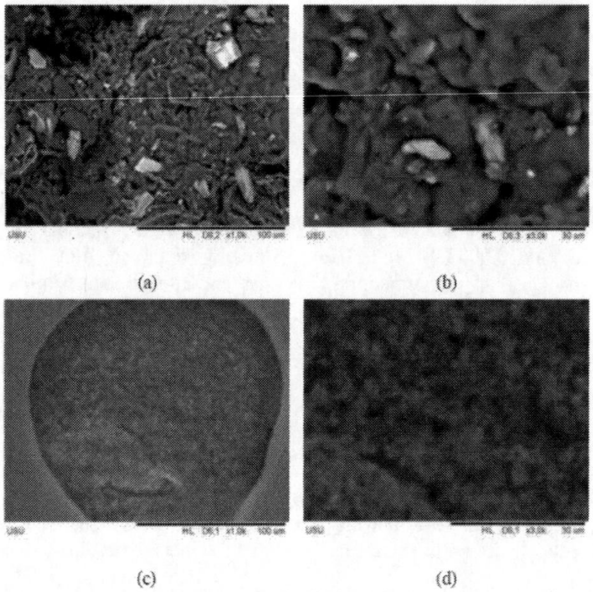

Figure 4. Scanning Electron Microscopy (SEM) Test Results for Edible Film Biomaterials Surface from 5% Mango Seed Starch Waste with 0% Lemongrass Extract (a) 1000x magnification (b) 3000x magnification; and Mango Seed Starch Waste 5% with Lemongrass Extract 9% (c) Magnification 1000x (d) Magnification 3000x

From Figure 4 it can be seen that there are differences in the morphology of edible films at the same starch concentration but with different concentrations of lemongrass extract. In edible film with a 0% lemongrass extract concentration, there are many protrusions and pores on the surface of the film. This is caused by the presence of bacteria on the film. Apart from that, the film surface also looks uneven and not smooth because of the large number of starch granules on the film surface. This means that the ingredients that make up the film are not mixed evenly.

Meanwhile, in edible film with a concentration of 9% lemongrass extract, the film surface was smoother and without bulges, which means there were no bacteria on the film. The SEM test results prove that the active compounds in lemongrass extract are proven to be able to kill bacteria. However, the addition of lemongrass extract to edible film causes pores to form on the surface of the film. This occurs due to weak bonds between amylose molecules in the film structure which causes a decrease in the density of the film matrix. As a result, the tensile strength value of the film will also be smaller. This is in accordance with the tensile strength test data which states that the tensile strength value of edible film for mango seed starch waste with a concentration of 5% and lemongrass extract with a concentration of 0% is greater than for lemongrass extract with a concentration of 9%.

3.5 The Effect of Coating on Weight Loss

Weight loss is a process that occurs when fruit experiences a decrease in weight due to transpiration, respiration and the activity of microorganisms. The reduction in weight loss for each variation after storage for 7 days can be seen in Figure 5.

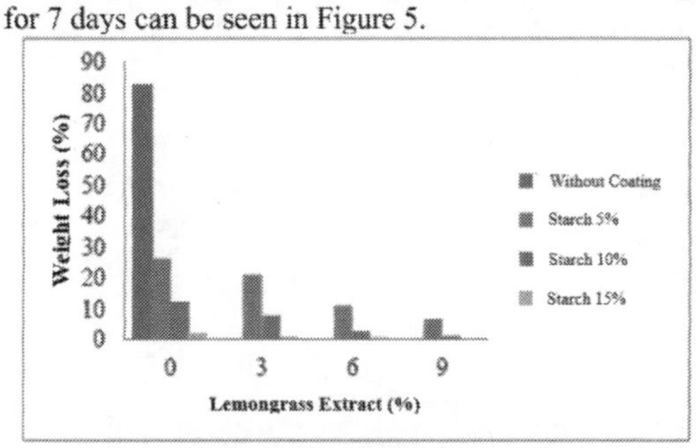

Figure 5. Effect of Coating Composition with Variations of Mango Seed Waste and Lemongrass Extract on Strawberry Weight Loss during 7 Days Storage

From Figure 5 it can be seen that there is a difference in the percentage increase in weight loss for strawberries without using a coating and using a coating. The highest weight loss was obtained in uncoated strawberries of 82.79% and the lowest weight loss in strawberries coated with a variety of starches was 15%; 9% lemongrass extract was 0.51%. This difference in weight loss is caused by the respiration and transpiration processes in uncoated strawberries so that the carbohydrates in the strawberries are broken down into CO_2 and H_2O which evaporate and cause the strawberries to experience a large weight loss. Meanwhile, for strawberries with coatings, weight loss can be minimized because of the coating which protects the fruit from gas exchange.

4 Conclusion

In this research, edible coating biomaterials was made from mango seed starch with the addition of lemongrass leaf extract as an antibacterial. Edible coating biomaterials will be applied to strawberry. Edible coating will be analyzed mechanical properties of the film will be analyzed in the form of tensile strength, elongation and young's modulus and SEM analysis. The results of the analysis are that the greater the tensile strength of the film, the better the film is at resisting mechanical damage. The best tensile strength value was obtained from a variation of 5% mango seed starch waste with 0% lemongrass extract. The greater the elongation of the film, the more elastic the film. The best film elongation value was obtained from a variation of 5% mango seed starch waste with 9% lemongrass extract. The smaller the elastic modulus value, the more elastic and flexible the film formed. Elastic modulus value The best elastic modulus value was obtained from a variation of 5% mango seed starch waste with 0% lemongrass extract. SEM test results show that there are cluster differences between films without lemongrass and lemongrass. Films with the addition of lemongrass extract have better morphology compared to films without lemongrass extract.

References

1. P. S. Awodi, J. C. Ogbonna, and T. N. Nwagu. *Heliyon*, 8 (2022)
2. J. L. F. Alves, J. C. G. da Silva, G. D. Mumbach, R. F. Alves, M. Di Domenico, and C. Marangoni. Journal *Thermochim. Acta*, 725 (2023)
3. C. Onyutha and N. Auma. *Heliyon*, 9 (2023)
4. A. O. Martínez-Olivo, M. U. Carlos-Murillo, S. G. Sáyago-Ayerdi, J. A. Sánchez-Burgos, and V. M. Zamora-Gasga. *Chem. Adv*, 3 (2023)
5. dan C. W. Miskiyah, Widaningrum. Jurnal *Hortikultura*, 21 (2011)
6. A. Tazi, A. Zinedine, and J. Miguel. *J. Pre-proof*, (2024)
7. M. de S. Alves *et al. Ind. Crops Prod*, 199 (2023)
8. M. A. Balti, B. Hadrich, and K. Kriaa. *Chem. Eng. Process. Process Intensif*, 124 (2018)
9. R. Cantoia Júnior *et al.*, "Lemongrass essential oil in sugarcane silage: Fermentative profile, losses, chemical composition, and aerobic stability," *Anim. Feed Sci. Technol.*, vol. 260, no. December, 2020, doi: 10.1016/j.anifeedsci.2019.114371.
10. R. Fokom *et al. Ind. Crops Prod*, 138 (2019)
11. J. Saha, M. I. H. Mondal, F. Ahmed, and M. Rahma. *Arab. J. Chem.*, 16 (2023)
12. dan K. T. Majewska, E., M. Kozlowska., E. Sekowska., D. Kowalska. *Polish J. Food Nutr. Sci*, 69 (2019)
13. O. Samuel Olugbenga, P. Goodness Adeleye, S. Blessing Oladipupo, A. Timothy Adeleye, and K. Igenepo John. *Waste Manag. Bull*, 1 (2024)
14. P. Chavan *et al. Food Chem*, 418 (2023)
15. P. . Roeswitawati, D., Santoso, U dan Lestari, Journal *Nabatia*, 7(2019)
16. K. D. Tafa, N. Satheesh, and W. Abera,. *Heliyon*, 9 (2023)
17. K. B. Lim, G.O., Jang, S.A. dan Song. *Journal Food Eng*, 98 (2010)
18. A. M. S. Jorge, M. C. Gaspar, M. H. F. Henriques, and M. E. M. Braga. *Innov. Food Sci. Emerg. Technol*, 88 (2023)
19. L. Fama, L., Rojas, A.M., Goyanes, S. dan Gerschenson. *J. LWT*, 38 (2005)

 https://doi.org/10.1051/e3sconf/202456002013

An Analytical Study on the Measurement of Agricultural Net Irrigation Water Requirements

Rongxiang Hu [1*], Xu Luo [2], Zheng Zheng [2]

[1]Zhejiang Institute of Hydraulics & Estuary (Zhejiang Institute of Marine Planning and Design) Hangzhou, Zhejiang, 310020, China
[2]Shaoxing Water Resources and Hydropower Survey and Design Institute Co., Ltd. Zhejiang Shaoxing, 312000, China

Abstract: The net irrigation water requirement (NIWR) in the irrigated area is the "terminal" of water-use efficiency and the basis for implementing quota management and water-saving incentives. Also, it is significant for farmland water management and water conservation in irrigated areas. In light of the average water consumption per mu and the crop area, which are two key indicators of the NIWR, this paper adopts the field selection measurement method and the statistical method based on the image patches achieved by China's third national land survey to acquire the data of water consumptions and crop areas. In this case, four principles of field selection are proposed, including "basic requirements", "moderate size", "controllable water inflow", and "measurement practicality", in an attempt to increase the number of fields selected for analysis. In addition, based on the image patches, the "complete investigation" and the "typical investigation" are suggested for irrigated areas at different scales to determine the crop areas, which are proven practicable and available for evaluating the water-use efficiency of irrigated areas at different scales.

1 Introduction

As the "terminal" of water-use efficiency, the net irrigation water requirement (NIWR) in the irrigated area that is committed to crop growth and stable yield and quality can be considered as the minimum water volume lifted and diverted from the canal headwork (i.e., the pump station), which is the data that is of great significant to water resource management [1]. At present, the NIWR is an important indicator in calculating the effective utilization coefficient of farmland irrigation water [2]. The solution recommended by the Ministry of Water Resources involves field selection and field measurement (e.g., direct measurement and observation-based analysis) to understand the net irrigation quota [3], where principles of field selection are provided. The quantity and quality of fields selected can greatly impact the determination of the NIWR, and more detailed farmland standards are required in practice.

Determining the crop areas in an irrigated area is necessary for the real-time optimization and distribution of water resources over the target area [4]. The two major determination methods for crop areas are the internal survey and the field survey, which are often used in combination. The internal survey involves the form-based questionnaire and the image identification (i.e., remote sensing analysis and manual interpretation) [5]. While remote sensing analysis applies to extensive survey regions, manual interpretation provides high-precision results but is labor-intensive. Currently, remote sensing technology has been widely applied in various scenarios [6] other than agricultural water management for it requires high-caliber experts, which appears to be a shortage in this field. For agricultural water management, the popularization of remote sensing technology is a long-term process. On the other hand, the form-based questionnaire is a convenient method to acquire the data for each village. However, the data relies mostly on farmers and their experience in the absence of scientific plotting. In practice, it is difficult to guarantee the questionnaire quality and review the data, due to the uncontrollability of respondents' sense of responsibility and experience. Thus, a crop statistical approach that is applicable to agricultural water management is needed.

In 2019, China carried out the third national land survey, which segmented the farmland according to remote sensing data and others to yield massive segmented images of farmland, laying a solid base for crop statistics. By virtue of the third national land survey's achievements, it is expected that crop survey and relevant research should be the focus of the field for an extended period.

*364601233@qq.com

2 Methodology

2.1 Calculation

According to the *National Technical Instructions for Measuring and Analyzing the Effective Utilization Coefficients of Farmland Irrigation Water*, an irrigated area's NIWR can be obtained by multiplying the average

NIWR per mu of each crop type with the actual irrigation area of the crop. The equation is as follows:

$$W_{net} = \sum_{n}^{j}\sum_{m}^{i} w_{ij} A_{ij} \quad (1)$$

where, W_{net} is the irrigated area's NIWR (10,000 m^3); w_{ij} is the average NIWR per mu of Crop i in the jth section in the irrigated area (m^3/mu); A_{ij} is the actual irrigation area of Crop i in the jth section in the irrigated area (mu); n is the number of the sections divided in the irrigated area (large and medium-scaled irrigated areas can be divided into an upper reach and a lower reach); m is the number of crop types.

In addition, the irrigated area's NIWR can be calculated using the equation below:

$$W_{net}=W_{gross}*\eta \quad (2)$$

where, W_{net} is the irrigated area's NIWR (10,000 m^3); W_{gross} is the NIWR per mu (10,000 m^3/mu); η is the effective utilization coefficient of farmland irrigation water.

In Equation 1, the average NIWRs per mu of crops and the actual irrigation areas of crops are two key factors. In contrast, Equation 2 emphasizes the gross NIWR and the effective utilization coefficient.

2.2 Field Selection and Water Measurement

The selection of typical fields should consider both measurement accuracy and practicality. To the best of our knowledge, "basic requirements", "moderate size", "controllable water inflow", and "measurement practicality" are proposed:

(1) Basic Requirements. Dominant crop types and irrigation methods should be considered. Paddy rice (early rice, late rice, and single-cropping rice), fruits, and vegetables are major water consumers in irrigated areas. Thus, rice fields, orchards, and vegetable fields, as well as canal irrigation, pipeline irrigation, and micro irrigation, should be taken into consideration. No survey fields are assigned to those crops planted at a small scale or consuming less water, and only relevant investigations are carried out on them;

(2) Moderate Size. According to the settings for irrigation experimental plots, a survey field should be within 0.5 to 2.5 mu and can be clearly distinguished from adjacent fields, which accords with the definition of the typical "large field". Oversized fields may suffer conveyance losses of irrigation canals and tend to involve a scenario where water flows from one field to another for irrigation and drainage, which deviates from the concept of the NIWR. On the other hand, undersized fields cannot afford the space the water-measuring facilities require. For mountains where the topography is a significant restriction, smaller values close to 0.5 are available. For plains, larger values are acceptable, but large fields should be divided into smaller, watertight fields;

(3) Controllable Water Inflow. Survey fields should be provided with gates, valves, or other controllers in the water inlets. If possible, only one inlet is set up for a single survey field. In this case, it allows for water management and precise measurement not only but also reduces the irrigation time and the times of observations. Also, irrigation-drainage synchronization and broad irrigation can be avoided;

(4) Measurement Practicality. Observation-based analysis, where a water-measuring facility is required in the inlet, is preferred because it is straightforward and easy to operate in practice. The standard section method, as well as measuring weirs (and flumes), recommended by the *Specifications for Water Measurement* do not apply to field surveys due to they require constant attention to the variations in water level and result in low-precision measurements. Water-measuring meters can automatically record the volume of water irrigated, which is friendly to farmers. In the case of straight stream, full pipe flow, and unobstructedness, these devices can realize higher precision. To solve the problem of straight stream, an additional section of pipe can be installed behind the gap of the existing pipe. It is recommended that this pipe should be laid on or along the field ridge to save land and avoid incurring troubles with farming. Moreover, it is advisable to place a sand sediment trap in front of the pipe to collect silt deposits, so as to minimize the impact of impurities in the water. To achieve full pipe flow, a discharging well should be placed on the outlet, and the outlet's height should be no lower than the pipe's top elevation, so as to realize jacking flow inside the pipe. If the water source has the ability to store water, a combination of the portable pump and the water-measuring meter is practicable. When irrigation is needed, the portable pump and the meter can be moved to the field to start measurement while irrigating. Given the high content of impurities in canals, the ultrasonic water meter, the electromagnetic flowmeter, and other large-diameter devices that are not prone to clogging are recommended. The mechanical water meter should be avoided.

2.3 Statistical Survey of Crops

Taking the image patches achieved by the third national land survey as the base map, we conducted crop surveys in the irrigated areas. The following conclusions are summed up:

(1) Survey Time. We analyzed the crops and the planting habits in the irrigated areas to understand the phenological phases of major water-consuming crops. In Zhejiang Province, the planting habits are rice-rape rotation and rice-wheat rotation, yielding two crops a year. Usually, early rice is from April to July, late rice is from July to November, and single-cropping rice is from June to November. In this case, two field surveys were conducted in mid-to-late April and October, respectively, to cover the growth of major crops;

(2) Survey Preparation. Appropriate survey methods were adopted for different crops. Specifically, a field investigation was used for rice and outdoor vegetables, a solution integrating the initial imagery identification and the on-site review was used for perennial forestland, and the independent investigation was used for larger farmers who planted commercial crops, which often can be

harvested for more than one time a year. The boundaries of the irrigated areas were nested with the image patches to generate the image patches of the irrigated areas, which were numbered in sequence to form a table of field patches;

(3) Survey Method. If the irrigated area has a size of no larger than 5,000 mu, complete investigation and review are acceptable. Otherwise, if the size exceeds 5,000 mu, on account of the massive workload of a complete investigation, typical random sampling is recommended to capture the data from the typical regions and use the data to extrapolate the whole irrigated area. In line with the image patches, farmers or technicians fill in the crop names on the forms or mark the crops on the image patches based on facts. When a patch's boundary is inconsistent with the fact, the patch and the information should be modified accordingly. Also, unmanned aerial vehicle remote sensing would be a preference for extensive irrigated areas;

(4) Area Statistics. The size of an irrigated area, where a complete investigation is conducted, can be obtained by adding the sizes of all patches together. For those subject to a typical investigation, it should first sum up the planting area of each crop in the irrigated area, then analyze the characteristics of the planting area of each crop on different surface features, and eventually deduce the irrigated area's planting area in proportion.

3 Materials

The research takes the Ximihu Irrigated Area and the Yangshuxia Irrigated Area as the subjects. As a medium-sized irrigated area, the Ximihu Irrigated Area has a designed irrigation area of 105 million mu, supported by a water-lifting pump station and conveyance canals. The Yangshuxia Irrigated Area is a small-sized irrigated area, which has a designed irrigation area of 400 mu, with a water-lifting pump station and conveyance canals. Both irrigated areas cultivate mainly single-cropping rice.

In this case, we selected 6 and 2 survey fields from the two irrigated areas, respectively. The field sizes were between 0.5 and 2.5 mu. By crop type, in the Ximihu Irrigated Area, 3 fields were single-cropping rice, and 3 fields were double-cropping rice; in the Yangshuxia Irrigated Area, 1 was single-cropping rice, and 1 was double-cropping rice. The ultrasonic flowmeter (Diameter: 100-150mm) was used for water measurement. The farmers were responsible for providing the observations. The irrigation time was from May 5 to October 30, 2023. By nesting the boundary with the image patches, it was determined that the Ximihu Irrigated Area had an agricultural land area of 11,440 mu, and the farmland area was 10,523 mu (9579 for paddy fields, 948 for dry fields). A total of 839 image patches were recognized. The irrigated area was too large to perform a complete investigation.

We divided the Ximihu Irrigated Area into three sections. The field survey was used for Sections 1 and 2 to determine the crop types and their planting areas. Based on the data from Sections 1 and 2, we derived the data for Section 3. The Yangshuxia Irrigated Area possessed an agricultural land area of 222 mu, of which 197 was farmland. A total of 28 patches were involved. The irrigated area was small enough for a complete investigation. For both irrigated areas, the investigation was performed on April 20, 2023 and October 20, 2023.

4 Result Analysis

4.1 Survey Results of Crop Areas

For the Ximihu Irrigated Area, Sections 1 and 2 involve 54 and 58 patches(Figure 1), respectively. The crops in each section are determined according to the actual field conditions(Table 1). As calculated, the multiple cropping index for paddy rice is 1.55, and the multiple cropping index for dry fields is 2.0, approximately. The area of paddy rape accounts for 6% of the total area of the paddy fields. The other crops are consistent with the surface feature data provided by the third national land survey. The measured and calculated results of crop areas in the Ximihu Irrigated Area are summed up in Table 2.

Table 1. Survey Results of Crop Areas in Sections 1 and 2 in the Ximihu Irrigated Area (Unit: Mu)

Region	Section 1	Section 2	Total
Area of Early Rice	398.16	388.26	786.42
Area of Late Rice	398.16	388.26	786.42
Area of Single-Cropping Rice	334.66	299.38	634.04
Area of Oilseed Rape	88.88	0.00	88.88
Area of Vegetables	10.81	56.22	67.03
Area of Orchards	0.00	1.68	1.68
Total	1230.67	1133.8	2364.47

Table 2. Crop Areas in the Ximihu Irrigated Area (Unit: Mu)

Crop	Early Rice	Late Rice	Single-Cropping Rice	Vegetable	Orchard	Oilseed Rape	Tea Garden	Total
Area (Mu)	5295	5295	4269	1896	912	599	3	18270

https://doi.org/10.1051/e3sconf/202456002013

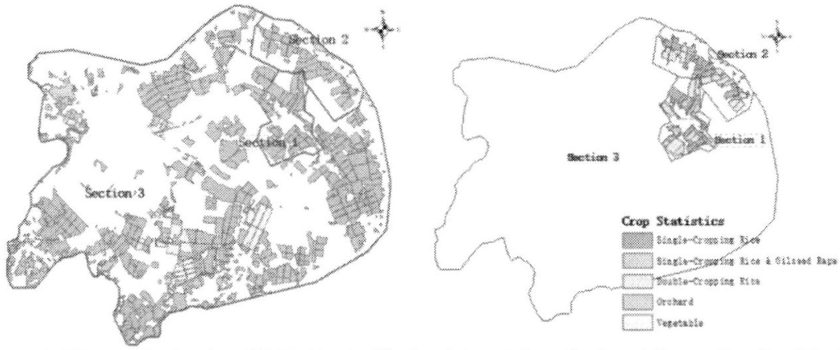

Figure 1. Schematic Diagram for Sections Divided in the Ximihu Irrigated Area (Left) and Survey Results of Typical Crop Areas (Right)

The small-sized Yangshuxia Irrigated Area is available for a complete investigation. The patches are marked according to actual field conditions(Figure 2). The crop areas are summed up in Table 3. In this area, some ponds are planted with rice to increase the rice area over the dry area.

Table 3. Crop Areas in the Yangshuxia Irrigated Area (Unit: Mu)

Crop	Vegetable	Early Rice	Late Rice	Oilseed Rape	Single-Cropping Rice	Total
Area	20.32	127.75	127.75	138.15	138.15	552.12

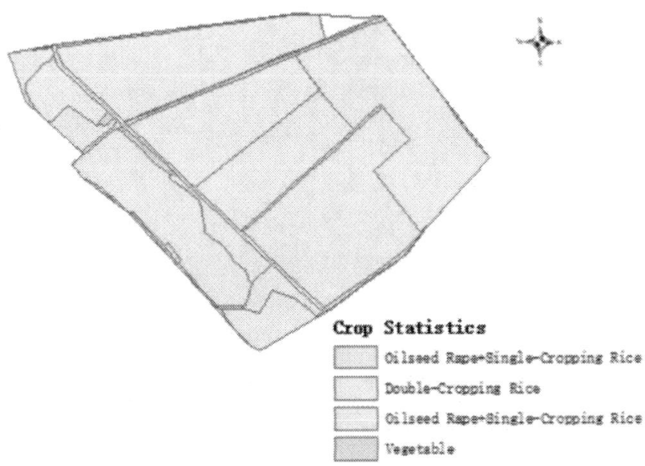

Figure 2. Crop Areas in the Yangshuxia Irrigated Area

4.2 Calculation of NIWRs

According to the measured data in the typical fields in 2023, in the Ximihu Irrigated Area and the Yangshuxia Irrigated Area, the average irrigation water consumptions for early rice, late rice, and single-cropping rice are 180, 250, and 320 m^3/mu, and the average number of irrigation times are 6, 7, and 10, respectively. The data of the other crops are based on the reference quotas or field surveys. According to the calculation method of NIWR above, the total NIWRs of the Ximihu Irrigated Area and the Yangshuxia Irrigated Area are 3.75 million and 0.100 million m^3, and the average NIWRs are 328 and 452m^3/mu.

On the other hand, using the "calculating water consumptions with electricity consumptions" method in the headwork pump station, it is calculated the volumes of water lifted for the Ximihu Irrigated Area and the Yangshuxia Irrigated Area are 6.250 million and 0.167 m^3. With the average effective utilization coefficient of farmland irrigation water (0.62) considered, the NIWRs of the Ximihu Irrigated Area and the Yangshuxia Irrigated Area are 3.88 million and 0.103 m^3. The deviation between the two methods is 3% approximately, and the performances are similar.

5 Conclusion & Prospect

5.1 Conclusion

The NIWR is of great importance to farmland water management. Since the NIWR relies on the average water consumption per mu and the crop area, this research proposes the field selection measurement method and the statistical method based on the image patches achieved by the third national land survey. According to the utilization coefficient of farmland irrigation water, the specifications for field selection are suggested. Applicable water-measuring devices are recommended as well. By virtue of the data achieved by the third national land survey, the research combines with field investigation to determine the sizes of the irrigated areas more conveniently. For irrigated areas at different scales, two corresponding investigation methods are proposed, expecting to provide a reference for future research on crop areas in irrigated areas.

5.2 Suggestion

Due to the complexity of field selection, the fields and crops selected in this research tend to be inadequate and monotonous. Also, the water consumption data of some crops were based on survey analysis instead of actual measurement. If possible, it is recommended to increase the number of fields to improve the representativeness.

Acknowledgment

Foundation Item: Zhejiang Province Basic Public Welfare Research Program (LG22E090001); Water Conservancy Science and Technology Program (RC2119); Zhejiang Institute of Marine Planning and Design (ZIHE23Y007).

Author

Hu Rongxiang (1986.7-); Male; Senior Engineer; Engaged in Farmland Irrigation and Water Conservancy. Email: 364601233@qq.com

References

1. Ren S.M., Li B., Countermeasure to Develop Efficient Irrigation for Limited Use Water in Hetao Irrigation Area [J]. Journal of China Agricultural University, 2000.
2. Han Z.Z. and Feng B.Q., Calculation and evaluation of efficiency of irrigated water use [J]. China Water Resources, 2016.23.
3. SL/Z 699-2015. Technical Guide for the Measurement and Calculation of Irrigation Efficiency [S].
4. Niel T G V , Mcvicar T R .Assessing positional accuracy and its effects on rice crop area measurement: an application at Coleambally Irrigation Area[J].Australian Journal of Experimental Agriculture, 2001.DOI:10.1071/EA00140.
5. Lin C., Improving the New Round of Crop Area Survey [J]. China Statistics, 2018(2):2.
6. Ray S S , Dadhwal V K .Estimation of crop evapotranspiration of irrigation command area using remote sensing and GIS[J].Agricultural Water Management, 2001, 49(3):239-249.DOI:10.1016/S0378-3774(00)00147-5.

Research on the impact of ecological literature on the environment and society—taking Chinese ecological literature as an example

Yuxi Chen*

School of Liberal Arts, Panzhihua University, 617000 Panzhihua City, Sichuan Province, China

Abstract. China's rapid economic development is accompanied by severe environmental problems. Chinese government has taken some measures to improve environmental conditions and protect diverse ecosystems, but these problems still exist. Science and technology are the main measures to improve the ecological environment, but the impact of ecological literature on the environment cannot be underestimated. As a medium for expressing opinions and emotions, ecological literature has become an important way to reflect the social environment, which has attracted widespread attention from ecological and environmental scholars and social people. Therefore, this article aims to explore the power of ecological narratives in ecological literary works and how this narrative affects people's awareness and social behavior of ecological and environmental. Taking China as an example, this article analyzes the ecological narrative elements, the influence of readers, the influence of social public opinion, and the influence of the Chinese government's environmental policies in many ecological literary works in China. Finally, it was discovered that ecological literature, as a literary form, shows the power to transcend time and space, profoundly affects the environmental awareness of contemporary society, and lays a solid foundation for the improvement of the ecological environment.

1. Introduction

China's rapid economic development is accompanied by serious environmental problems, such as ecological environment damage, soil pollution, water pollution, etc., as shown in Figure 1. Chinese government has taken some measures to improve environmental conditions and protect diverse ecosystems, but these problems still exist and more comprehensive and long-term efforts are needed to achieve a sustainable ecological environment[1].Science and technology are the main measures for environmental improvement. At the same time, people's awareness of environmental issues and their social impact cannot be underestimated. As a medium for expressing opinions and emotions, literature has become an important way to reflect the social environment, which has attracted widespread attention from ecological and environmental scholars and social people[2]. Therefore, this article takes Chinese ecological literature as the research object, aiming to explore how literary works express awareness of environmental issues and analyze the impact of this awareness in society.

As a unique literary field, ecological and environmental literature explores the relationship between people and the environment through literary expression, which has become an important way to protect the environment. Therefore, scholars have a very rich research results and related works on this issue. Craig A[3], Hans P[4], Werner I[5] and others analyzed the interrelationship between literature, science and the public in Silent Spring, a classic work describing the ecological environment, making people realize the importance of ecological literature as a medium for the advancement of climate awareness. Sandip K[6] selected ecocritical literary works in the world for analysis, and obtained environmentally oriented literary works that can bring ecological literacy to readers and have a higher ecological awareness, so as to take good care of nature. Adam Trexler[7] explores the complexity of climate change as a scientific and literary phenomenon, and obtains methods for writing characters and plots in literary works in the context of harsh ecological environments. Yang Yu[8] stated that the state of the ecological environment and its changes affect contemporary Chinese literature, giving rise to "ecological literature" and "ecocriticism" in modern Chinese literature, as well as the "three types" of literary creation. Long Q L[9] said that ecological writers adhere to a humanistic stance and criticize the huge damage caused by industrialization to the ecological environment and its impact on simple human nature. Chen Tingting[10], Zhu

* Corresponding author: 914347668@qq.com

 https://doi.org/10.1051/e3sconf/202456002014

Yali[11], Han Xiaoxiao[12] and others have analyzed and summarized the impact of environmental awareness and ecological criticism on China's ecological environment in the works of Li Juan, a famous Chinese ecological and environmental prose writer. They have always agreed with the ecological environment. Literary works have played a vital impact on China's ecological environment.

Fig. 1. Famous ecological environment damage in China (a: Environmental pollution in the Tengger Desert in Inner Mongolia Autonomous Region; b: Major water pollution accident in Tuojiang River, Sichuan; c: Dead fish incident in Baiyangdian, Hebei; d: Blue algae outbreak in Dianchi Lake, Yunnan)

Scholars' research on ecological literature mainly focuses on the impact of a single literary work on the ecological environment (such as Silent Spring), or scholars' research focuses on the impact of multiple literary works on a single ecological environment, such as industrial pollution. Existing published research results lack a multi-faceted and systematic analysis of the impact of many works on various aspects of the local ecological environment (such as various elements of ecological narratives, various impacts on readers, literary works and public opinion, ecological literature and environmental policies, etc.) .To sum up, scholars around the world have rich research results on ecological and environmental literature, but there is a lack of comprehensive and systematic postgraduate literature works on the impact of ecological environment, especially in China. As the earth's ecological and environmental problems become more and more serious, ecological and environmental narrative literary works can improve the public's awareness of environmental protection and enhance social influence. Therefore, as the largest developing country, China needs to conduct research on ecological narrative literature and improve the ecological and environmental protection awareness of its 1.4 billion people.

2. Overview of Chinese ecological literature

2.1. Definition and evolution of ecological literature

Ecocriticism originated in the early 20th century. Early ecological literature mainly focused on natural description and environmental protection themes. However, it later gradually evolved into an interdisciplinary subject involving literature, culture, ecology and environmental philosophy[13]. The development stages of ecological literature are shown in Figure 2.

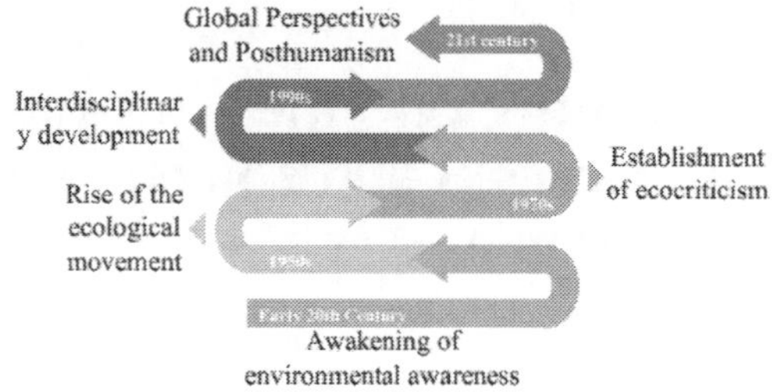

Fig. 2. Development process of ecological literature

- Awakening of environmental awareness

Early naturalistic literature and the environmental movement helped to draw attention to the natural environment.

- Rise of the ecological movement

The rise of environmentalism and ecological movements prompted the literary world to pay attention to the relationship between humans and nature, which in turn gave birth to ecological literature.

- Establishment of ecocriticism

Ecological literature gradually evolved into ecocriticism, which is no longer just a description of environmental themes, but also examines the deep-seated issues of the relationship between man and nature through literary works.

- Interdisciplinary development

Ecocriticism gradually expanded into other disciplines, including philosophy, sociology, cultural studies, etc., forming a more complex theoretical system.

- Global Perspectives and Posthumanism

Ecological literature gradually integrates into a global perspective, paying attention to global environmental issues, and the idea of posthumanism appears in some theoretical schools, emphasizing the interaction between humans and nature.

2.2. Special status of Chinese ecological literature

The development of ecological literature in China is relatively new, but as environmental issues become increasingly prominent, it has gradually attracted the attention of scholars and writers. This article lists ten classic Chinese ecological literature writers and their representative works as shown in Table 1[14].

Table 1. Ten classic Chinese ecological literature writers and their representative works

·	Author	Representative works
1	Xu G	*Loggers, wake up!*, *Snken land*, *Rvers are not the eternal flow*
2	We an	*Things on earth*, *Ater the sun rises*, *Bekeepers*
3	Zhang W	*Integrate into the wilderness*, *Stories of the fields*, *Your trees*

4	Han S G	*Mountains south and rivers north*
5	Yang W F	*Nature notes, Why butterflies are so beautiful, Sick bonsai*
6	Li C B	*Garland under the mountain, Green heavenly book, Wet dream of the river*
7	Li Q S	*Mikania, Land ethics, Spirit*
8	Gu Y	*Worry about the source of rivers, Who will repent for mankind, Love letter to the sources of the three rivers*
9	Li J	*Winter pasture, Sheep path, Corner of altay*
10	Zhou X F	Birds, Spots, Collection

To summarize, the special status of Chinese ecological literature is mainly reflected in the following points:

- Naturalistic tradition

Chinese literature has a long tradition of describing nature, such as classical poetry, landscape poetry, etc. These traditions laid the foundation for China's ecological literature.

- Contemporary environmental literature

As environmental issues have attracted increasing attention, Chinese writers have begun to create literary works involving ecological themes, exploring the relationship between humans and nature and the impact of environmental changes on human society.

- Ecological poetry and prose

Some contemporary poets and essayists focus on ecological themes to express concerns about the beauty of nature and concerns about environmental issues.

- Social concerns and academic research

In the academic field, Chinese researchers have also begun to pay attention to ecological literature and ecocriticism, and explore ecological elements and environmental ethics in Chinese literature.

Generally speaking, while retaining traditional naturalistic elements, China's ecological literature has also developed new forms driven by modern environmental awareness, becoming a literary genre that reflects social concerns and raises thoughts on environmental issues.

3. Ecological narrative and environmental awareness

3.1. Ecological narrative elements in literary works

There are many ecological narratives in Chinese literary works. This article lists some classic representative writers to analyze the elements of ecological narratives.

Li Juan is known for her depictions of nature, ecology and rural life. Her "Winter Pasture" is a collection of essays set in the western mountains, reflecting the author's unique observations and perceptions of natural landscapes, rural life and the relationship between man and nature. "Sheep Path" series of essays expresses the yearning for a simple and natural life through the description of natural scenery and rural life. The book "A Corner of Altay" mainly contains the author's essays and prose in rural life, covering tea culture, natural scenery and every detail of life, showing the author's yearning for a simple and peaceful life. Li Juan's works are loved by readers for their delicate descriptions and sensitive observations of the relationship between nature and humanity. She conveys a deep understanding of nature, ecology and traditional culture through literary means.

Chen Zhongshi is famous for his descriptions of nature, ecology and rural life in works such as "White Deer Plain". The novel "White Deer Plain" is considered one of the classics of contemporary Chinese literature. Through the rise and fall of a family in Inner Mongolia over the past hundred years, it shows the contradictions and conflicts between human beings and nature, tradition and modernity. The work has been praised for its profound observations of the ecological environment and rural life. Through literary means, Chen Zhongshi reflects the impact of social development on nature and traditional lifestyles, and expresses his unique understanding of nature, land and life. Xu Gang has landmark significance in the development of contemporary ecological prose in China. He published "Lumberjack, Wake Up!" " is one of the most influential works of contemporary ecological prose in China and is considered China's "Silent Spring". In "Lumberjack, Wake Up!" "", Xu Gang uses a sad and indignant tone, field investigation data, and a comparative perspective from China and foreign countries to describe the heavy ecological price China has paid in the process of modernization. Xu Gang has a strong sense of urgency and has persisted in traveling around the country for decades to gain an in-depth understanding of natural ecology and reflect deeply on the shortcomings of modern lifestyles, creating a new world for Chinese ecological prose creation.

Han Shaogong's "Tianya" magazine launched "Nanshan Minutes" focusing on ecological and environmental issues. The publication of Han Shaogong's collection of ecological essays "Mountain South and Water North" caused a sensation in the reading world. The experience of nature and the observation of the earth have become the main thread running through "Mountain South and Water North". In Han Shaogong's view, working in nature is of great significance to human beings. It is not only a way for writers to observe the earth and understand nature, but also has a hidden effect on people's mental health. Han Shaogong re-lived his daily life in the agricultural era, reflected on modern civilization and ecological crises through his experience of natural culture, and proved the possibility of returning to nature and rebuilding a harmonious state between man and nature through the practice of rural seclusion. Han Shaogong put forward a new understanding of the term original sin, believing that the loss of other lives caused by human beings in order to maintain their own lives is a kind of original sin.

In addition to the representative works of the above representative writers, China also has many ecological literary works. Summarizing many ecological literary works, it is not difficult to find that Chinese ecological narrative elements are usually reflected in the description and reflection of the natural environment, animals and plants, the relationship between humans and nature, and environmental changes. Commonly used ecological narrative elements in Chinese ecological literature include the following points.

- Description of natural scenery

The writer uses vivid language to describe the beauty of nature, such as mountains, rivers, forests, lakes, etc., to convey his unique perception of the natural environment.

- Description of flora and fauna

Through vivid descriptions of plants and animals, it shows the interrelationship of various elements in the ecosystem and emphasizes the importance of ecological balance.

- Description of rural life

Some works reflect the close connection between man and nature by depicting rural life, expressing nostalgia for traditional agricultural lifestyles or concerns about changes in the rural environment.

- Environmental changes

Literary works often describe processes such as urbanization and industrialization to show the impact of environmental changes on humans and nature, as well as the impact on traditional lifestyles.

- Disasters and struggles

Describe the impact of natural disasters (such as floods, earthquakes, sandstorms, etc.) on human society and ecological systems, as well as people's struggle and repair against disasters.

- Ecological philosophy and ethics

Some works express concerns about ecological ethics and sustainable development through the characters' thinking or dialogue, and think about the harmonious coexistence between humans and nature.

- Awakening of ecological consciousness

Describe the characters' gradual awakening and awareness of the importance of environmental issues in the story, and their positive actions to protect nature.

3.2. Impact of ecological narratives on readers

China's ecological narrative literary works, with their profound cultural heritage and unique narrative methods, are permeated with the love of nature and thinking about the relationship between man and nature. Through literary expression, these works not only show China's rich natural resources and cultural traditions, but also reveal the seriousness of environmental problems.

Chinese ecological narrative literary works show diverse natural landscapes through vivid descriptions, from magnificent mountains and rivers to vast grasslands to deep forests, making readers feel as if they are immersed in these natural environments. "A Yurt in the Grassland" allows readers to feel the vastness and vitality of the prairie in Inner Mongolia with its vivid description. Such descriptions not only make people appreciate the beauty of nature, but also inspire unique emotions towards this land.

Ecological narrative literary works often arouse readers' concerns about environmental protection through the depiction of environmental issues. Taking "Evolution from Polluting the World" as an example, the author triggered readers to reflect on the urban development model by describing the damage caused by urbanization to the natural environment. The focus on pollution, deforestation and ecological balance in the work guides readers to think about the responsibilities of individuals and society in environmental protection.

Chinese ecological narrative literary works enable readers to have a deeper understanding and appreciation of the beauty of nature through meticulous descriptions. "The Classic of Mountains and Seas" leads readers into the mysterious natural world with its fantastic narrative. The descriptions of landscapes, flowers and plants in the works allow readers to have a more intuitive and emotional experience of the beauty of nature. Such appreciation helps cultivate people's environmental awareness and makes them cherish natural resources more.

Ecological narrative literary works not only focus on nature, but also deeply explore the relationship between man and nature. By telling the relationship between humans and plants, "Human Plants" reveals humans' dependence on nature and the blessings nature bestows on humans. The concept of symbiosis and co-prosperity between man and nature in the work guides readers to re-examine the interaction between man and nature, inspiring them to be more cautious about the use of natural resources.

Ecological narrative literary works arouse emotional resonance in readers by depicting the emotions and thoughts of characters. "Jiangnan Water Town", with its delicate character descriptions, enables readers to deeply experience the blend of characters and nature. Readers emotionally resonate with the characters in the literary works and become more involved in the environmental narrative, as if they are experiencing it themselves.

Some ecological narrative literary works inspire readers' creativity and desire for artistic expression with their unique literary styles and creative techniques. "Notes on a Mountain Dwelling" enables readers to feel the infinite possibilities of literature through its unique narrative structure and use of imagery. Such inspiration helps readers find more ways of expression outside of literature, and express their love for nature and thinking about environmental issues through their own creativity and artistic techniques.

Chinese ecological narrative literary works have profoundly influenced readers through descriptions of the natural environment, concerns about environmental issues, and thinking about the relationship between man and nature. These works not only arouse readers' environmental awareness, but also allow them to have a deeper understanding and appreciation of the beauty of nature and re-examine the relationship between humans and nature. At the same time, through emotional resonance and unique artistic expression, these literary works inspire readers' creativity and thinking about environmental issues. In general, Chinese ecological narrative literary works play an important role in promoting environmental awareness and prompting people to rethink the relationship between man and nature.

4. Social impact of ecological literature

4.1. Literary works and public opinion

China's ecological literary works play an increasingly important role in today's society. They reflect and explore environmental issues through literary expressions and have a profound impact on public environmental concerns and attitudes. The mutual influence between ecological literary works and public opinion is mainly reflected in the following points.

- Consciousness awakening and environmental concern

Ecological literary works arouse the public's awareness of environmental issues through descriptions of the natural environment and reflections on the relationship between man and nature. For example, Wang Huaiyu's novel "Blood Grassland" deeply demonstrates the impact of human activities on nature through the description of grassland ecology, and triggers readers to think deeply about environmental issues. The awakening of this awareness is the starting point for public attention to environmental protection, making people gradually pay attention and care about changes in the natural environment.

- Triggering of emotional resonance

Ecological literary works often resonate with readers through delicate emotional descriptions, making them feel the urgency of environmental issues more deeply. For example, Yu Hua's novel "Beyond the Castle" uses the description of ecological disasters and the protagonist's experiences to arouse readers' emotional resonance with environmental changes. This kind of emotional resonance can make the public more actively participate in environmental protection actions and form a more positive attitude towards environmental issues.

- Deep reflection on the relationship between man and nature

Ecological literary works prompt the public to re-examine the relationship between man and nature by deeply reflecting on the impact of human activities on the natural environment. For example, Chi Zijian's novel "The Right Bank of the Ergun River" guides readers to reflect on the damage caused to nature by human development models through the description of ecological destruction. This profound reflection helps the public form a more responsible and sustainable lifestyle, thereby affecting the overall society's environmental attitude.

- Promotion of social topics

Some ecological literary works have attracted widespread social attention and become hot topics in society. The environmental issues reflected in literary works have guided the public to pay attention to environmental policies, supervision and sustainable development. For example, Ma Boyong's novel "The Twelve Hours of Chang'an" aroused public discussion on water resources management through the description of water conservancy projects in history. This promotion of social topics can help guide the public to form a more rational and positive attitude towards environmental issues.

- Diversity of literary creation

Ecological literary works show diversity in literary creation, including novels, poems, essays and other forms. This diversity enables readers with different literary tastes and aesthetic concepts to find works that suit them, thereby increasing public attention to environmental issues. Diversified literary creation not only expands the scope of influence of ecological literature, but also enables a wider range of readers to participate in environmental concerns and form a broader public opinion.

- Interaction between media and public opinion

With the development of social media, ecological literary works can more easily interact with the public and form a larger public opinion through sharing, commenting, etc. Readers share their insights on ecological literary works on social media, forming interactions and discussions, which expands the scope of dissemination of environmental information, thereby influencing the public's attitude towards environmental protection. The interactive mechanism of social media provides a wider communication channel for the influence of ecological literary works, helping to form public opinion.

- Continued attention and calls for action

Some ecological literary works not only arouse public attention to environmental issues in their works, but also guide readers to take practical environmental protection actions through the end of the work or the author's call. For example, Liu Cixin's science fiction novel "The Three-Body Problem" made an urgent call for environmental protection by depicting the earth's environmental crisis. This direct call to action makes the impact of ecological literature more far-reaching and can inspire the public to become more actively involved in environmental protection.

Ecological literary works have had a profound impact on public opinion through awareness awakening, emotional resonance, reflection on the relationship between man and nature, promotion of social topics, diversity of literary creation, media interaction and calls for action.

4.2. Ecological Literature and Environmental Policy

As a country facing increasingly severe environmental challenges, China has tried to balance economic development and ecological protection by increasing the formulation and implementation of environmental policies in recent years. In this context, ecological literary works, as a form of literary expression, are gradually becoming an important factor in guiding and influencing environmental policies through profound depictions of nature, the environment, and the relationship between humans and nature. The potential impact of ecological literature on environmental policy making in China and how literary works influence public decision-making are analyzed below.

- Realistic reflections of ecological literary works

Ecological literary works often reflect the real environmental problems faced by society through profound descriptions of natural landscapes, environmental problems, and the relationship between man and nature. These works provide powerful social feedback for presenting environmental issues to the public and policy makers. This reflection of reality helps policymakers better understand the environmental challenges society faces.

- Guidance of ecological literary works on environmental protection concepts

Ecological literary works often use emotional narrative methods to guide readers to identify with environmental protection concepts. The unique depiction of the natural environment and the profound reflection on the relationship between man and nature in the work can inspire readers to resonate with environmental protection. This resonance helps shape public attitudes toward environmental policies and promotes a more positive environmental culture in society.

- Impact of ecological literary works on public awareness of environmental protection

Through detailed descriptions that deeply reflect natural ecology, ecological literary works help shape public awareness of environmental issues. The ecological destruction, resource depletion and other issues shown in the works can arouse the public's attention to environmental issues and make them realize that environmental protection is an urgent and important task. The consciousness-shaping effect of ecological literary works can prompt the public to more actively support and cooperate with the implementation of environmental protection policies, and promote the formation of more environmentally friendly values in society.

- Enlightenment of ecological literature works on policy formulation

The profound reflection on environmental problems and discussion of solutions in ecological literature can help inspire policy makers. The works may include suggestions on strengthening environmental protection supervision, promoting corporate responsibility, promoting green lifestyles, etc., providing reference and thinking directions for policy makers. This revelation helps policies to be closer to social reality and problem pain points, and to solve environmental challenges in a more targeted manner.

- Interaction between ecological literature works and policy implementation

There is an interactive relationship between ecological literature and policy. The implementation of the policy may have inspired writers to pay attention to social issues and promoted more literary creation on environmental issues. At the same time, through its influence, ecological literary works may also influence the thinking of policymakers to some extent, making them more deeply aware of society's concerns about environmental protection issues. This two-way interaction makes ecological literary works not only a reflection of reality, but also a catalyst for social change and policy formulation.

- Ecological literary works condense social consciousness

Ecological literary works help to gather environmental awareness in society through in-depth reflection on environmental issues. This gathering of awareness can encourage the public to participate more actively in environmental protection activities and form a social consensus. And this consensus will affect the decision-making direction of policymakers to a certain extent, because the implementation of policies requires broad support from society.

- Support for public participation in policy implementation

Ecological literary works help to form broader public participation by guiding the public's understanding and attitude towards environmental issues. Public attention and support for environmental protection have played a positive role in promoting the implementation of policies. The concerns and appeals for environmental protection issues in the works can stimulate the public's sense of environmental responsibility, make them more actively participate in environmental protection causes, and provide strong support for the implementation of policies. To sum up, there is a profound mutual influence between Chinese ecological literary works and environmental policies. Ecological literary works contribute to China by reflecting reality, guiding environmental protection concepts, shaping public environmental awareness, inspiring policy formulation, interacting with policy implementation, condensing social awareness, and supporting public participation. Provides useful reference and support for the formulation and implementation of environmental policies. In the future, policymakers and literary creators should strengthen exchanges and cooperation to jointly promote the sustainable development of environmental protection. With the collaboration of literature and policy, China is more likely to usher in a fresher and greener future.

5. Interdisciplinary value of ecological literature

Ecological literature not only focuses on the protection and restoration of the natural environment, but also deeply explores the relationship between humans and nature and the impact of human behavior on the natural environment. This field involves multiple disciplines such as literature, culture, ecology, and environmental philosophy, providing us with new perspectives and methods for understanding and solving environmental problems.

As an interdisciplinary field, ecological literature is characterized by the organic integration of literature with other disciplines. First of all, ecological literature uses literature as a carrier to vividly show the beauty and fragility of the natural environment through narrative, description, lyrical and other techniques, arousing readers' attention and thinking about the natural environment. Secondly, ecological literature also involves cultural studies, exploring human cognition and attitudes towards the natural environment under different cultural backgrounds, and how these cognitions and attitudes affect human behavior. In addition, ecology and environmental philosophy also provide important theoretical support for ecological literature, allowing it to more deeply analyze the relationship between humans and nature, as well as the causes and solutions to environmental problems.

The unique value of ecological literature from different disciplinary perspectives mainly includes the following contents:

- Literary perspective

Ecological literature uses literary techniques to show the beauty and fragility of the natural environment, allowing readers to intuitively feel the seriousness of environmental problems. At the same time, ecological literature also reflects the relationship between humans and nature through character creation and plot arrangements, and triggers readers to think and act on environmental protection.

- Cultural perspective

Ecological literature focuses on human cognition and attitudes towards the natural environment under different cultural backgrounds, helping us to understand environmental issues more comprehensively. By comparing and analyzing different cultures, we can discover the diversity and complexity of the impact of human behavior on the natural environment, thereby formulating more effective environmental policies.

- Ecological perspective

Ecology provides rich practical experience and theoretical support for ecological literature. Through the research and analysis of ecological systems, ecological literature can explore the roots and solutions of environmental problems more deeply, and provide scientific basis for environmental protection.

- Environmental philosophy perspective

Environmental philosophy is concerned with the relationship between humans and nature and the ethical and moral dimensions of environmental issues. Ecological literature shows the harmonious and symbiotic relationship between humans and nature through literary techniques, guides people to establish correct environmental values, and provides spiritual motivation for environmental protection.

6. Conclusion

Chinese ecological literature shows outstanding potential in cultivating environmental awareness. Through writers' descriptions of natural scenery and deep reflections on environmental issues, ecological literature has successfully aroused readers' sensitivity to the environment. The ecological ethics contained in the works guide people to rethink the relationship between man and nature, and promote positive changes in society's understanding of ecological issues. As a literary form, ecological literature shows the power to transcend time and space, profoundly affects the environmental awareness of contemporary society, and lays a solid foundation for the sustainable development of society. In the future, we should further pay attention to the development potential of ecological literature and promote it to have a far-reaching impact on a wider scale and contribute more importantly to the sustainable development of global society.

References

1. Xia B., Zheng L. (2022) Ecological environmental effects and their driving factors of land use/cover change: the case study of Baiyangdian Basin, China. Processes, 10(12): 2648-2648.
2. Russotto M. (2023) Birds: ecological awareness and literature. Gragoatauff, 28(61): 816-831.
3. Craig A. (2021) Taking lessons from silent spring: using environmental literature for climate change. Literature, 2021, 1(1): 2-13.
4. Hans P., Dagny A., Emma L., Emma L. (2023) Avoiding the next silent spring: our chemical past, present, and future. Environmental science & technology, 57(16): 6355–6359.
5. Werner I., Hitzfeld B. (2012) 50 Years of ecotoxicology since silent spring–a review. Ecological perspectives for science and society, 21(3): 217-224.
6. Sandip K. (2016) Ecocriticism: a study of environmental issues in literature. Journal of educational research, 6(4): 168-170.
7. Adam T., Adeline J. (2011) Climate change in literature and literary criticism. Wires climate change, 2(2): 185-200.
8. Yang Y. (2023) A brief discussion on the relationship between ecological civilization and Chinese contemporary literature. Economist, 8: 254-255.
9. Long Q. L. (2023) Earth discourse and planetary vision in Western ecocriticism. Journal of Hunan university of technology (social science edition), 28(04): 36-41.
10. Chen T. T. (2023) A study on ecological consciousness of Li Juan's prose. Guangxi normal university.
11. Zhu L. Y. (2018) Study on Xinjiang prose in the new period from the perspective of ecological aesthetic education. Shihezi University.
12. Han X. X. (2022) Ecological thought in Li Juan's prose from the perspective of ecocriticism. Hunan normal university.
13. Phillips D. (1999) Ecocriticism, literary theory, and the truth of ecology. New literary history, 30(3): 577-602.
14. Long Q. L. (2022) A brief review of ten contemporary Chinese ecological prose writers. Southwest literature, 1(1): 1-3.

Effect of Sterilization on the Change of Available Phosphorus During the Wetting and Drying Process

Jiayao Jin[a], Gang Dai[b*]

College of Chemistry and Environmental Science, Inner Mongolia Normal University, 010022 Hohhot Inner Mongolia Autonomous Region, China

Abstract. Wetting and drying are typical natural events in arid and semiarid areas. In order to explore the contribution of microorganisms to the change in soil phosphorus components during wetting and drying processes, soil incubation experiments were conducted on calcareous soil from Inner Mongolia grasslands. By comparing the change in various phosphorus fractions between sterilized and unsterilized soils, the microbial effect on the change in available phosphorus with soil moisture changes was studied. The results indicated that there was no significant difference in APi ($NaHCO_3$-extractable inorganic phosphate) content between the two treatments. Sterilization caused the content of APo ($NaHCO_3$-extractable inorganic phosphate) to significantly increase. These results suggested that soil microorganisms have a negligible influence on changes of soil available inorganic phosphorus (APi) in the present soil sample during the wetting and drying process. The change in APo may be related to the high-temperature and high-pressure sterilization processes. These results enrich our knowledge about the microbial influence on dynamics of soil available phosphorus during wetting and drying events.

1 Introduction

Phosphorus (P) is an essential nutrient for plant growth[1-3]. The content of phosphorus can constrain primary productivity in terrestrial ecosystems[4]. Phosphorus fractions in soil are complex, and the bioavailability of different phosphorus fractions is different. Water soluble phosphorus is most easily absorbed and utilized by plants, while phosphorus that binds to metal minerals is relatively stable[5]. Phosphorus in soil can be divided into inorganic phosphorus and organic phosphorus according to their existing forms. The inorganic phosphorus/organic phosphorus ratios in soil solid samples range from 0.1 to 3[6].

Intermittent precipitation and artificial irrigation can cause soil to undergo repeated wetting and drying cycles. Soil moisture undergoes drastic changes during these processes. Changes in soil properties due to wetting and drying can affect phosphorus cycle in soil. The occurance of wetting and drying events will become more frequently in the future due to the global climate change. Numerous studies have shown that wetting and drying can lead to a transient increase in soil phosphorus[7-11]. But Zhang[12] found that the change of phosphorus in soil was slight during various wetting and drying process. A study on 29 grassland soils showed that most of the water-soluble phosphorus released after rewetting of the dried soil is organic phosphorus, and the dissolved organic phosphorus directly come from microorganisms because microorganisms undergo osmotic shock and cell rupture during the wetting[13]. In another study, Wang[14] found that micomial biomass phosphorus was the cause of the increase of phosphorus after wetting and drying cycle. However, some studies suggest that microorganisms do not play a major role in it. Bunemann[15] found that physical and chemical processes might be more important for the increase in phosphorus caused by wetting and drying in a grassland soil with low phosphorus bioavailability but high microbial phosphorus content. In summary, there is still controversy over whether microorganisms are the source of increased phosphorus during the wetting and drying process.

While most study concentrated on the transient changes of soil phosphorus after wetting of the soil that is air-dried, little attention is paid on dynamics of phosphorus during drying process in soil. The soil will undergo a drying process after natural precipitation and irrigation. The change of available phosphorus during the drying process after wetting is poorly unsterstood. And it is necessary to figure out the effect of microbial during this process. Present study examined the change of readily available P during wetting and slow drying process of a soil collected from Inner Mongolia where prolonged draughts are frequently observed. A comparison was made between sterilized and unsterilized soil to elucidate the role of active microorganisms during this process.

* Corresponding author: [b]E-mail: daigang@imnu.edu.cn
[a]E-mail: 2068539609@qq.com

2 Materials and Methods

2.1 Soil sampling

We used a grassland soil that was collected from Sonid Zuoqi, Xilingol League, Inner Mongolia Autonomous Region, China (112.843694, 44.671538) on May 28, 2023. Single point sampling was used to collect 0-20 cm of surface soil. After removing plant roots, these collected soils were air dried at ambient temperature, subsequently, these soils were stored for future use. these soils are calcareous with a natural moisture content of 6.9%. The pH is 6.65; soil organic matter (SOM) is 19.94 g/kg; total nitrogen (TN) content is 1548 mg/kg; the total phosphorus (TP) content is 978 mg/kg; and the particle content<50-2000 μm reaches 85%, this soil is loamy sandy soil.

2.2 Experimental design

Fifty samples of soil that have passed through a 10 mesh sieve were weighed in 50 crucible filters (height and diameter were about 3.00 cm and 3.50 cm, respectively). Twenty-five of them were sterilized, and the others remained untreated. The sterilization treatment was conducted twice: the crucible filters containing soil were put into the autoclave for the first sterilization (121°C, 1 hour and 30 minutes), and the second sterilization was carried out at an interval of three days after the first sterilization. After two times of sterilization, the soil sample was placed for an appropriate time for drying before soil incubation.

The design of soil incubation was carried out as follows: The crucible filter containing soil samples was placed in a plastic tray containing water in order to slow down the drying rate of soil samples (the level of water in the tray was kept below the soil to avoid their direct contact). Fifteen milliliters of deionized water was added to each soil sample so that the initial water content (gravimetric water content, GWC) of soil samples was 33%. All soil samples were incubated at room temperature. During incubation, these soil samples dried naturally at ambient temperature and moisture. The soil samples were sampled at 0 (air-dried), 1, 2, 4, and 7 days after wetting to determine soil GWC, pH and various phosphorus fractions. Five replicates were set for each sampling time.

2.3 Extraction and determination of aviailable phosphorus

We used sodium bicarbonate to extract available phosphorus in soil samples. The specific operation is as follows: We used 0.25 mol · L^{-1} $NaHCO_3$ to extract available phosphorus, including available inorganic phosphorus (APi) and available organic phosphorus (APo). The weight of each soil was 5.00 g and the ratio of soil to extractant was 1:4. It should be noted that the volume of the extractant used needed to be deducted from GWC of the soil itself. The APi content in extractant was determined using Mo-Sb Anti-spectrophotometer methord[16]. Subsequently, the extractants were digested with 50 g·L^{-1} $K_2S_2O_8$ in an autoclave (120°C, 30 min, 1.1kPa) to determine the total phosphorus (TP) content. The TP content deducted the APi content was the content of available organic phosphorus (APo). Available inorganic phosphorus (mg·kg^{-1}) was caculated as:

$$APi = C \times V \times a / W \quad (1)$$

where C (mg · L^{-1}) is the phosphate concentration obtained from the standard curve; V (mL) is the volume of chromogenic solution; a is total volume of leachate/ the volume of extract leachate; W (g) is soil quality.

Available organic phosphorus (mg·kg^{-1}) was caculated as,

$$APo = TP - APi \quad (2)$$

The determation of microbial biomass phosphorus (MBP) used hexanol fumigation method. Briefly, weighed 10 samples (2.00 g per sample) at each sampling day. Five of them were fumigated: Soil samples were placed in 50 mL centrifuge tube, respectively. Ultrapure water (20 mL), 2-hexanol (2 mL), and an anion exchange resin strip (AEM, VWR/BDH, 1×2cm) that preloaded with $NaHCO_3$ were added in the centrifugation tube, respectively. Then these centrifuge tubes were put in a constant temperature shaker to shake for 16 hours. The AEM was taken out and eluted with 0.25 mol·L^{-1} H_2SO_4 to exchange the phosphate. The content of phosphate in the eluate was determined using Mo-Sb Anti-spectrophotometer methord. The other samples were also extracted with ultrapure water but did not fumigate with hexanol. The phosphorus content in fumigated sample should subtracting the phosphorus content in nonfumigated sample and this difference was divided by 0.4 to calculate the MBP[17].The calculation of phosphate content in fumigation and unfumigation is shown in formula (1), MBP (mg·kg^{-1}) was cacluated as,

$$MBP = (Pf - Pu)/0.4 \quad (3)$$

where Pf is the phosphate in fumigation treatment; Pu is the phosphate in unfumigation treatment.

2.4 Determination of soil properties

GWC was obtained by weighing. 5.00 g soil was extracted with 12.5 mL deionized water to determine soil pH. SOM, TN, TP, and mechanical composition were determined by the Hangzhou Yanqu Information Technology Co., Ltd.

2.5 Statistics

The data in this study were analysed by Excel 2021 and SPSS Statistics 26 and drawn by Origin 2022.

3 Results and discussion

3.1 Effect of sterilization on soil GWC and pH

GWC decreased gradually in both conditions throughout the entire incubation process and remained at around 5% on day 7 (Fig. 1.).

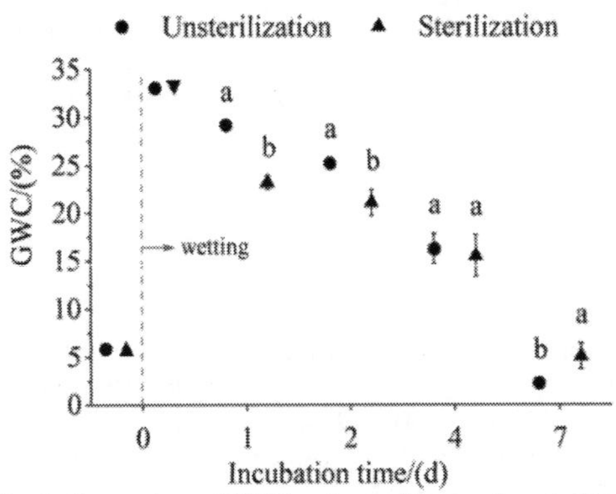

Fig. 1. Comparison of GWC under sterilized and unsterilized conditions.

The soil pH increased after 2 days of wetting and decreased in the remaining days in both treatments (Fig. 2.). The maximum pH value occurred at 1 day and 2 days of incubation in unsterilized and sterilized treatments, respectively. Overall, sterilization treatment did not have a significant impact on pH; but the pH in unsterilized treatment was smaller than that in sterilized treatment on the fourth day of incubation. The pH in sterilization treatent and unsterilization treatment were 6.96 and 6.74, respectively, on the fourth day of incubation.

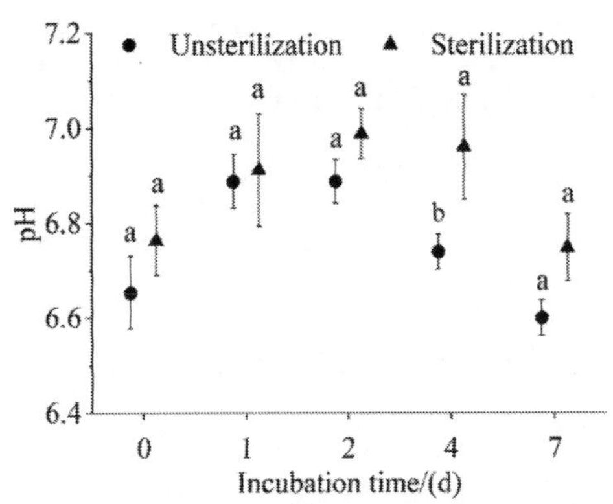

Fig. 2. Comparison of pH under sterilized and unsterilized conditions.

3.2 Effect of sterilization on available inorganic and organic phosphorus

The sodium bicarbonate extracted fraction has the highest bioavailability and is considered available phosphorus in soil. The average values of APi in air-dried soil were 2.84 and 2.64 mg · kg^{-1} in unsterilization treatment and sterilization treatment, respectively, the APi content in these two treatments had no significant difference (Fig. 3). Sterilization also did not bring any significant difference in APi content between sterilized and unsterilized soil in other sampling days. The APi in two treatments still increased on the first and second days of incubation compared with air-dried soil. The mean APi content increased 0.32 and 0.46 mg · kg^{-1} in unsterilization treatment and sterilization treatment, respectively. Both the unsterilization treatment and sterilizatin treatment decreased after two days of incubation. Wetting of the air-dried soil still induced a pulse in APi under sterilization conditions, indicating that microorganisms are not the main cause of the increase in soil available phosphorus content. Butterfly[18] also had the same conclusion. A long-term incubation also indicated that weting and drying cycle can decrease biologically-mediated phosphorus cycling and increase the change of inorganic phosphorus[19].

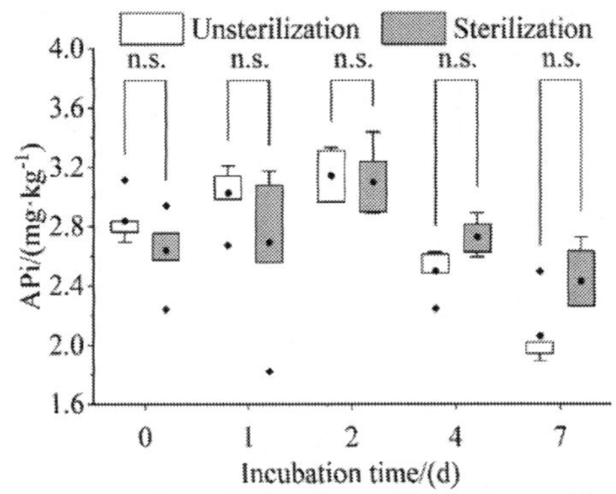

Fig. 3. Comparison of APi under sterilized and unsterilized conditions.

APo is also a highly bioavailable form of phosphorus. Sterilization had a significant impact on APo.The APo content in the unsterilized treatment was extremely low, to the extent that some sampling days were under the detection limit. However, the APo content reached a maximum of 5.91 mg·kg^{-1} in the sterilization treatment on day 7. (Fig. 4.) The APo content in sterilization treatment was 32-231 times than that in unsterilization treatment during the incubation. The increase in APo content may not originate from microbial cell lysis during sterilization. Because the content of water-soluble inorganic phosphorus is comparable to the organic phosphorus in microbial cells[17], if sterilization caused a significant release of microbial organic phosphorus, a similar increase in water-soluble inorganic phosphorus would be observed. Therefore, a likely reason for the increase in APo content is that the high-temperature and high-pressure sterilization process may increase the solubility of inherent organic phosphorus in the soil. Sterilization can alter soil organic matter chemistry[20]. Autoclaving may alter the structure of soil components and release soil organic matter that can not to be accessed before. Berns[21] found that autoclaving induced dissolved organic matter (DOM) increased compared with untreated soil. The increase of DOM may directly induce the high APo content in sterilization treatment in our study.

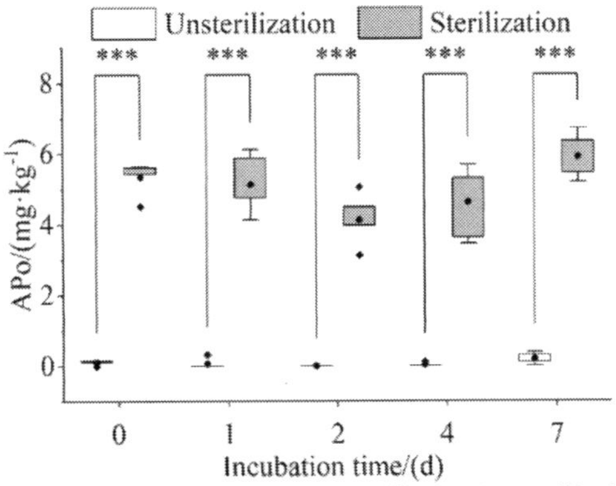

Fig. 4. Comparison of APo under sterilized and unsterilized conditions.

The content of MBP did not have significant difference in unsterifization treatment and sterilization treatment (Fig. 5). However, the mean MBP in sterilization soil was lower than that in unsterilized soil in all sampling days. Sterilization may not have caused all microbial cells to rupture, and there were still some microorganisms that maintained their cell structure. MBP was extracted from this portion of unruptured cells after fumigation with hexanol.

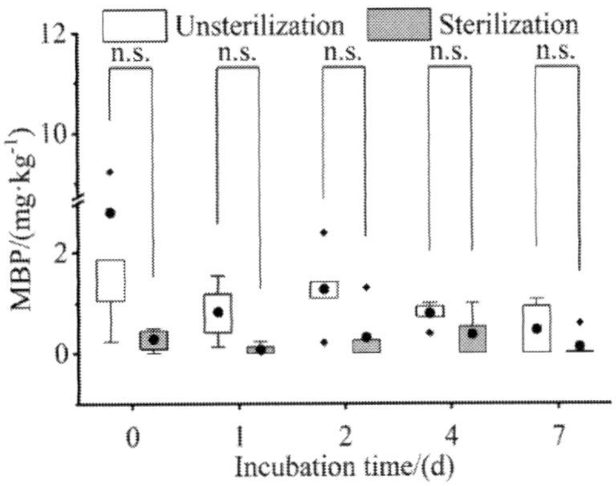

Fig. 5. Comparison of MBP under sterilized and unsterilized conditions.

4 Conclusion

Sterilization did not have a significant influence on APi content compared with the unsterilization treatment during wetting and the following slow drying process, indicating that soil microorganisms have a negligible effect on changes in APi in the present soil sample during the wetting and drying process. The change in APo may be related to the high-temperature and high-pressure sterilization processes. Sterilization had no significant effect on MBP. These results enrich our knowledge about the microbial influencec on dynamics of soil available phosphorus during wetting and drying events. The incubation time is relatively short in this study, it is necessary to carry out incubation for several months to further unsterstand the role of microorganisms in soil phosphorus cycle during wetting and drying process.

References

1. Ahmad M., Ahmad M., El-Naggar A.H., et al. (2018) Aging effects of organic and inorganic fertilizers on phosphorus fractionation in a calcareous sandy loam soil. Pedosphere, 28: 873-883.
2. De-Bashan L.E., Magallon-Servin P., Lopez B.R., et al. (2022) Biological activities affect the dynamic of P in dryland soils. Biol. Fert. Soils, 58: 105-119.
3. Hou E., Luo Y., Kuang Y., et al. (2020) Global meta-analysis shows pervasive phosphorus limitation of aboveground plant production in natural terrestrial ecosystems. Nat Commun, 11: 637.
4. Hosokawa N., Ozawa Y., Hayakawa A., et al. (2022) Effect of active aluminum on soil phosphorus forms in a forested watershed in Akita, Japan. Geoderma, 416: 115800.
5. He X., Augusto L., Goll D.S., et al. (2023) Global patterns and drivers of phosphorus fractions in natural soils. Biogeosciences, 20: 4147-4163.
6. Kruse J., Abraham M., Amelung W., et al. (2015) Innovative methods in soil phosphorus research: A review. J. Plant Nutr. Soil Sc., 178: 43-88.
7. Chen H., Jarosch K.A., Mészáros É., et al. (2021) Repeated drying and rewetting differently affect abiotic and biotic soil phosphorus (P) dynamics in a sandy soil: A ^{33}P soil incubation study. Soil Biol. Biochem., 153: 108079.
8. Marschner P., Zheng B. (2022) Direction and magnitude of the change in water content between two periods influence soil respiration, microbial biomass and nutrient availability which can be modified by intermittent air-drying. Soil Bio. Biochem., 166: 108559.
9. Khan S.U., Hooda P.S., Blackwell M.S.A., et al. (2022) Effects of drying and simulated flooding on soil phosphorus dynamics from two contrasting UK grassland soils. Eur. J. Soil Sci., 73.
10. Lisboa M.S., Schneider R.L., Sullivan P.J., et al. (2020) Drought and post-drought rain effect on stream phosphorus and other nutrient losses in the Northeastern USA. J. Hydrol-Reg Stud., 28: 100672.
11. Gao D., Bai E., Li M., et al. (2020) Responses of soil nitrogen and phosphorus cycling to drying and rewetting cycles: A meta-analysis. Soil Biol. Biochem., 148: 107896.
12. Zhang Z., Wang D., Li M. (2022) Soil respiration, aggregate stability and nutrient availability affected by drying duration and drying-rewetting frequency. Geoderma, 413: 115743.
13. Turner B.L., Haygarth P.M. (2001) Phosphorus solubilization in rewetted soils. Nature, 411: 258.
14. Wang C., Guo J., Zhang W., et al. (2021) Drying-rewetting changes soil phosphorus status and enzymatically hydrolysable organic phosphorus fractions in the water-level fluctuation zone of Three Gorges reservoir. Catena, 204: 105416.
15. Bünemann E.K., Keller B., Hoop D., et al. (2013)

Increased availability of phosphorus after drying and rewetting of a grassland soil: processes and plant use. Plant Soil, 370: 511-526.

16. Murphy J., Riley J.P. (1962) A modified single solution method for the determination of phosphate in natural waters. Anal. Chim. Acta, 27: 31-36.
17. Brookes P.C., Powlson D.S., Jenkinson D.S. (1982) Measurement of microbial biomass phosphorus in soil. Soil Biol. Biochem., 14: 319-329.
18. Butterly C.R., Bünemann E.K., Mcneill A.M., et al. (2009) Carbon pulses but not phosphorus pulses are related to decreases in microbial biomass during repeated drying and rewetting of soils. Soil Biol. Biochem., 41: 1406-1416.
19. Helfenstein J., Frossard E., Pistocchi C., et al. (2021) Soil phosphorus exchange as affected by drying-rewetting of three soils from a Hawaiian climatic gradient. Front. Soil Sci., 1: 1-23.
20. Kelsey J.W., Slizovskiy I.B., Peters R.D., et al. (2010) Sterilization affects soil organic matter chemistry and bioaccumulation of spiked p,p'-DDE and anthracene by earthworms. Environ. Pollut., 158: 2251-2257.
21. Berns A.E., Philipp H., Narres H.D., et al. (2008) Effect of gamma-sterilization and autoclaving on soil organic matter structure as studied by solid state NMR, UV and fluorescence spectroscopy. Eur. J. Soil Sci., 59: 540-550.

Molybdenum and its alloys in advanced engine applications: from material selection to surface optimization

Zhiyu Kuang*

Nanchang University, Nanchang, Jiangxi, 330031, China

Abstract. This article explores the potential of molybdenum and its alloys for use in advanced aerospace engines, focusing on their high-temperature capabilities and corrosion resistance. It highlights the importance of enhancing oxidation resistance through methods such as alloying and the application of specialized coatings. The paper provides a detailed overview of high-temperature alloys, especially molybdenum-based alloys, discussing the benefits of various alloying elements and the types of oxidation-resistant coatings like aluminide and silicide coatings that protect against high-temperature degradation. The review of coating technologies includes modern methods like slurry sintering, atmospheric plasma spraying, and chemical vapor deposition, which improve the performance and longevity of these coatings. The paper concludes by identifying future research directions, emphasizing the need to boost the high-temperature oxidation resistance of coatings to meet the evolving demands of aerospace engine technologies.

1.Introduction

The advancement of aerospace technology is closely linked to the capabilities of aviation engines, which are pivotal in defining the upper limits of various types of aircraft. As the aerospace field evolves, there is an increasing demand for engines that can withstand higher temperatures and exhibit superior specific strength. High-temperature alloys, known for their high melting points and excellent high-temperature strength, are extensively used in the aerospace and nuclear-powered industries attributable to their ability to form stable, protective oxide films under the high-temperature oxidative conditions found in engines[1-3].

Refractory metal molybdenum, with its exceptionally high melting point, robust high-temperature strength, corrosion resistance, and lower thermal expansion, offers significant compositional design flexibility due to its substantial solubility in various alloys. The use of alloying and coating techniques to enhance its oxidation resistance at medium to high temperature positions molybdenum as a potential replacement for traditional high-temperature alloys such as nickel, cobalt, and titanium-based alloys in the development of advanced aircraft.

This paper provides an overview of high-temperature materials used in aeroplane engines, highlighting the advantages of molybdenum-based alloys. It reviews techniques to enhance the oxidation resistance of molybdenum-based alloys, focusing on the types of oxidation-resistant coatings, their fabrication technologies, and their effectiveness. Furthermore, it analyses the current state and future trends of oxidation resistance technology for molybdenum-based alloys.

2.Overview of high-temperature alloys

High-temperature alloys are metal materials designed specifically for operation in high-temperature environments, extensively used in industries such as aviation, aerospace, power plant, and chemicals. These alloys are primarily based on nickel, cobalt, or iron, including TiAl alloys and refractory metals and their alloys. The addition of other alloying elements enhances their strength, creep resistance, and corrosion resistance at high-temperatures. They are commonly employed in critical components such as turbine blades and combustion chambers of gas turbines, where they must withstand high temperatures and significant mechanical loads while resisting severe corrosion.

2.1 Nickel-based high-temperature alloys

Nickel-based high-temperature alloys offer outstanding high-temperature oxidation and corrosion resistance, along with high-temperature strength and creep strength. As nickel-based alloys can dissolve a variety of alloying elements, alloying helps maintain structural stability and form intermetallic compounds as strengthening phases. For example, the addition of Al and Cr enhances strength and high-temperature oxidation and corrosion resistance, respectively, making these alloys widely used in the aerospace sector. However, their surface chemistry in oxidative environments is unstable, and the service temperature of nickel-based single crystal high-temperature alloys used in engine blades has reached

* Corresponding author: ytx641608175@vip.qq.com

1150°C, close to 85% of their melting point, making further increases in working temperature challenging [4,5] .

2.2 Cobalt-based high-temperature alloys

Cobalt-based high-temperature alloys maintain good wear resistance and corrosion resistance above 1000°C. Precipitation-hardened cobalt-based high-temperature alloys, compared to traditional solid solution and carbide strengthened alloys, show better strengthening effects. New ternary Co-Al-W alloys with an L12 structure offer better mechanical properties and higher melting points than traditional nickel-based high-temperature alloys, but their γ' phase has lower high-temperature stability, larger alloy density, prone to precipitation of harmful phases, and poorer high-temperature mechanical properties[6,7].

2.3 High-temperature titanium alloys

Titanium alloys with service temperatures above 400°C are known as high-temperature titanium alloys. Compared to ordinary titanium alloys, they possess higher high-temperature strength and creep resistance, and are extensively used as structural materials in the aviation sector. The stable service temperature of high-temperature Ti-Al-based materials reaches 700-900°C, but their high-temperature oxidation resistance remains poor, rapidly oxidizing above the "thermal barrier" temperature of 600°C, thus limiting their development in aviation engine applications [8-10]

2.4 Molybdenum and molybdenum alloys

Molybdenum, with a high melting point (2620°C), moderate density (10.23g/cm^3), outstanding high-temperature mechanical properties, and low coefficient of thermal expansion, maintains outstanding mechanical properties at 1500°C. The addition of trace elements like Ti, Zr, and C in titanium-zirconium molybdenum alloys significantly enhances their high-temperature strength, hardness, and mechanical properties. High-temperature molybdenum, developed by adding Al, Si, K, can produce "potassium bubbles" that strengthen the matrix at high temperatures[11]. The adding of rare earth elements or rare earth oxides for dispersion strengthening improves its high-temperature performance, creep resistance, and oxidation resistance, making molybdenum-based high-temperature alloys promising in the aerospace and aeronautics fields.

3.Oxidation resistance of molybdenum and its alloys

Molybdenum undergoes oxidation reactions at high temperatures as indicated by equations (1) and (2). It begins to oxidize at about 300°C, forming a green oxide film on the surface. As the temperature rises to 600°C, a dense dark green oxide layer form. Between 600°C and 700°C, some oxides start to volatilize; with further temperature increases, molybdenum oxidizes extensively, forming highly volatile MoO_3. This volatilization intensifies with rising temperatures, leading to a rapid reduction in the mass of molybdenum alloys, which consequently results in failure[12].

$$Mo + O_2 = MoO_2(300\sim475°C) \quad (1)$$

$$2MoO_2 + O_2 = 2MoO_3(475\sim875°C) \quad (2)$$

3.1 Alloying

Alloying involves combining two or more metal elements to form a new material with improved physical and chemical properties. Alloying can somewhat advance the high-temperature oxidation resistance of molybdenum, and applying oxidation-resistant coatings on molybdenum and its alloys is also considered an effective measure.

3.1.1 Mo-Si-B alloy.

Molybdenum-silicon alloys, with their high melting points, can be used at high temperatures. However, their weak high-temperature creep resistance and poor room-temperature fracture toughness limit their application as high-temperature structural materials. Additionally, Mo-Si alloys undergo severe oxidation through the Pesting phenomenon between 600°C and 800°C. Adding a small amount of boron can mitigate mid-temperature Pesting, thereby enhancing high-temperature oxidation resistance[13,14].

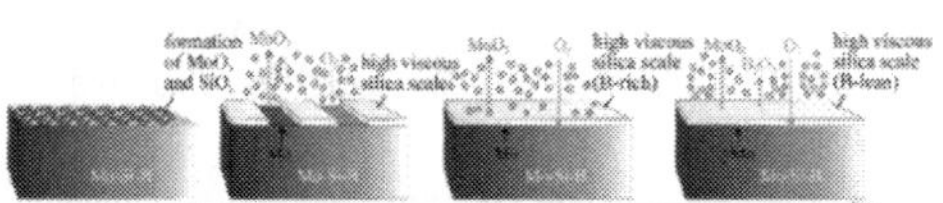

Figure 1 Schematic oxidation mechanism of Mo–Si–B alloys at elevated temperatures[15]

In Mo-Si-B alloys comprising the three phases α-Mo, Mo_3Si, and Mo_5SiB_2, α-Mo oxidizes to MoO_3 at 350°C. MoO_3 begins to volatilize at 500°C and rapidly volatilizes between 700°C and 800°C, showing poor oxidation resistance; molybdenum silicide like Mo_3Si can form SiO_2 glass phases, which, while somewhat oxidation-resistant, generally have limited flowability and tend to crack and form pores. Figure 1 shows the oxidation mechanism of Mo-Si-B alloy, and it can be seen that B_2O_3 can play a role in reducing the viscosity of the oxide layer. Because of this, the Mo_5SiB_2 intermetallic can better cover the alloy surface by dint of a borosilicate glass phase with lower viscosity than SiO_2, which provides excellent high-temperature oxidation resistance. Increasing the boron content in Mo-Si-B alloys enhances the volume fraction of Mo_5SiB_2, reducing the α-Mo and Mo_3Si phases, with the best oxidation resistance observed in the Mo-12Si-17B alloy, demonstrating good high-temperature oxidation resistance between 1000°C and 1300°C[16].

Although alloying can improve the oxidation resistance of molybdenum to some extent, the reduction in mass loss of molybdenum silicon boron alloys is minimal, and the enhancement in high-temperature oxidation resistance is limited. Molybdenum's alloying capacity is not high, and adding slightly more alloying elements can reduce the processability of molybdenum alloys.

Therefore, to further enhance the stability and oxidation resistance of molybdenum and its alloys under high-temperature conditions, it is still necessary to protect the molybdenum matrix through the application of coatings.

3.2 Oxidation-resistant coatings

Oxidation-resistant coatings can enhance the high-temperature oxidation resistance of molybdenum and its alloys while maintaining their mechanical and processing properties. Currently, the types of high-temperature oxidation-resistant coatings primarily include the following[17]: (1) alloy coatings that can form a dense oxide layer on the substrate surface; (2) intermetallic compounds that can form a dense oxide layer on the substrate surface; (3) intermetallic compounds capable of forming a dense, continuous glassy oxide film; (4) inert oxides that do not react with the substrate material; (5) metals or alloys that form volatile oxides which react very slowly or not at all with the substrate material. The main types of high-temperature oxidation-resistant coatings used on molybdenum and its alloys include aluminide coatings, silicide coatings, and other types of high-temperature oxidation-resistant coatings.

3.2.1 Aluminide coatings.

Aluminide coatings are a type of thermal diffusion coating that reacts with the substrate material to produce a dense alumina ceramic layer, effectively inhibiting oxygen diffusion. These coatings are cost-effective and simple to produce, making them primarily used under static load conditions for oxidation resistance. Chakraborty et al. [18]used a pack cementation technique to prepare uniform and adherent aluminide coatings on the surface of TZM alloy. Active aluminium diffused into the substrate forming several aluminides (such as $Al_{17}Mo_4$, $Al_{12}Mo$, Al_7Mo_4, Al_5Mo, $Al_2(MoO_4)_3$), and the coatings were found to possess good oxidation resistance and wear resistance. However, at temperatures exceeding 1400°C, the performance of aluminide coatings deteriorates rapidly, and their mechanical properties significantly decline. Additionally, under stress conditions, these coatings are more susceptible to damage and even detachment.

3.2.2 Silicide coatings.

Silicide coatings are high-melting, moderate-density coatings with good thermal stability, usable up to 1700°C. The primary mechanism involves the formation of a dense, continuous SiO_2 glass phase oxide film on the surface at high temperatures, which has some fluidity. This fluidity can fill cracks that develop due to high temperatures, endowing the coating with self-healing properties.

Single-component coatings are simpler in composition and structure. Alam et al. [19] prepared a single $MoSi_2$ coating on a molybdenum substrate, which after 13 thermal cycles at 1100°C, showed extensive rapid oxidation. After two thermal cycles at 1500°C for 1 hour, the coating detached from the surface, resulting in a mass loss of 2.9mg/cm^2. The mismatch in thermal expansion coefficients between the coating and the substrate led to through-coating cracks, thereby reducing the oxidation resistance of the coating. Figure 2 illustrates bare and Coated Mo samples before and after cyclic oxidation at 1100 °C.

Composite coatings utilize silicide and other compounds to form coatings with complex phase compositions. Xu et al. [20]co-deposited a ternary Mo-Si-Al coating on a molybdenum substrate using the pack cementation method. After high-temperature oxidation tests at 1100°C, the Mo-Si-Al coating, which formed mullite ($3Al_2O_3{\cdot}2SiO_2$), exhibited better oxidation resistance, which acted protectively and beneficially compared to the single $MoSi_2$ coating without Al. Yoon et al. [21]prepared a $MoSi_2$-Si_3N_4 nanocomposite coating on a molybdenum substrate using a two-step deposition method. Through 1850 low-temperature cycling oxidation tests, the coating showed good cyclic oxidation resistance as Si_3N_4 oxidized preferentially over $MoSi_2$, thereby inhibiting the formation of big amounts of MoO_3. However, as the number of cycles increased, the low-temperature oxidation rate of the coating accelerated due to cracks caused by the evaporation of N_2 during oxidation.

Preparing multilayer coatings with a gradient distribution of thermal expansion coefficients on a molybdenum substrate can mitigate crack caused by the mismatch in the coefficients of thermal expansion between the coating and the substrate and thus enhance the oxidation resistance of the coating. Edward et al. [22]used CVD and MOCVD to prepare $MoSi_2$-SiO_2 coatings with uniform thickness and excellent adhesion on a molybdenum substrate. These coatings showed exceptional performance in isothermal oxidation tests. The low diffusion coefficient of O_2 in SiO_2 caused a delay in oxygen transmission, resulting in a final mass loss of only 4 mg/cm^2, significantly lower than the 20 mg/cm^2 of the single $MoSi_2$ coating. Additionally, Zhang et al. [23]used a liquid siliconizing method to prepare dense Si-$MoSi_2$ functional gradient coatings on a molybdenum substrate, consisting of a surface layer (Si-$MoSi_2$ layer), a middle layer ($MoSi_2$ layer), and a transition layer (Mo, Mo_5Si_3, and Mo_3Si layers). The coating's mass gain after oxidizing at 1600°C for 70 hours was only 0.17%, showing superior oxidation resistance compared to traditional silicide coatings.

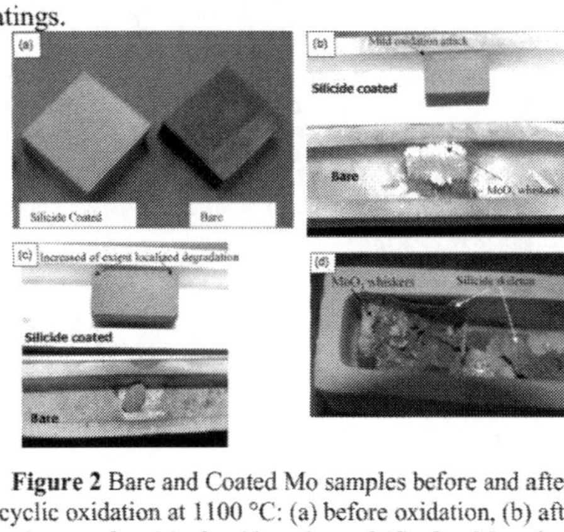

Figure 2 Bare and Coated Mo samples before and after cyclic oxidation at 1100 °C: (a) before oxidation, (b) after three cycles, (c) after 13 cycles and (d) after 35 cycles showing the coated sample in disintegrated condition.[19]

3.2.3 Other oxidation-resistant coatings.

Apart from the aforementioned aluminide and silicide coatings, the oxidation-resistant coating systems also include oxide coatings and rare earth-modified coatings.

Oxide coatings, which involve inert oxides that do not react with the substrate, primarily function by increasing the path for oxygen diffusion, thereby mechanically blocking oxidation. These coatings are often quite thick. When applied to molybdenum and its alloys, the thickness of the coating and the difference in thermal expansion coefficients between the coating and the substrate material must be carefully considered to avoid large cracks within the coating.

Rare earth-modified coatings involve incorporating rare earth elements into the coating to modify the surface. For instance, the best-performing thermal barrier coating, YSZ, is created by doping ZrO_2 with Y_2O_3 to produce YSZ. YSZ features a low thermal conductivity (2.1 $Wm^{-1}K^{-1}$), high expansion coefficient (10.7×10-6 K^{-1}), and excellent high-temperature stability, with 8wt% YSZ being the most commonly used, also noted for its excellent high-temperature mechanical properties[24].

4.Coating preparation methods

Currently, the main methods used for preparing coatings on molybdenum and its alloys include slurry sintering, thermal spraying, chemical vapor deposition, and pack cementation.

4.1 Slurry sintering

Slurry sintering involves mixing a specific ratio of raw materials, activators, and binders to create a suspension, which is then uniformly applied to the surface of the substrate material. The coated substrate is dried and sintered at specific temperatures and atmospheres. Figure 3 shows the process method of preparing the layer by the slurry method. Cai et al. [25]used slurry sintering to prepare Si-Mo coatings on TZM alloy surfaces. The coatings included layers of $MoSi_2$ and Mo_5Si_3 and exhibited a minimal mass gain of only 0.816 mg/cm^2 at 1650°C, with a protection duration of 14 hours. This method is straightforward, produces uniform coatings, and ensures good adhesion with the substrate material. However, improper selection of binders can lead to porosity during sintering.

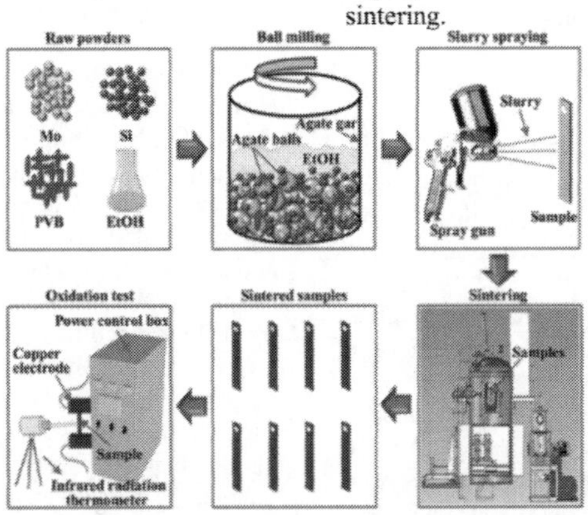

Figure 3 Diagram of coating process prepared by slurry method.25

4.2 Atmospheric plasma spraying

Atmospheric plasma spraying heats coating materials using a plasma arc and then uses a high-speed air stream to spray the molten materials onto the substrate surface, forming various functional coatings. Figure 4 illustrates the process of atmospheric plasma spraying. Huang Yan et al. [26]25 used this method to prepare $MoSi_2$ 40wt.% boride fully enveloped composite coatings. After 20 hours of oxidation at 1500°C, the mass reduction was only 1.22%. Traditional plasma spraying may lead to particle adhesion causing cracks in the coating, but the addition of boron enhanced the density of the coating. This method is efficient in film formation and can locally repair coatings, but it is prone to impurities such as oxides during spraying, which can weaken the bond between the coating and the substrate and even cause cracking.

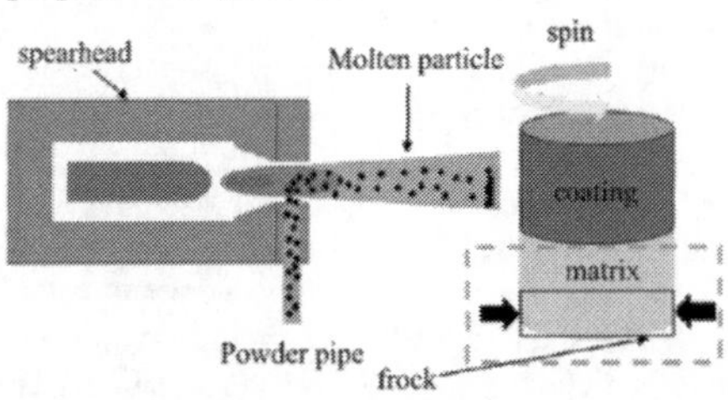

Figure 4 Schematic diagram of spraying process.[26]

4.3 Chemical vapor deposition (CVD)

Chemical vapor deposition (CVD) is a process that heats coating materials to a gaseous precursor state and diffuses them onto the substrate surface, where they undergo decomposition, synthesis, diffusion, adsorption, and desorption to form a dense coating. Figure 5 illustrates the CVD process. Yoon et al. [27]used CVD to prepare molybdenum disilicide coatings on a molybdenum substrate. The research found that the growth rate of the coating was directly proportional to the input amount of C_i/H, but once this input exceeded a critical value, the growth rate would gradually decrease, displaying a parabolic pattern. This method allows for precise adjustment of the coating's composition and structure, and produces coatings of high purity, though it has low processing efficiency and high costs.

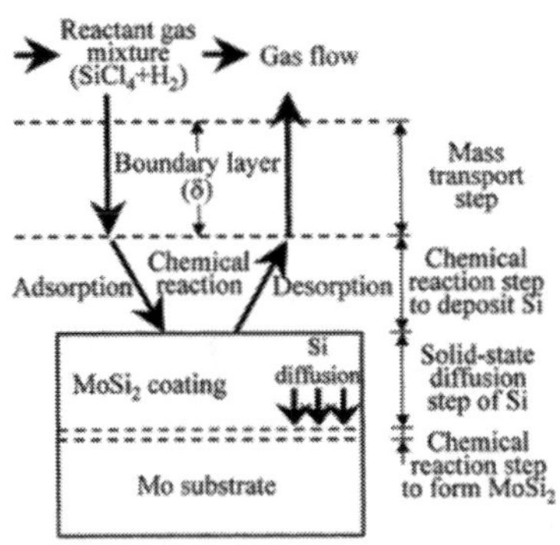

Figure 5 Schematic illustration of the CVD process for Si deposition on Mo substrate.[27]

4.4 Pack cementation

Pack cementation is a method that involves burying the substrate in a powder of raw materials and then causing diffusant elements to diffuse onto the substrate surface under high temperatures to form a coating. Figure 6 illustrates the process flow of packing cementation. Huang Yan et al. [26]used a two-step pack cementation method to prepare a $MoSi_2/SiC\text{-}Mo_2C$ composite coating on a pure molybdenum substrate. After oxidation at 1300°C for 30 hours, the coating had only a slight mass increase of 0.61 mg/cm^2, with an average oxidation life of 56.67 hours. Sun et al. [28] utilized a halide-activated pack cementation method with Al_2O_3, SiC, and SiO_2 as fillers, and NH_4F as an activator to prepare a $MoSi_2$ coating on a molybdenum substrate. The presence of SiC in the filler accelerated the deposition process and enhanced the oxidation resistance of the coating, with the best-performing coating showing only a 0.05% weight gain after 110 hours of oxidation at 1473K. Compared to other coating preparation methods, pack cementation is straightforward, produces a coating with strong adhesion to the substrate, and allows for thickness uniformity. The thickness of the diffusion layer can be adjusted by varying the duration of the pack cementation. Multi-step pack cementation can create gradient composite coatings, which closely bond the layers together, eliminating interlayer stress, and can be modified with rare earth elements to further enhance coating quality or be tailored for specific functional requirements depending on the application.

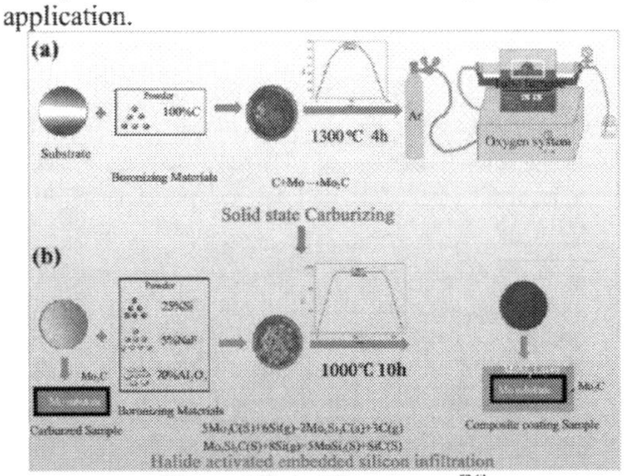

Figure 6 Embedding process.[26]

5.Conclusions

Despite considerable advancements in high-temperature oxidation-resistant coatings for molybdenum and its alloys, the continuously escalating demands of advanced engines in the aerospace sector necessitate further development of existing coating systems. To extend the service life of high-temperature oxidation-resistant coatings at higher temperatures, future research could focus on the following areas:

1)Research on silicide-based high-temperature defensive coatings. The study of silicide-based high-temperature protective coatings remains at the forefront of technological advancement. In terms of composition, it is crucial to ensure that the coatings can form stable glassy oxide films with low oxygen transmission rates during oxidation while maintaining the integrity of essential components within the coating. Structurally, attention should be given to the bond strength between the coating and the substrate.

2)Enhancing performance Above 1700°C. Silicide-based high-temperature protective coatings are primarily effective up to 1700°C. Beyond this temperature, the volatility of SiO_2 oxide films increases, making it difficult to maintain the coating's protective performance. Therefore, improving the oxidation resistance of silicide-based coatings at higher temperatures is an urgent technical challenge that needs addressing. Incorporating ultra-high-temperature ceramic materials such as ZrB_2, which produces high-melting-point ZrO_2 and low-viscosity B_2O_3, could play a crucial role in enhancing the high-temperature oxidation resistance of silicide-based coatings.

3)Behavior of multi-component Multi-layer composite protective coatings. The behavior of multi-component, multi-layer composite protective coatings in high-temperature oxidative environments is significantly more complex, involving the interdiffusion of elements such as B, Si, and O, and the formation of various compounds or phases. A deeper understanding of these complex chemical and physical processes can be achieved through combined thermodynamic and oxidation kinetic analyses. This approach will enable the development of more effective and durable high-temperature protective solutions.

References

1. Fu H. (1998) Challenge and development trends to future aero engine materials. https://qikan.cqvip.com/Qikan/Article/Detail?id=3317893&from=Qikan_Search_Index
2. Yi Z, Tan D and Tang Y, et al. (2022) Microstructure and oxidation resistance of in-situ silicide composite coating on Mo-2.5 Zr-2.5 ZrO2-xW (x= 0, 20, 30) substrate. J. Surface and Coatings Technology, 451: 129038. https://doi.org/10.1016/j.surfcoat.2022.129038.
3. Cai Z, Shen H and Liu S, et al. (2020) Review and prospect of refractory metal alloys and high temperature oxidation resistance coatings. J. The Chinese Journal of Nonferrous Metals, 30(09): 1991-2010. http://www.ysxbcn.com/paper/paperview.aspx?id=paper_321802
4. Tang Z, Guo T and Fu Y, et al. (2014) Research present situation and the development prospect of nickel-based superalloy. J. Metal World, 1: 36-40. https://xueshu.baidu.com/usercenter/paper/show?paperid=9fe8595d4be07ed34615de66c9cceb65&site=xueshu_se
5. Wang H, An Y and Li C, et al. (2011) Research progress of Ni-based superalloys. J. Materials Review, 25 (S2): 482-486. https://xueshu.baidu.com/usercenter/paper/show?paperid=ef84e33f52ddf066834ebe8208daa4dc
6. Sato J, Omori T and Oikawa K, et al. (2006) Cobalt-base high-temperature alloys. J. Science, 312(5770): 90-91. https://doi.org/10.1126/science.1121738.
7. Zhang X, Ma Q and Zhang H, et al. (2024) Research progress on composition design of novel cobalt based superalloy. J. Materials China, 43(03): 230-237. http://www.mat-china.com/oa/darticle.aspx?type=view&id=202107062
8. Liu Z, Xin S and Zhao Y. (2023) Research progress on the creep resistance of high-temperature titanium alloys: a review. J. Metals, 13(12): 1975. https://doi.org/10.3390/met13121975.
9. Zhang T, Yue K and Yu J, et al. (2023) Research overview and application of high-temperature titanium alloys and titanium-based (Ti-Al) materials for aerospace. In: The 14th CSM Steel Congress. Chongqing. 6. https://www.yongyiti.com/ziliao/516.html
10. Li X, Zhao J and Liu S, et al. (2020) Research progress of high temperature titanium alloys for aviation. In: 2020 China Foundry Congress. 109-113. https://cpfd.cnki.com.cn/Article/CPFDTOTAL-ZJXZ202011001020.htm
11. Gao J, Wang Y and Chen G, et al. (2007) Microstructure and properties of Si, Al, K-doped molybdenum wire. J. Powder Metallurgy Technology, 25(1): 7-12, 16. https://en.cnki.com.cn/Article_en/CJFDTOTAL-FMYJ200701001.htm
12. Fu M, Zhang Y and Zhang G, et al. (2023) Research progress of high temperature antioxidant modified silicide coatings of molybdenum and its alloys. J. Materials Reports, 37(03): 177-184. http://www.mater-rep.com/EN/10.11896/cldb.21030219
13. Karahan T, Ouyang G and Ray P, et al. (2017) Oxidation mechanism of W substituted Mo-Si-B alloys. J. Intermetallics, 87: 38-44. https://doi.org/10.1016/j.intermet.2017.04.005.
14. Kong Y, Wang C and Chen X, et al. (2023) Review of research progress on Mo–Si–B alloys. J. Materials, 16(15): 5495. https://doi.org/10.3390/ma16155495.
15. Gatzen C, Smokovych I and Scheffler M, et al. (2021) Oxidation-resistant environmental barrier coatings for Mo-based alloys: a review. J. Advanced Engineering Materials, 23(04): 2001016. https://doi.org/10.1002/adem.202001016.
16. Li B, Lin X and Li R, et al. (2018) High-temperature oxidation resistance of Mo-Si-B alloys with different B contents. J. Acta Metallurgica Sinica, 54(12): 1792-1800. https://en.cnki.com.cn/Article_en/CJFDTotal-JSXB201812009.htm
17. Wang Y, Wang D and Sun A, et al. (2012) The advance on the oxidation resistance coatings of molybdenum and its alloys. J. Materials Review, 26(01): 137-141. https://en.cnki.com.cn/Article_en/CJFDTOTAL-CLDB201201031.htm
18. Chakraborty S, Banerjee S and, Singh K, et al. (2008) Studies on the development of protective coating on TZM alloy and its subsequent characterization. J. Journal of Materials Processing Technology, 207(1-3): 240-247. https://doi.org/10.1016/j.jmatprotec.2007.12.075.
19. Alam M, Venkataraman B and Sarma B, et al. (2009) $MoSi_2$ coating on Mo substrate for short-term oxidation protection in air. J. Journal of alloys and compounds, 487(1-2): 335-340. https://doi.org/10.1016/j.jallcom.2009.07.141.
20. Xu J, Zhou C and Gong S, et al. (2002) Codeposition of aluminum and silicon on pure molybdenum substrate using halide activated pack cementation treatments. J. Acta Metallurgica Sinica(English Edition), 15 (2): 167-171. https://www.cnki.com.cn/Article/CJFDTotal-JSXY200202004.htm
21. Yoon J, Kim G and Han J, et al. (2005) Low-temperature cyclic oxidation behavior of $MoSi_2/Si_3N_4$ nanocomposite coating formed on Mo substrate at 773 K. J. Surface and Coatings Technology, 200(7): 2537-2546. https://doi.org/10.1016/j.surfcoat.2005.01.035.
22. Nyutu E, Kmetz M and Suib S. (2006) Formation of $MoSi_2$–SiO_2 coatings on molybdenum substrates by CVD/MOCVD. J. Surface and Coatings Technology, 200(12-13): 3980-3986. https://doi.org/10.1016/j.surfcoat.2005.02.212.
23. Zhang Y, Wang Q and Wang D, et al. (2023) Effect of reaction temperature on the molten salt electrodepo

sition of silicide coating on molybdenum substrate an d its oxidation behavior. J. Journal of The Electroche mical Society, 170(8): 082503. https://doi.org/10.114 9/1945-7111/acf076.

24. Cheng T, Wang Z and Dai S, et al. (2022) Effect of S iC whiskers on thermal cycling performance of YSZ-based sealing coating. J. Ceramics International, 48 (5): 6443-6452. https://doi.org/10.1016/j.ceramint.20 2111.188.
25. Cai Z, Liu S and Xiao L, et al. (2017) Oxidation beha viour and microstructural evolution of a slurry sintere d Si-Mo coating on Mo alloy at 1650° C. J. Surface a nd Coatings Technology, 324: 182-189. https://doi.or g/10.1016/j.surfcoat.2017.05.054.
26. Huang Y. (2023) Preparation of MoSi_2 composite c oatings and research on oxidation resistance and abla tion properties. https://cdmd.cnki.com.cn/Article/CD MD-10674-1023701172.htm.
27. Yoon J, Kim G and Byun J, et al. (2003) Effect of Cl /H input ratio on the growth rate of $MoSi_2$ coatings fo rmed by chemical vapor deposition of Si on Mo subs trates from $SiCl_4$–H_2 precursor gases. J. Surface and Coatings Technology, 172(2-3): 176-183. https://doi. org/10.1016/s0257-8972(03)00428-6.
28. Sun J, Fu Q and, Guo L, et al. (2016) Effect of filler on the oxidation protective ability of $MoSi_2$ coating f or Mo substrate by halide activated pack cementation. J. Materials & Design, 92: 602-609. https://doi.org/1 0.1016/j.matdes.201512.079.

Spatial and Temporal Variation of Satellite-Derived Chlorophyll-a in Bohai Sea over the Past 20 Years

Xue Wang [1,*], Dekui Yuan [2]

[1]Department of Mechanics, Tianjin University, Tianjin 300350, China
[2]State Key Laboratory of Hydraulic Engineering Simulation and Safety, Tianjin University, Tianjin 300350, China

Abstract. The Chlorophyll-a (Chl-a) concentration in the surface layer of the Bohai Sea from 2003 to 2022 was obtained after processing the MODIS data with the improved shortwave infrared (SWIR) atmospheric correction algorithm and the Clack three-channel method. Its spatial and temporal variation were analyzed. The results show that it has obvious interannual spatial distribution characteristics, showing a trend of being high in nearshore waters and low in offshore waters, high in the estuarine bays and low in other sea areas. The interannual Chl-a variation shows an inverted U-shape. The year-by-year seasonal average spatial characteristics are basically consistent, with high in nearshore waters and low in other waters. The Chl-a in spring and summer was higher than that in autumn and winter. The temporal trend of the four seasons was similar to the interannual trend, showing an inverted U-shaped trend of first increasing, then stabilizing and then decreasing.

1 Introduction

In recent years, marine environmental issues have been receiving much attention in line with globalisation. As an important component of the marine ecosystem, phytoplankton is an important indicator of seawater quality. Bohai Sea, the northernmost offshore sea of China, is the only semi-enclosed inland sea among the four major seas of China, also an important marine aquaculture area. it is of great significance to study the phytoplankton biomass in the Bohai Sea.

Phytoplankton are major producer in the water ecosystem, and Chl-a can reflect their biomass well. The study of Chl-a can effectively monitor the change of seawater quality. The traditional methods of observing Chl-a, station observation and ship measurement are limited by space and time, which makes it difficult to obtain long time and large-scale data. Satellite remote sensing observation can well make up for the shortcomings of traditional methods. With its long observation time and wide coverage, it has provided more and more effective data for monitoring Chl-a in recent years [1]. Scholars at home and abroad have carried out a large number of studies using satellite remote sensing data for the distribution characteristics of Chl-a in the long time series of typical sea areas. Yu Kun and Lu Dengmei studied the Chl-a of the Bohai Sea in December of each year from 2000 to 2006 using MODIS data, and the results showed that the Chl-a tended to increase during the whole period of time [2]. Dongyan Liu and Yueqi Wang investigated the distribution of Chl-a in the surface waters of the Bohai Sea and the Yellow Sea from 1997 to 2011 using the SeaWiFS and MODIS data. The spatial and temporal variations of Chl-a were analyzed, and it was pointed out that the decadal trend of Chl-a was decreasing in the offshore area and increasing in the nearshore area [3]. Xu Shiguo and Fu Yanzhao analyzed the spatial, interannual and seasonal distribution characteristics of surface Chl-a in the Bohai Sea in 2003 - 2013 by applying MODIS level 3 data, and the results showed that the highest value was distributed in the three inland bays, including Laizhou Bay, Bohai Bay and Liaodong Bay, and that except for the Laizhou Bay,it showed a yearly trend of increasing [4]. Tian Hongzheng, Liu Qinping et al. studied the distribution of Chl-a in the Bohai Sea from spatial and temporal variations angle for nearly 20 years from 1997 to 2018, and the results showed that the annual Chl-a increased significantly in 20 years and showed an increasing trend in all seasons; the increase was more pronounced from the coastline of the Luanhe River estuary to the Bohai Strait [5]. Zhai Fangguo and Wu Wenfan et al. studied the interannual variation of Chl-a in the Bohai Sea in 2002 - 2019, and pointed out that it showed an increasing trend in 2003 - 2011 and a decreasing trend in 2012 - 2018, and that these two decadal trends were more pronounced in the near-shore than in the off-shore waters [6].

Atmospheric correction is a key step in the inverse performance of Chl-a concentration using satellite remote sensing data [7]. For the Bohai Sea, which is a typical case II water body, the SWIR atmospheric correction algorithm is a more commonly used method [8]. We use the improved SWIR atmospheric correction algorithm and the Clack three-channel method to extract Chl-a concentration from the L1B data of the MODIS in

* Corresponding author: xwang19951111@tju.edu.cn

the Bohai Sea for the period of 2003 - 2022 [9], and analyse the spatial and temporal variations in the 20-year period, with the aim of contributing to the further improvement of the Bohai Sea's environmental problems.

2 Data and methodology

2.1 Study area

The Bohai Sea (37°7′~40°59′N, 117°35′~122°16′E) is surrounded by land on three sides, with Liaodong Bay in the north neighbouring Liaoning, Bohai Bay in the west connecting Hebei and Tianjin, and Laizhou Bay in the south connecting Shandong, around the central basin of the Bohai Sea, and the eastern part connecting to the Yellow Sea through the Bohai Strait (Fig.1). It is China's largest inland sea. It covers an area of about 80,000 square kilometers , with an average depth of about 18 meters, making it an important marine aquaculture area. Along the coast, there are more than 40 rivers of various sizes, such as the Yellow River, Liaohe River, Haihe River, Luanhe River, etc., converging into it, and there are many harbors. The Bohai Sea is an important part of Bohai Economic Circle. The convergence of rivers and the increase of human activities make most of the Bohai Sea water body turbidity remains high. The Bohai Sea is a typical case II water body.

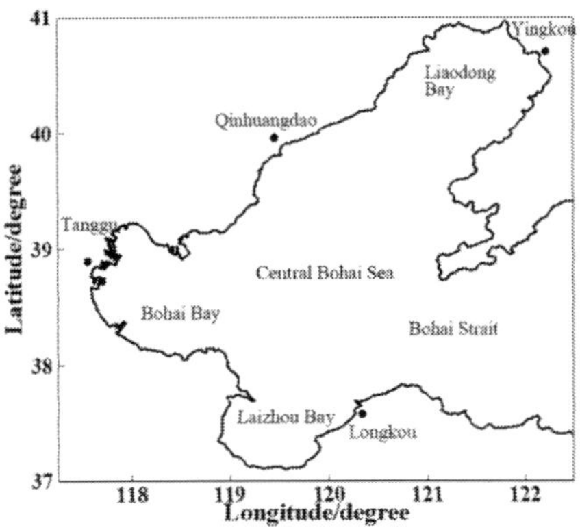

Fig. 1. Schematic of the study area.

2.2 Data source

The data were downloaded from the MODIS sensor on board the Terra satellite of the National Aeronautics and Space Administration (NASA) of the United States, with data type L1B, a temporal resolution of 1 d and a spatial resolution of 1 KM.

The wind and barometric field data used in the atmospheric correction step were downloaded from the National Centers for Environmental Prediction (NCEP) with a temporal resolution of 6 h and a spatial resolution of 0.312°*0.312°.

The temporal range of the data is from 1 January 2003 to 31 December 2022, and the spatial range is to cover the whole Bohai Sea area.

2.3 The improved SWIR atmospheric correction algorithm

The total radiance received by satellite radiometers in case II waters is the sum of the radiance generated during atmospheric transport, the water-leaving radiance and the reflection from the white crown of the waves:

$$L_t(\lambda_i) = L_{path}(\lambda_i) + t(\lambda_i)L_w(\lambda_i) + t(\lambda_i)L_{wc}(\lambda_i) \quad (1)$$

where, i is the sensor's i channel, λ_i is the corresponding wavelength of the i channel; $L_{path}(\lambda_i)$ is the radiance generated during the transmission of sea surface radiation in the atmosphere; $L_w(\lambda_i)$ is water-leaving radiance; $L_{wc}(\lambda_i)$ is the reflection from the white crown of the waves; $t(\lambda_i)$ is the atmospheric transmittance of water to the sensor.

Reflections from the white crowns of the waves can be calculated by a relatively precise empirical formula. The radiance during atmospheric transport includes atmospheric Rayleigh scattering and atmospheric aerosol scattering; Rayleigh scattering has an accurate method for calculation, and aerosol scattering is the computational difficulty in obtaining the water-leaving radiance. Using the blackbody assumption in the near-infrared band in clean water, the water-leaving radiance is approximated to be 0 [10], the following extrapolation relation is established:

$$\varepsilon(\lambda_i, \lambda_j) = \frac{L_a(\lambda_i)F_0'(\lambda_j)}{L_a(\lambda_j)F_0'(\lambda_i)} \quad (2)$$

$$\varepsilon(\lambda_i, \lambda_j) = \exp\left[c(\lambda_i - \lambda_j)\right] \quad (3)$$

where, λ_j is the reference band, λ_i is the extrapolated band; $\varepsilon(\lambda_i, \lambda_j)$ is the atmospheric correction factor of the i band, c is constant, and is used to extrapolate the atmospheric correction factor of other bands; $F_0(\lambda_i)$, $F_0(\lambda_j)$ is the luminous flux at wavelength λ_i , λ_j of sunlight incident from outside the atmosphere. $F_0'(\lambda_i)$, $F_0'(\lambda_j)$ is the value of $F_0(\lambda_i)$, $F_0(\lambda_j)$ after it has been attenuated twice through the atmosphere.

For MODIS, the reference bands i , j are fixed, and the aerosol scattering corresponding to i , j is used to calculate the constant c by using (2), (3). Further fixe j , the atmospheric correction factors $\varepsilon(\lambda_i, \lambda_j)$ for different bands are calculated by using (3), and then the water-leaving radiance is obtained by using (2), (1). This is the SWIR atmospheric correction algorithm [11].

In view of the fact that the atmospheric correction factor is not a strictly (logarithmic) linear function between the extrapolated and reference bands, segmentation is performed and 869 nm is used as the cut-off point:

$$\lambda_i > 869nm \ , \ \varepsilon(\lambda_i, \lambda_j) = \exp[c_1(\lambda_j - \lambda_i)] \quad (4)$$

$$\lambda_i < 869nm \ , \ \varepsilon(\lambda_i, \lambda_j) = \exp[c_2(\lambda_i)(\lambda_j - \lambda_i)] \quad (5)$$

where c_1 is a constant; $c_2(\lambda_i)$ is a variable that varies with wavelength. The rest of the methods are the same as the SWIR correction algorithm. This is an improved SWIR atmospheric correction algorithm.

Steps such as preprocessing MODIS data, calculating by the improved SWIR atmospheric correction algorithm, and then Chl-a concentration extraction by the Clack three-channel method, which is suitable for the Bohai Sea region [12], were used to obtain the daily data of Chl-a concentration over the past 20 years (2003-2022). The daily data were then used to calculate the average seasonal and annual values. Winter is December, January and February, spring is March, April and May, summer is June, July and August, and autumn is September, October and November.

3 Results and discussion

3.1 Interannual spatial and temporal variability of Chl-a

3.1.1 Characteristics of interannual spatial distribution

From 2003 to 2022, the Bohai's Chl-a has obvious interannual spatial distribution characteristics, showing a trend of gradual decrease from the nearshore sea area to the central sea area (Fig.2). The value of the nearshore sea area is higher than 5 mg/m^3, and the central area of the Bohai Sea is obviously lower. The Bohai Strait, which is bordered by the Yellow Sea, is the lowest. Larger values were concentrated in the Liaodong Bay, Bohai Bay and Laizhou Bay near the mouths of the rivers and ports, and significantly high values were found near the ports of Dalian, Yingkou, Tangshan, Caofeidian, Tianjin and Huanghua, while the mouths of the Fuzhou River, Daqing River, Liaohe River, Daling River, Haihe River, Zhangweixin River, Tuhai River, Yellow River and Xiaoqing River also showed significantly high values, with the maximum value occurring near the port of Huanghua port.

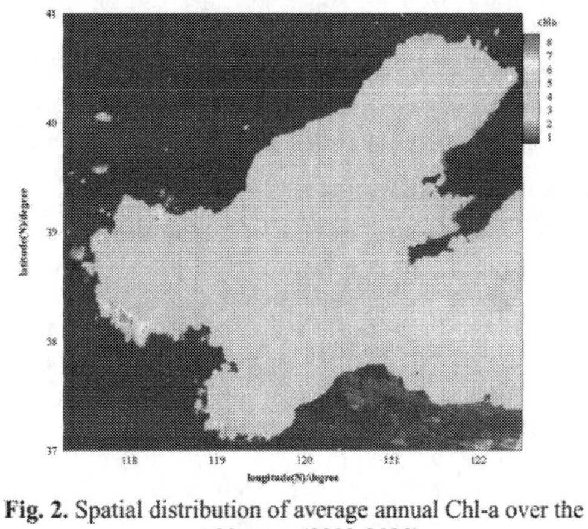

Fig. 2. Spatial distribution of average annual Chl-a over the past 20 years (2003-2022)

The spatial distribution of Chl-a is basically consistent from year to year (Fig.3), showing the distribution pattern of high in the nearshore sea area and low in the central sea area. Chl-a in the near-shore sea area changes significantly from 2003 to 2022. In particular, the high values of Chl-a along the coasts of Liaodong Bay, Bohai Bay and Laizhou Bay gradually increased from 2003 to 2007, stabilized at high values from 2007 to 2016, and gradually decreased from 2016 to 2022.

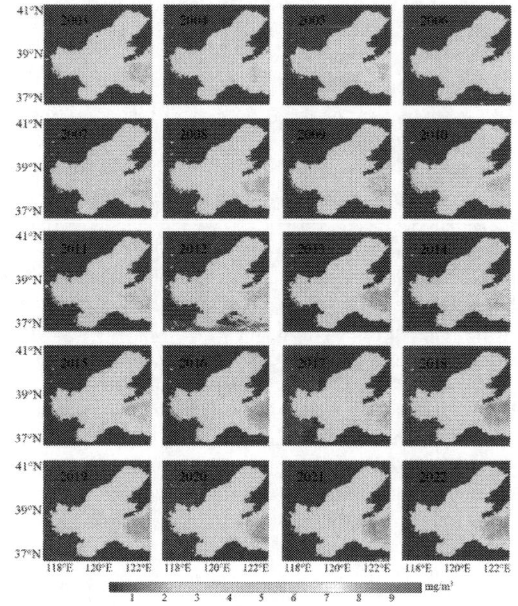

Fig. 3. Spatial variation of Chl-a over the past 20 years (2003-2022)

3.1.2 Characteristics of interannual temporal distribution

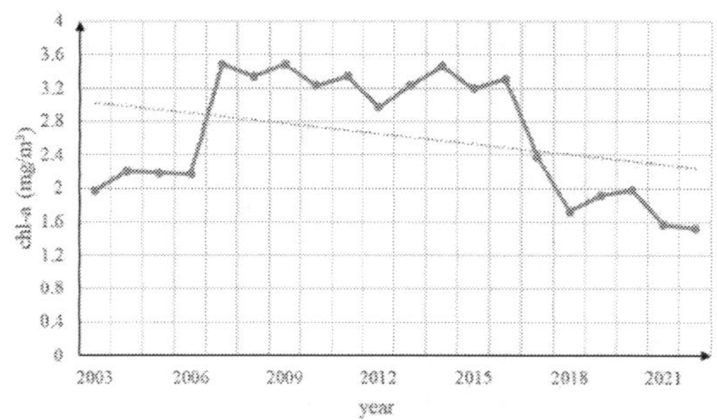

Fig. 4. Temporal variation of Chl-a over the past 20 years (2003-2022)

Temporally, the interannual Chl-a from 2003 to 2022 (Fig.4) showed a trend of first increasing, then stabilising and then decreasing, with the overall change showing an inverted U-shape. 2003-2006 showed a fluctuating increasing trend; 2007 increased steeply, and 2007-2016 fluctuated slightly but was basically stable and maintained above 3 mg/m^3. The period from 2016 to 2022 showed a fluctuating decreasing trend. The maximum value is 3.48 mg/m^3 in 2007 and the minimum value is 1.53 mg/m^3 in 2022.

3.2 Seasonal spatial and temporal variability of Chl-a

3.2.1 Characteristics of seasonal spatial distribution

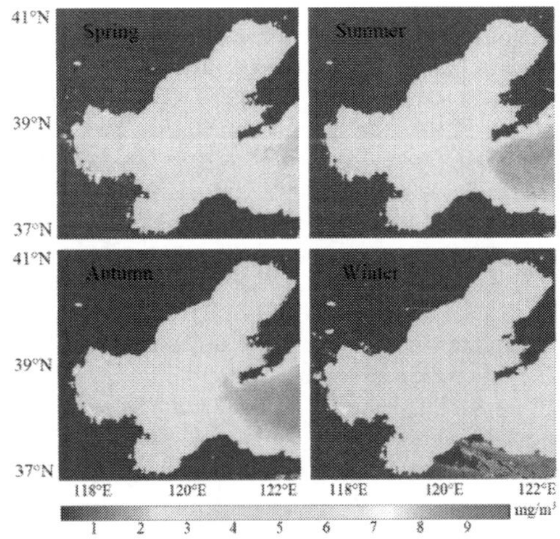

Fig. 5. Spatial distribution of average seasonal Chl-a over the past 20 years (2003-2022)

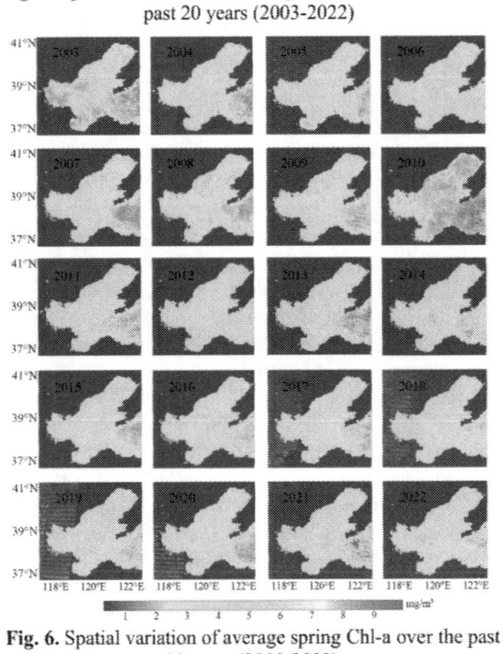

Fig. 6. Spatial variation of average spring Chl-a over the past 20 years (2003-2022)

The spatial characteristics of Chl-a in the four seasons from 2003 to 2022 (Fig.5) were basically the same. The value in the nearshore sea area is significantly higher than that in the central sea area. It is the lowest in the Bohai Strait, and in Liaodong Bay, Bohai Bay and Laizhou Bay is higher, with significantly high values occurring in the coastal harbors and river inlets. The value tends to peak in spring,surpassing those in autumn and winter,and reaches its nadir during autumn. Especially, the nearshore area of Bohai Bay showed obvious performance, which coincided with the time period of red tide in the nearshore area in spring and summer.

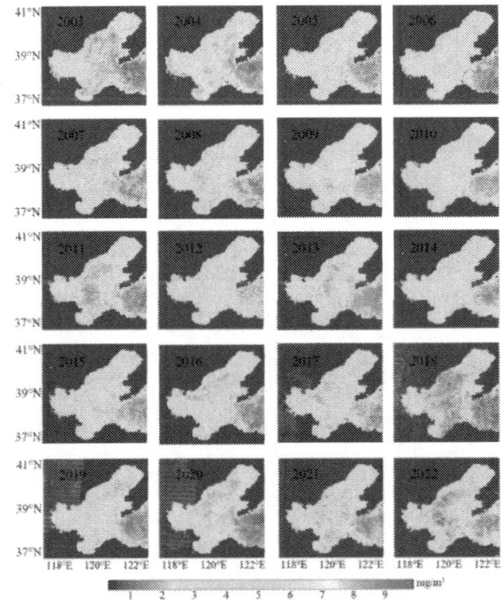

Fig. 7. Spatial variation of average summer Chl-a over the past 20 years (2003-2022)

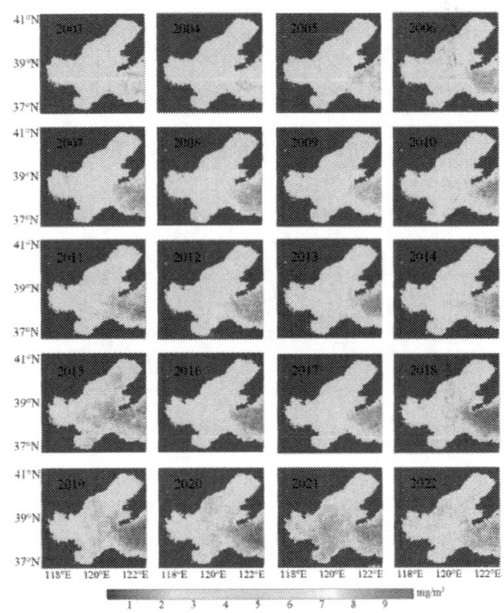

Fig. 8. Spatial variation of average autumn Chl-a over the past 20 years (2003-2022)

The spatial distribution patterns of Chl-a in spring, summer, autumn and winter are more or less the same (Fig.6-9), showing a trend of gradual increase, stabilization and then gradual decrease. The changes in the nearshore waters are evident in all four seasons. In particular, the distribution area of high Chl-a concentrations along Liaodong Bay, Bohai Bay and Laizhou Bay showed a gradual increase from 2003 to 2007, a stabilization of high value from 2007 to 2016, and a decrease in fluctuations from 2016 to 2022.

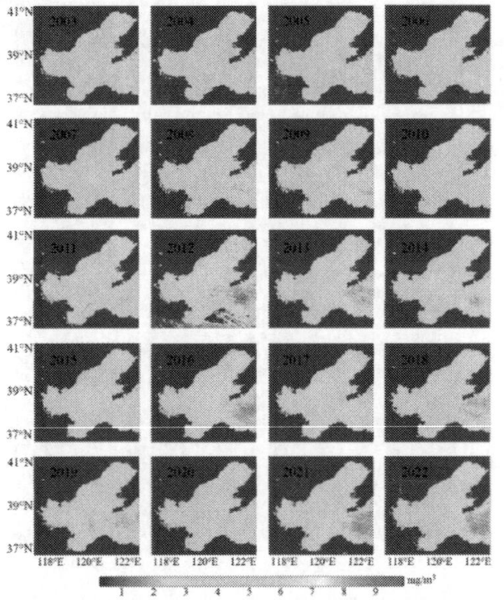

Fig. 9. Spatial variation of average winter Chl-a over the past 20 years (2003-2022)

3.2.2 Characteristics of seasonal temporal distribution

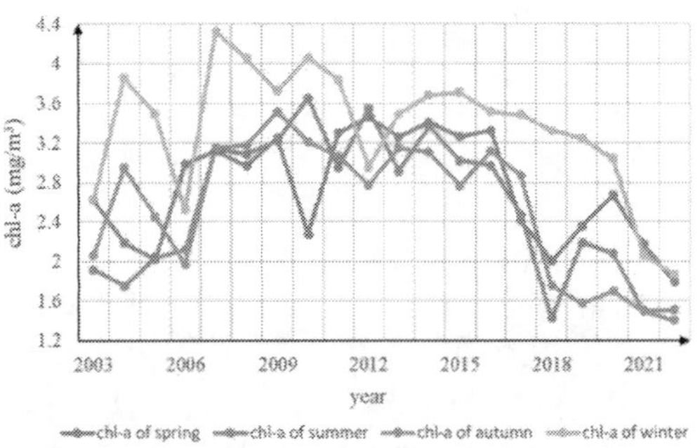

Fig. 10. Seasonal Chl-a variations in the Bohai Sea over the last 20 years (2003-2022)

The temporal trends of Chl-a in the four seasons from 2003-2022 (Fig.10) are different, but they were similar to the interannual trends, showing an inverted U-shaped trend. Spring and summer showed an increasing trend in the longer time series, while autumn and winter showed a decreasing trend. 2003-2006 showed fluctuating increasing trends in all seasons, with a significant relative low value in spring in 2010, a significant decrease in spring, summer and autumn from 2017-2022 compared with 2007-2016, and a basic stability in winter from 2013-2020. The spring maximum occurred in 2012 at 3.46 mg/m^3 and the minimum in 2022 at 1.79 mg/m^3. The summer maximum occurred in 2010 at 3.66 mg/m^3 and the minimum in 2018 at 1.43 mg/m^3. The autumn maximum occurred in 2009 at 3.52 mg/m^3 and the minimum in 2022 at 1.40 mg/m^3. The winter maximum occurred in 2007 at 4.32 mg/m^3 and the minimum occurred in 2022 at 1.86 mg/m^3.

4 Summary

Based on the MODIS L1B data of the Bohai Sea from 2003 to 2022, the Chl-a concentration was calculated with the improved SWIR atmospheric correction algorithm and a series of methods. The seasonal and annual average data of Chl-a concentration during the period of 2003-2022 were obtained, and the characteristics of the spatial and temporal variation were analyzed. The following main conclusions were obtained:

(1) The Chl-a over the past 20 years has obvious interannual spatial distribution characteristics. It shows a trend of high values in nearshore sea areas and low values in offshore sea areas, high in estuaries and bays and low in other sea areas. The larger values appeared centrally in Liaodong Bay, Bohai Bay, and Laizhou Bay near the mouths of rivers and harbors, with the largest value appearing near Huanghua port.

Temporally, the overall variation of interannual Chl-a concentration showed an inverted U-shape. 2003-2006 showed a fluctuating increasing trend; 2007 increased steeply, and 2007-2016 fluctuated slightly but basically stable; 2016-2022 showed a fluctuating decreasing trend.

(2) The spatial characteristics of Chl-a in the four seasons from 2003 to 2022 were basically the same. The value in the nearshore sea area is significantly higher than that in the central sea area. It is the lowest in the Bohai Strait, and in Liaodong Bay, Bohai Bay and Laizhou Bay is higher, with significantly high values occurring in the coastal harbors and river inlets. The value is higher in spring and summer than in autumn and winter, the highest in spring and the lowest in autumn. Especially, the nearshore area of Bohai Bay showed obvious performance.

The temporal trends of Chl-a in all seasons from 2003-2022 showed an inverted U-shaped trend. Spring and summer showed an increasing trend in the longer time series, while autumn and winter showed a decreasing trend. 2003-2006 showed fluctuating increasing trends in all seasons, with a significant relative low value in spring in 2010, a significant decrease in spring, summer and autumn from 2017-2022 compared with 2007-2016, and a basic stability in winter from 2013-2020.

In recent years, the impact of human activities has made the eutrophication of the Bohai Sea waters serious, and red tides and other disasters have occurred frequently, bringing huge economic losses. The environmental management of the Bohai Sea has always been a top priority. The Action Plan for Bohai Sea Comprehensive Management Battle was issued and implemented in 2018, and Chl-a concentration of the Bohai Sea improved significantly and stabilized gradually in 2018, which further corroborated that the spatial and temporal distribution of the Chl-a basically reflects the distribution characteristics of the Bohai Sea phytoplankton. The Chl-a concentration is affected by various factors such as light, temperature, wind speed, precipitation and nutrient salts, etc., and a refined analysis of the influencing factors of Chl-a concentration needs to be continued.

References

1. J. O'Reilly, S. Maritorena, et al. Ocean color chlorophyll algorithms for SeaWiFS. Journal of Geophysical Research: Oceans 103.C11(1998).
2. K. Yu, D. Lu, and H. Xiong. Remote sensing inversion of winter surface seawater chlorophyll concentration in the Bohai Sea sea area and its change analysis in the last 7 years. Remote Sensing Information 6 (2009): 8.
3. D. Liu, Y. Wang. Trends of satellite derived chlorophyll-a (1997-2011) in the Bohai and Yellow Seas, China: Effects of bathymetry on seasonal and inter-annual patterns. Progress in Oceanography 116(2013):154-166.
4. S. X, Y. Fu, P. Kang. Seasonal and interannual variations of Chlorophyll a in Bohai Sea. Marine Environmental Science (2015).
5. H. Tian, Q. Liu, G. Joaquim, et al. Spatial and temporal variations of chlorophyll a concentration in the Bohai Sea over the last 20 years. Journal of Oceanography 41.8 (2019): 10.
6. F. Zhai, W. Wu, et al. Interannual-decadal variation in satellite-derived surface chlorophyll-a concentration in the Bohai Sea over the past 16years. Journal of Marine Systems 215(2020).
7. J. Chen, M. Zhang, et al. A Review of Some Important Technical Problems in Respect of Satellite Remote Sensing of Chlorophyll-a Concentration in Coastal Waters. IEEE Journal of Selected Topics in Applied Earth Observations & Remote Sensing 6.5(2013):2275-2289.
8. M. Wang, S. H. Son, W. Shi. Evaluation of MODIS SWIR and NIR-SWIR atmospheric correction algorithms using SeaBASS data. Remote Sensing of Environment 113.3(2009):635-644.
9. S. Li, D. Yuan, X. Wang. Improvement of SWIR atmospheric correction algorithm for case II water. (2017).
10. H. Gordon, M. Wang. Surface-roughness considerations for atmospheric correction of ocean color sensors 1: The Rayleigh-scattering component. Applied Optics (1992).
11. M. Wang. Remote sensing of the ocean contributions from ultraviolet to near-infrared using the shortwave infrared bands: simulations. Applied Optics (2007).
12. S. Li. Characteristics of chlorophyll distribution in the Bohai Sea based on MODIS data and hydrodynamic modelling[D]. Tianjin University (2016)

Research on the Technical Model of Resource Utilization of Rural Domestic Sewage

Xinyu Yang*

Resources and Environmental Science, Northwest A&F University, Xianyang, Shaanxi, China

Abstract: At present, China's rural sewage treatment technology is mostly used in municipal sewage treatment methods, with high energy consumption and complicated operation and maintenance. Through analyzing the problem of rural sewage management, we discussed the necessity and feasibility of rural sewage resource utilization, presented the principles that should be followed in the resource utilization according to local conditions, promoting the treatment, economic rationality, and ecological safety, and Summarized the technical models of irrigation utilization, ecological dissipation and decentralized on-site utilization for the resource utilization of rural domestic wastewater. The expansion of the resource utilization model will be more conducive to reducing the cost of rural sewage management and enhancing the quality of the rural environment and human settlements.

1 Introduction

With the economic development of rural regions and the improvement of the living standards of the population, the amount of sewage discharged from rural life has been constantly increasing.[1] Rural domestic sewage mainly consists of human, and poultry feces and urine, as well as bathing, washing, and kitchen wastewater, which contains high levels of organic matter, nitrogen, phosphorus, and other pollutants.[2] Due to the characteristics of rural domestic sewage, which is small in volume, dispersed, and far away from the sewage network, it is normally discharged directly in the vicinity without any treatment, resulting in the aggravation of black odor and eutrophication of water bodies in rural areas, and adversely affecting the rural ecological environment and the quality of life of the inhabitants.

China attaches great importance to the management of rural sewage and is now vigorously building rural sewage treatment facilities throughout the country, which has achieved definite results.[3] However, most of the existing rural sewage treatment facilities are characterized by difficulties in promoting their application due to high investment and operating costs, complicated operation and maintenance, and difficulties in achieving a long-term stabilization of the treated effluent to meet the standards. Meanwhile, with the continuous promotion of rural revitalization and the construction of new countryside, China's requirements for the discharge of treated rural sewage effluent have gradually increased, and the available rural sewage treatment facilities are facing enormous challenges. Therefore, the development of high-efficiency, low-investment, low-energy consumption, simple operation, and maintenance of rural domestic wastewater process and equipment plays an essential role in improving the efficacy of rural domestic wastewater treatment and contributing to the construction of new rural villages.[4]

Rural domestic wastewater is highly characterized by organic pollutants and differs from urban industrial wastewater because it involves the introduction of toxic and hazardous substances, and therefore has excellent potential for resource utilization. Sewage resource utilization refers to the sewage after harmless treatment to meet specific water quality standards, as renewable water instead of conventional water sources for industrial production, municipal miscellaneous, ecological water replenishment agricultural irrigation, and so on, which is of vital significance in promoting water resources intensive conservation, energy saving and carbon reduction, and promoting the high-quality development of China's economy. Nine ministries and commissions, including the National Development and Reform Commission (NDRC), explicitly requested in the Guiding Opinions on Promoting the Resourceful Utilization of Sewage issued in January 2021 to steadily promote the resourceful utilization of agricultural and rural sewage, popularize the combination of planting and raising, and promote the way of using to facilitate treatment to promote the solution to the problems of water resource shortages, water environment pollution, and water ecological damages.[5] Resource utilization has been made an important mode of agricultural sewage management, striving to achieve significant development in the resource utilization of rural domestic sewage by 2030, to manage rural domestic sewage in a natural, green, and ecological approach, contributing to the realization of the national " Double Carbon" goal.

*Corresponding author: 17622822665@163.com

2 Major problems in rural sewage management

2.1 Rural Wastewater Treatment Technological Limitations

At present, the vast majority of rural villages use the village committee's wells to supply water at regular intervals, most of the rural washing and kitchen wastewater is used to suppress dust in courtyards or for landscaping, and the designed water volume of the completed sewage treatment facilities is high while the actual emissions are small, which leads to their inability to operate normally.[6] In contrast, the northern part of the country is highly seasonal, with the cold climate in winter, rural water consumption and wastewater generation are greatly reduced and cannot be discharged continuously, and the wastewater discharge process is more likely to freeze into ice, resulting in the treatment facilities not being able to operate properly.

2.2 Inconsistency between the implementation of emission standards and the actual situation

So far, China has not published a standard for the discharge of rural domestic sewage, mostly adopting the Pollutant Discharge Standards for Urban Sewage Treatment Plants (GB18918-2002) and local standards issued by the respective provinces. Generally, rural domestic sewage effluent is divided into resource utilization and direct discharge. [7]Currently, most districts and counties have not yet established complete mechanisms and systems for the operation and maintenance of rural domestic sewage treatment facilities. The actual situation in rural areas varies from province to province, which makes it impossible to guarantee that the treatment facilities meet the discharge standards.[8]

2.3 Inadequate costs of construction and operation of rural domestic wastewater

The amount of investment in rural domestic sewage treatment efforts was large, and the ability of local governments to shoulder the financial burden was limited. Difficulties in operation and maintenance at the grass-roots level have resulted in insufficient environmental infrastructure. Most villages have collective financial challenges, resulting in the physical operation of most treatment facilities and the inability to ensure that effluent water quality meets standards and that treatment facilities are operated and maintained effectively over the long term.[9]

2.4 Operation and maintenance system need for perfection

There are more types and numbers of rural sewage facilities, which are scattered and difficult to maintain, and over a long period, some of the equipment is gradually deteriorating, facility failures are occurring, and there are also problems such as broken pipe networks, and the treatment efficiency of rainwater and sewage diversion has been greatly reduced. There are common rural sewage treatment processes at this stage, mainly mobile integrated facilities, artificial wetlands, and traditional activity sludge system A^2/O technology, etc. Owing to the lack of funds and professional and technical personnel, operation and maintenance are not in place promptly, and some of the sewage treatment projects have been idling, without achieving the effectiveness of the treatment. [10]Because of the limited local finances and the pressure to raise matching funds, it is extremely difficult to effectively guarantee funds for operation and maintenance at a later stage, especially for sewage treatment facilities that require power maintenance. The operation and maintenance of treatment facilities require a certain degree of expertise and specialized personnel, but at present, rural areas lack management personnel or fail to meet the technical requirements, resulting in sewage treatment facilities not being able to operate normally or not being able to achieve optimal operating conditions.

2.5 Awareness of pollution prevention among the villagers is weak, and the phenomenon of direct discharge of sewage is prevalent

Presently, in the rural areas of China, due to the limitations of economic conditions, rural residents' to protect the environment and pollution prevention awareness is weak, most areas lack perfect sewage collection systems and discharge sewage at random, resulting in the rural sewage discharged directly into the river or septic tanks after a simple treatment of infiltration into the ground, seriously polluting the river water and groundwater, which leads to the increase of nitrogen and phosphorus content in the water, triggering eutrophication of the water body, seriously disrupting the balance of rural ecosystems, resulting in the loss of the original self-purification of the body of water, so that the rural residents of the deterioration in the living environment, and greatly enhanced the difficulty of the management of rural domestic wastewater.[11]

3 Feasibility analysis of the utilization of rural domestic sewage resources

Initially, China has proposed the point target of achieving " Carbon Peak and Carbon Neutrality", which is a major decision for China to promote high-quality economic development and comprehensively build a strong modern socialist country. In the context of the national " Double Carbon" goal, the concept of rural sewage resource utilization and local consumption has been deepening, and national and local policies to promote and encourage the use of rural sewage resources have been launched one after another. [12]In July 2019, the Central Agricultural Office, the Ministry of Agriculture and Rural Development Ministry of Ecology and Environment, and other 9 departments jointly issued the "Guiding Opinions on Promoting the Treatment of Rural Domestic Sewage", which emphasized that the basic idea of rural domestic sewage treatment is to "suit the local conditions and respect the habits, curing as much as possible, utilizing as the first step, locally close to the place and ecological recycling, advancing gradually, constructing and managing with equal importance, mobilizing farmers, and having a long-term effect". Oriented on wastewater reduction, classification and local treatment, and recycling. In June 2022, the Ministry of Ecology and the Environment and other ministries issued the "Implementation Plan for Reducing Pollution, Reducing Carbon, and Synergizing Efficiency", which put forward the strategy of "promoting the centralized or decentralized treatment of rural domestic wastewater and nearby reuse, based on local conditions and with the guidance of resources, ecology, and sustainability" regarding rural domestic wastewater. The strategy is based on resource utilization, decolonization, and sustainability. Therefore, the realization of rural sewage resources is a key measure and an inevitable requirement to promote the successful realization of the " Double Carbon" goal and the green and coordinated development of China's economy.

Furthermore, rural sewage is characterized by high nitrogen and phosphorus content and high organic matter content, if directly discharged into a water body, there will be a risk of causing eutrophication of the water body, resulting in black smelly water, whereas if treated appropriately, it can become a good water resource for field irrigation, and the nitrogen and phosphorus elements it contains can be a source of nutrients for the growth of crops. Rural areas are rich in land resources, with powerful nitrogen and phosphorus acceptance capacity, and rural sewage water volume is small, decentralized discharge, with good biochemical properties. At the same time, differing from municipal sewage, the content of heavy metals and other toxic and harmful substances in rural sewage is extremely small, so in terms of safety considerations, rural sewage also has a huge potential for resource utilization. If the treatment purpose is changed from standard discharge to ecological use, it not only greatly alleviate the problem of nitrogen and phosphorus in the effluent of rural sewage treatment facilities which is difficult to meet the standard stably, and promote its long-term operation, but also save the cost of construction of sewage treatment facilities, operating costs, but also save the water resources and various types of fertilizers and other additional costs, and at the same time can also reduce the secondary pollution of the environment brought about by various types of wastewater conditioning agents.

Eventually, although the rate of rural sewage treatment has been increasing, there are mismatches in the identification of treatment modes, the selection of process technologies, and so on, that are not ecologically friendly and green enough and not quite in line with the actual needs of the countryside, coupled with the lack of professionalism in operation and maintenance and the small amount of funding, which leaves a gap with the actual effectiveness of operation and maintenance. For facilities in normal operation, feasibility surveys on resource utilization can be carried out to comprehensively assess the economic and environmental benefits of the facilities in terms of service life, operational energy consumption, maintenance costs, and inputs for resource utilization and renovation, to provide a basis for decision-making on the management of agricultural wastewater.

4 Basic Principles for Utilizing Rural Domestic Sewage Resources

4.1 Economically rational and ecologically safe-oriented principles

Considering the limited economic conditions in rural areas, the optimization of economic benefits should be observed in the process of resource utilization of rural domestic wastewater, which should not produce negative benefits, selecting technological measures involving relatively inexpensive construction, stable operation, low carbon, and energy saving, and simplified maintenance, which encourage the combination of the upgrading of old and low-scalar facilities with the renovation of the standard to achieve the eco-cycling of wastewater to ensure its economic reasonableness and sustainability.[13] Simultaneously, sewage treatment and resource utilization need to always grasp people-oriented thinking, all efforts need to improve the people's quality of life as the starting point and finishing point, any resource utilization of rural sewage must meet the requirements of the regional ecological environment objectives, shall not appear in the water, soil and agricultural products, suffering from pollution and endangering human health and other phenomena.

4.2 The principle of utilization for governance and site-specific challenges

Renewal and reconstruction of existing rural domestic wastewater treatment facilities, the effective use of potential physical energy resources in rural domestic wastewater as far as possible, through the recovery of raw materials, harmless treatment, can be achieved through the use of low-function or biological nutrient utilization and water resources back to the use of resource utilization pathways, but also several modes of utilization can be combined. Concerning local water resource endowment, carrying capacity of the water environment, development needs, economic and technological level, farmers' production and living habits, conditions for the utilization of sewage resources, and other factors, we will carry out the utilization of sewage resources, implement differentiated measures, and reasonably determine the program for the utilization of sewage resources in rural areas, sewage treatment process, and construction and operation modes.[14]

5 Main technological processes for rural wastewater resource utilization

5.1 Ecological treatment technologies

5.1.1 Artificial Wetlands

Artificial wetland treatment system originates from the simulation of natural wetlands, mainly utilizing the synergistic effect of plants, substrate, and microorganisms of natural ecosystems to realize water purification. The main body of the artificial wetland consists of soil and filler filled at a certain level, with aquatic plants planted on the surface of the bed to form a unique ecosystem.[15]

5.1.2 Stabilization Ponds

Stabilization ponds are sewage ponds that are artificially and appropriately manicured with an embankment and impermeable layer. Its purification principle is similar to the self-purification mechanism of natural water bodies, through the metabolic activities of microorganisms (bacteria, fungi, algae, protozoa, etc.), and accompanying physical, chemical, and physical-chemical processes, multi-stage conversion, degradation, and removal of pollutants in sewage.[16]

5.1.3 Land Disposal

Land disposal technology is a treatment method that utilizes a soil-plant-microbe composite ecosystem under artificial control to purify sewage through a series of physical, chemical, and biological actions and to realize the recycling of water and nutrients in sewage. According to the different flow rates and trajectories of water movement, land treatment systems can be categorized into four types: slow percolation systems, rapid percolation systems, surface diffusion systems, and subsurface percolation systems (i.e., capillary land percolation treatment technology).[17]

5.2 Biofilm technologies

The biofilm method is an artificial treatment technology mainly applied in decentralized domestic wastewater treatment, including two types of anaerobic and aerobic biofilm. Anaerobic or aerobic microorganisms attach to the surface of the carrier to form a biofilm to adsorb and degrade the pollutants in the sewage to achieve purification purposes. This method has simple equipment, lower operating costs, and high treatment efficiency. The reactor is generally composed of three parts: filler (carrier), water distribution device, and drainage system, and the fillers used are inorganic (ceramic, slag, activated carbon, etc.) and organic (PVC, PP, plastic, fiber, etc.).[18]

5.3 Physical treatment technologies

Physical and chemical treatment methods for sewage mainly include coagulation, air flotation, adsorption, ion exchange, electrodialysis, reverse osmosis, and ultrafiltration. In a variety of physical and chemical treatment technologies, only coagulation technology is relatively in line with the rural requirements, and its greatest advantage is that it can be based on the nature of the pollutants in the sewage, selecting the appropriate flocculant, which guarantees the efficient removal of pollutants. The coagulation technology is effective in removing suspended solids, metal ions, colloidal substances, and inorganic phosphorus, but relatively weak in the removal of organic matter and nitrogen, and the operation process requires continuous addition of chemicals, so operating costs are comparatively high.

6 Summary

Against the backdrop of the promotion of the rural revitalization strategy and the intertwining of carbon-neutral and forced transformation, rural areas have emerged as a new front in China's environmental remediation as environmental pollution intensifies.[19] Rural sewage has already been a major source of pollution in the rural environment, which is why our country attaches great importance to the research of rural sewage collection and treatment technology, and has successively developed many technical methods and processes, and popularized their application in actual projects, providing important technical support for the improvement of the quality of the rural environment.

During future work, the future direction of sustainable development of rural sewage treatment in China has the following points: (1) At all levels of research, application, and management should continue to follow the principle of " Local conditions, Economical application, Simple maintenance", according to the natural, economic and technological conditions of each place, and effectively establish a technical system with regional characteristics.[20](2)The use of intelligent dosages of pharmaceuticals to reduce drug consumption, or utilizing solar energy instead of electricity to reduce electricity consumption, thus reducing indirect emissions in the wastewater treatment process. (3) Emphasize sewage treatment while taking into account the problem of sewage sludge treatment and disposal in rural areas to prevent secondary pollution. (4) Anaerobic digestion of sludge is currently the main way of organic matter resource utilization, and rural sludge can be synergized with agricultural straw, livestock and poultry manure, and other agricultural waste for anaerobic fermentation to produce methane, realizing the recovery of resources and energy. [21]

Rural habitat improvement is an important part of realizing rural ecological revitalization and promoting the construction of beautiful and livable villages, as well as promoting the construction of rural ecological civilization and enhancing the ecological well-being of rural residents. The CPC Central Committee and the State Council have paid great attention to the improvement of the rural habitat environment and have adopted a series of policy measures; the Outline of the Fourteenth Five-Year Plan for the National Economic and Social Development of the People's Republic of China and the Visionary Goals for the

Year 2035 has made the " significant improvement of the urban habitat environment" as one of the goals of economic and social development during the Fourteenth Five-Year Plan period, and has clearly pointed out that the goal of the construction of a beautiful China will be basically achieved by the year 2035. [22]However, facing the vision of 2035, the upgrading of rural human settlements requires a problem-oriented approach and the adoption of effective countermeasures to promote it in terms of organization, mode, funding, talents, system, and supervision:

(1) Strengthening the overall leadership of improving and upgrading rural human settlements, enhancing the role of grass-roots party organizations as a battleground, and giving full expression to the pioneering and exemplary role of party members. (2) Focusing on model innovation for the improvement and upgrading of rural human settlements. (3) Scientifically accounting for the funding gap and exploring diversified funding channels: increasing financial inputs from governments at all levels, ensuring that land appreciation proceeds are earmarked for this purpose, progressively improving diversified financing mechanisms. (4) Strengthening the cultivation of talents for upgrading rural human settlements: scientific talent, precise talent cultivation, flexible talent attraction, efficient talent utilization and long-term talent retention. (5) Continuously increase policies to support rural development, systematize and establish an effective institutional system, and develop comprehensive standards and rule of law safeguards for improvement and upgrading. (6) Improve long-term supervision and care mechanisms to strengthen supervision and inspection. [23]

Reference

1. Yongkang Shen, Di Zhang, Ying Zhao et al., " Progress on Rural Drainage Water for Rural Revitalization in Northwest China", JOURNAL OF COASTAL RESEARCH, pg. 23-27, (2021); doi: 10.3964/j.issn.1000-0593(2022)07-2113-07
2. WANG Junneng, ZHAO Xuetao, CAI Nan, et al.Pollution Discharge and Environmental Treatment Efficiency of Rural Domestic Sewage in China[J]. Research of Environmental Sciences,2020,33(12):2665-2674.DOI:10.13198/j.issn.1001-6929.2020.11.08
3. WANG Qi, HUANG Yi, ZHANG Yuanke, et al. Study of Typical Treatment Technology of Domestic Sewage in Mountainous Rural Areas[J/OL]. Environmental Science & Technology:1-7[2024-05-22]. (in Chinese)
4. LI Shi-feng, HUANG Kai, ZHAO Rong, et al. Research on the Systematic Treatment Mode and Technology of Rural Domestic Sewage in a New Area of Qingdao[J]. China Water & Wastewater,2023,39(18):15-21.DOI:10.19853/j.zgjsps.1000-4602.2023.18.003.
5. Yu Fawen, Hu Meimei, Wang Guangliang. Research on the improvement path and countermeasures of rural human settlement environment towards the vision goal of 2035 [J]. China Soft Science,2022(07):17-27.
6. Yu Fawen, Yu Ting. Study on the Model and Countermeasure of Rural Domestic Sewage Treatment[J]. Chongqing Social Sciences, 2019(03):6-17+2.DOI:10.19631/j.cnki.css.2019.03.001. (in Chinese)
7. GUO Fang, CHEN Yong, WANG Guotian, ZHANG Wenhui, et al. Present situation, problems, and development suggestions of rural domestic sewage treatment in China[J]. Water & Wastewater Engineering, 2022,58(S1):68-72.DOI:10.13789/j.cnki.wwe1964.2022.03.30.0018.
8. Peng Bin, Hu Siyuan, Wang Zhu, et al. Current situation and problems of decentralized treatment of rural domestic wastewater [J]. Research of Agricultural Modernization, 2021,42(02):242-253. (in Chinese)
9. JIA Xiao-mei, YU Qi, WANG Wen-yi, ZHAO Fang, DONG Xu-hui, et al. Considerations for rural domestic sewage treatment during the 14th Five-Year Plan[J].Journal of Agricultural Resources and Environment,2020,37(05):623-626.DOI:10.13254/j.jare.2020.0133. (in Chinese)
10. GAO Shengwang, Huang Zhiping, Xia Xunfeng, et al. Investigation and suggestions on the treatment of rural domestic sewage [J]. Journal of Agricultural Resources and Environment,2022,39(02):276-282. (in Chinese)
11. Li Fa Station, Lu Jiaxing. Problems and countermeasures in rural domestic sewage treatment [J]. Journal of North China University of Water Resources and Electric Power (Natural Science Edition),2018,39(04):23-28.
12. Li Yun, Xia Xunfeng, Chen Sheng, et al. Current situation, problems and countermeasures of local standards for domestic sewage treatment in rural China [J]. Journal of Environmental Engineering Technology,2022,12(01):293-300.
13. Li Huaizheng, Jin Wei, Zhang Wencan, et al. Research on comprehensive treatment of domestic sewage in rural China [J]. Engineering Science,2022,24(05):154-160.
14. Dong Liwei, Zhang Wei, Bai Lu, et al. Current situation and model analysis of resource utilization of rural domestic wastewater in China [J]. Journal of Environmental Engineering Technology,2022,12(06):2089-2094.
15. HU Xiao-bo, LUO Hui, JING Zhao-qian, ZHANG Ze-yu, et al., Research progress of rural domestic sewage treatment technology [J]. Applied Chemical Industry, 2020,49(11):2871-2876.DOI:10.16581/j.cnki.issn1671-3206.2020.11.016. (in Chinese)
16. WANG Chang, WANG Li, ZENG Ming, HAO Linlin, SUN Juanjuan, et al., Status and countermeasure for the treatment of rural domestic sewage in China[J]. Journal of Agricultural Resoureces and Environment,

2022,39(02):283-292.DOI:10.13254/j.jare.2021.0615.v (in Chinese)

17. Daud S M , Noor Z Z , Mutamim N S A .Microbial Electrochemical Systems and Membrane Bioreactor Technology for Wastewater Treatment[J].Chemical Engineering & Technology: Industrial Chemistry -Plant Equipment -Process Engineering - Biotechnology, 2023.
18. Toth A J , Fozer D , Mizsey S K J J .Physicochemical methods for process wastewater treatment: powerful tools for circular economy in the chemical industry[J].Reviews in Chemical Engineering, 2023, 39(7):1123-1151.DOI:10.1515/revce-2021-0094.
19. SHI Wei, HUANG Senwei, LI Xueyuan. Efficiency Evaluation of Rural Domestic Wastewater Treatment Based on SBM-DEA and Cloud Model: An Analysis of Panel Data from 27 Provinces in China(2011-2020)[J/OL].Ecological Economy:1-20[2024-05-22].
20. Li Huaizheng, Jin Wei, Zhang Wencan, Hu Feng, Ye Jianfeng, et al., Comprehensive Treatment of Rural Domestic Sewage in China[J].Strategic Study of CAE,2022,24(05):154-160. (in Chinese)
21. María Molinos-Senante, Poch M , Rosso D ,et al.From wastewater treatment plants to decentralized resource factories[J].npj Clean Water, 2024, 7(1).DOI:10.1038/s41545-024-00343-4.
22. YU Fawen, HU Meimei, WANG Guangliang. Paths and Countermeasures of Rural Living Environment Improvement for the Long-Range Objectives through the Year 2035[J].China Soft Science,2022(07):17-27.
23. WANG Bin, YU Fa-wen. The Strategic Task of Improving Rural Residential Environment Renovation in the 14th Five-Year Plan Period. [J]. Reform,2021(03):111-120. (in Chinese)

Research on key issues in promoting the green transformation and development of the power grid

Qian Zhang[a*], Bingxin Zeng[b], Guangrui Tang[c]

State Grid Energy Research Institute Company Limited, Beijing, 102209, China

Abstract: The green transformation of the power grid is an important component of energy transition. It primarily involves promoting carbon peaking and carbon neutrality as the driving force, establishing the concept of "energy transformation and green development," enhancing the inter-provincial power resource allocation capacity and power system flexibility, and accelerating the shift from high-carbon to low-carbon and from fossil fuel-based energy to clean energy in order to accelerate the adoption of green production and consumption patterns. This paper summarizes the characteristics of the green transformation of power grid development and puts forth relevant suggestions in terms of defining standards, enhancing the flexibility of the power grid system, and driving energy-saving and emissions reduction in various industries.

1 Introduction

In recent years, resource scarcity and environmental pollution have become two major global problems hindering human development and social progress. The resulting climate change anomalies, global warming and other problems have received widespread attention from the international community, mainly because of the large amount of greenhouse gas emissions brought about by economic development, especially carbon dioxide emissions from the use of fossil energy[1].

As the largest developing country, China has become the second largest economy after more than 30 years of development, while at the same time carbon emissions from energy consumption have climbed dramatically. Figures 1 and 2 show the statistics of China's carbon emissions from 1980 to 2020 and the proportion of China's carbon emissions in the world's total emissions, respectively. As can be seen from the figure, in 1980 China's carbon emissions level is still relatively low, only accounted for about 7% of the world, by the end of 2020 China's carbon emissions proportion reached 30.7%. China's CO_2 emissions from only 1500 million tons in 1980 to 9,899 million tons in 2020, CO_2 emissions during the 30 years has increased by five times. The past 40 years have been a period of rapid economic growth in China, but also a period of rapid growth in CO_2 emissions. China has become a major carbon emitting country, and as a responsible country in the international community, China's emission reduction task cannot be delayed, not only need to strictly control the total growth of carbon emissions, increase the use of clean energy, but also need to actively adjust the way and mode of energy utilization, improve energy efficiency, and take the road of sustainable energy development[2].

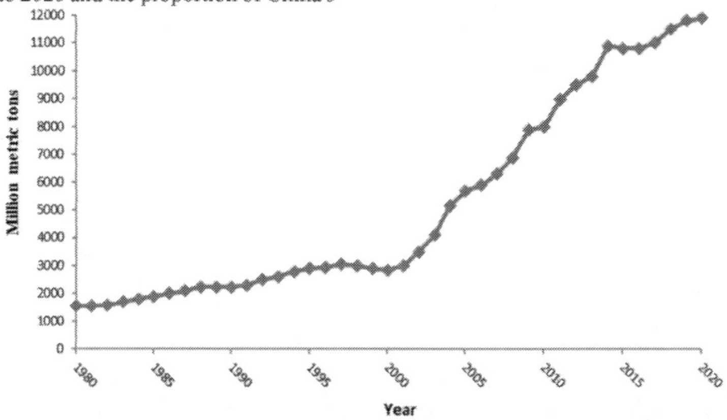

Figure 1 Statistical chart of China's carbon emissions from 1980 to 2020

Email of all the authors: [a]zhangqian881111@163.com; [b]zengbingxin@sgeri.sgcc.com.cn; [c]tangguangrui@sgeri.sgcc.com.cn;
*Corresponding Author: Zhang Qian; zhangqian881111@163.com; 18810316307; 010-66603486

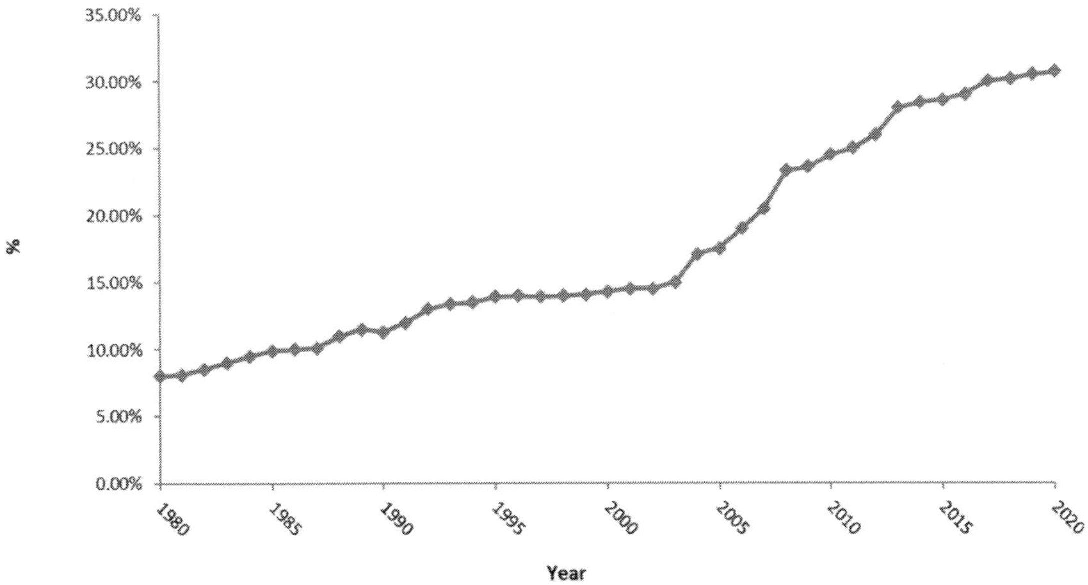

Figure 2 Proportion of China's carbon emissions to the world's total emissions from 1980 to 2020

Since the end of 2020, China has mentioned the peak carbon emissions and carbon neutrality on several important occasions and in policy documents. Peak carbon emissions refer to the peak value of carbon emissions in China before 2030, followed by a continuous decline; carbon neutrality refers to the balance between carbon emissions and absorption in China before 2060, achieving a relative zero emission status[3]. Scientific research has shown that excessive carbon emissions can lead to global warming, greenhouse effects, and other environmental issues.

Under the "dual-carbon" goals, the clean and low-carbon transformation of China's energy and power system is both significant and challenging. Relevant data shows that China's energy combustion emits approximately 9.8 billion tons of carbon dioxide, with the electricity industry contributing around 3.9 billion tons. It is evident that to achieve the goals of peaking carbon emissions and achieving carbon neutrality, the power sector is one of the main battlegrounds. The power grid serves as a vital platform connecting energy supply and consumption, and driving the green transformation of the power grid will be a key initiative to help achieve the "dual carbon" goals[4].

2 The main characteristics of the green transformation of the power grid development mode

It has a clear green standard system for the power grid. The green standard system is an important means to guide the green transformation of production and lifestyle[5]. For the power grid, it means having a comprehensive and clear green, low-carbon product and service standard system, realizing the gradual development of power grid equipment towards high-end, intelligent, and green direction, establishing a "dual carbon" standard, and promoting the gradual transition of energy consumption control from total amount and intensity to carbon emission total amount and intensity control system.

Energy production is gradually transitioning towards clean sources. This is mainly manifested in the large-scale development and utilization of clean energy such as new energy. Since 2020, the newly added installed capacity of wind power and solar power in China has exceeded 100 million kilowatts for two consecutive years, accounting for about 60% of the total newly added power generation capacity. The high proportion of new energy integration poses higher requirements for the stability of the power grid system. It is necessary to conduct in-depth research and application of the characteristics and control strategies of the large-scale power grid, and continuously improve the level of system safety operation[6].

Energy consumption continues to increase its level of electrification. This is mainly manifested in the rising share of electricity in final energy consumption. From 2000 to 2021, the global share of electricity in final energy consumption increased from 15.4% to 18.1%, while in China, it increased from 10.9% to 27%. In 2021, the per capita electricity consumption in China reached approximately 5,900 kilowatt-hours, which is 1.6 times higher than that of 2012 and surpasses the global average.

3 Analysis of key issues in the green transformation of the power grid development mode

First,the standard system for the green transformation of the power grid needs further clarification. Unlike abroad, in China, the focus is still mainly on the scope of smart grids, lacking supporting infrastructure for asset activities related to transmission and distribution, as well as equipment projects for grid system monitoring and control.

Moreover, unlike the detailed definition of benchmark and reference levels for different indicators such as energy consumption, material consumption, water consumption, pollutant emissions, and greenhouse gas emissions in clean and efficient coal utilization tools, there is still a lack of specific indicators and parameter definitions for the construction and operation of smart grids in improving the efficiency of clean energy consumption and reducing wind and solar curtailment.

Second, the insufficient flexible regulation capacity of the power system to support the integration of clean energy leads to significant pressure for low-carbon transformation. Clean energy, such as renewable energy, relies on the efficient absorption of electricity, which relies on the flexibility of the power system. In the past decade, comprehensive measures have been taken, including flexible retrofitting of thermal power plants and increasing interconnection of power grids, to promote the high-level integration of renewable energy. However, there is a limited total amount of available resources, leading to a "ceiling" in terms of integration capacity[7].

Third, Guiding the entire society in the task of energy conservation and efficiency is crucial. On one hand, With the increased electrification level, the demand for electricity will continue to grow. The power industry not only needs to accommodate the energy consumption and emissions transferred from transportation, buildings, and industries but also needs to make greater contributions in clean substitution of existing fossil energy resources. On the other hand, the power grid is the core hub connecting power production and consumption, playing a crucial role in carbon emission reduction within the power system. It has to ensure the large-scale development and efficient utilization of renewable energy while meeting the electricity demand of economic and social development. Power grid companies face significant pressure to ensure security, supply, and cost reduction, as well as the heavy responsibility of their own energy conservation and emission reduction tasks.

4 The green transformation and guarantee mechanisms for the development of the power grid

First, accelerate the improvement of the green transformation standards system for the power grid. Incorporate more infrastructure and services related to the power grid into the Green Industry Guidance Catalogue, such as supporting infrastructure for transmission and distribution assets, and power grid system monitoring and control equipment. Establish specific indicators and parameters for projects like smart grids to enhance the efficiency of clean energy accommodation and reduce wind and light abandonment, providing clear performance benchmarks and standards. Regularly update the Green Industry Guidance Catalogue and the Green Bond Support Project Catalogue to ensure they reflect current technological advancements and industry development needs[8]. Establish a comprehensive evaluation system to assess the environmental, economic, and social benefits of power grid projects. Ensure policy coordination among different government departments and industry associations to form a joint force in promoting the green transformation of the power grid.

Second, establish an effective incentivized mechanism for the development of shared energy storage to promote the early deployment of large-scale energy storage. Focus on constructing large-scale centralized shared energy storage in regions with abundant renewable energy resources, consolidating dispersed energy storage resources from renewable energy plants. Encourage various power generation enterprises and private capital to actively participate, while grid-related enterprises coordinate all stakeholders. Energy storage systems with excellent regulation capabilities will be dispatched by power grid companies with priority and settled according to ancillary service market rules, enabling the sharing of energy storage regulation capacity and auxiliary benefits. Following the principle of "those who benefit share the costs," the construction and leasing costs of energy storage will be shared by renewable energy enterprises, all industrial and commercial users, and other beneficiaries[9].

Third, establish a sound market mechanism that accurately reflects the value of power system capacity and allows the market to play a decisive role in resource allocation. Currently, expand the scope and variety of peak-regulation ancillary service markets, gradually integrating provincial markets into regional and even national markets, while introducing transaction types such as ramping services when appropriate. In the medium to long term, establish a capacity market through auctions and contracts to ensure that capacity market revenue covers the fixed costs of capacity resources, while the energy market covers "variable costs + reasonable returns + fixed costs not covered by the capacity market."

Fourth, implement an energy-saving management mechanism for the power grid to comprehensively promote electrification and energy efficiency. Optimize the grid structure, promote energy-saving conductors and transformers, strengthen energy-saving dispatch, and improve the energy-saving level of the power grid. Enhance green and low-carbon technology research and development in various stages of grid planning, design, construction, operation, and maintenance to achieve energy, water, material, land, and environmental protection throughout the entire process. Strengthen the recycling, reuse, and environmentally friendly disposal of sulfur hexafluoride gas and grid waste to protect the ecological environment[10]. Enhance building energy efficiency, promote energy-saving renovations for existing buildings and energy-efficient designs for new buildings, encourage the use of efficient energy-saving equipment, and fully utilize clean energy to meet energy demands.

5 Conclusion

The green transformation of the power grid development should prioritize meeting the electricity demand for economic and social development, while ensuring energy and power security as the bottom line. Regular reports should be made to government departments on the basic progress of promoting the green transformation of the power grid development. Efforts should be made to seek government policy support in areas such as distributed energy consumption and supporting energy storage. Cooperation with relevant departments should be actively sought to establish corresponding standards and regulatory systems for new market entities, such as energy storage. In line with the goal of green transformation of the power grid, major national scientific and technological programs should be implemented to promote innovation in basic theories and technological equipment.

Reference

1. Wu Kecheng, Zhao Dan. Research on Environmental Protection Management System for Green Development Oriented Grid Enterprise. Value Engineering, 2020, 39 (20): 12-13.
2. Zhang Xuefang. Research on Economic impact assessment of highway construction in Heilongjiang Province based on CGE model. 2015, Harbin Institute of Technology.
3. Yang Honglei, Xu Ouyang, Wang Jinli. Forward and backward Correlation effect analysis of electric power industry based on Input-output Method. Journal of Electric Power Science and Technology, 2015, 30(4):113-118.
4. Wang Ronghua, Du Ying, Gou Quanfeng. Evaluation on efficiency and benefit of provincial power grid planning investment considering transmission and distribution price reform. Electric Power Construction, 2020, 41(11):135-144.
5. Zheng Xinye, Wu You. Price Mechanism Innovation for Green Development of Energy System. Price Theory and Practice, 2018, (4): 12-16.
6. Song Zongyun. Study on power demand side management optimization under the background of air pollution control. 2018, North China Electric Power University .
7. Li Liang, Yang Yajuan. Research on Industrial capital investment benefit in Jiangsu Province from the perspective of input-output. Modern Management Science, 2021, (04):56-68
8. K. Adam. Energy Transformation and the UAE green economy: trade exchange and relations with Three Seas initiative countries. Energies, 2022, 15: 841-850.
9. Qiaochu Li, Peng Zhang, Junhua Chen. Does energy transformation improve green development efficiency? Evidence from Sichuan Province, China. Energy Reports, 2024, 12: 861-871.
10. L. Battisti. Energy, power, and greenhouse gas emissions for future transition scenarios. Energy Policy, 2023, 179: 113-115.

Simulation Analysis of Urban Drainage System Utilizing StormDesk Software

Shuyu Li *

Xi'an University of Architecture and Technology, School of Environmental and Municipal Engineering, China

Abstract. Employing the data from comprehensive drainage system surveys and pipeline defect diagnostics, the StormDesk software was utilized to construct a SWMM model. This model facilitates the simulation of operational scenarios within the pipe network, both with the existing network defects and following their rectification, and provides crucial insight for the repair and renovation of the drainage infrastructure. The results indicate that the current drainage system in the study area is inadequate, with a high incidence of pipeline defects. This poses a significant flood risk during intense rainfall events. Given the projected increase in rainfall intensity, it is imperative to substantially enhance the drainage capacity of the existing system.

1 Foreword

As urbanization rapidly progresses in China today, the impermeable surfaces in cities have expanded swiftly, leading to significant changes in the hydrological environment on both local and regional scales[1]. The initial construction of the drainage systems have often been inadequate, with minimal ongoing maintenance. To address the resultant disasters such as urban waterlogging, the use of various modeling software to simulate urban surface runoff and drainage systems has become increasingly common[2]. Geographic Information Systems (GIS) are employed to analyze environmental conditions and facilitate spatial analysis and management of drainage facilities[3]. Stormwater management models, represented by SWMM (Storm Water Management Model) and Info works ICM, assist users in modeling and simulating urban hydrology, sewage, and stormwater networks, analyzing the capacity of drainage systems to handle risks[4-6]. However, many studies using these models often overlook the existing defects in the pipeline network. This paper, based on the results of a network survey, simulates the drainage system in an old urban area with a high incidence of defects, evaluates the operational efficiency of the current network, and investigates the impact of these defects.

2 Data Collection

2.1 Overview of research area

Located in the western inland of China, this region experiences a subtropical continental monsoon climate. It receives an average annual precipitation of 815 mm, predominantly between June to September every year. The region's soil, primarily yellow-brown earth with a thickness ranging from 30 to 60 cm, exhibits moderate water retention capabilities.

The area spans 319 hm^2, and includes 35,450 meters of verified municipal drainage pipelines, with 40.4% comprising combined rainwater and sewage combined pipelines. In the case of heavy rain, occurrences of overflow and water accumulation are common[7].

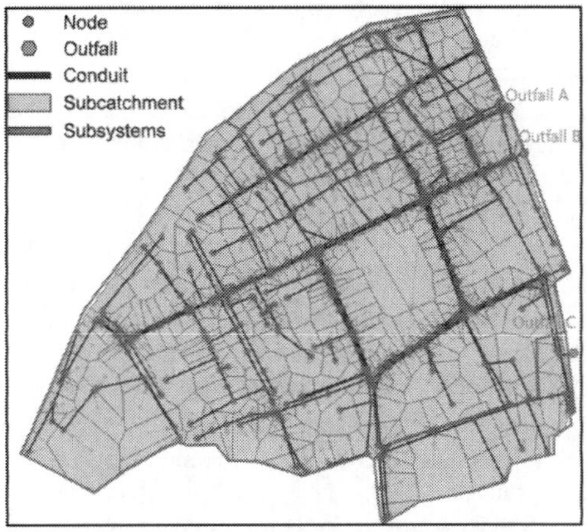

Figure. 1. Pipe network model diagram of the research area.

A comprehensive pipe network model with a complete topological structure was constructed based on measured data. The network comprises 578 pipe segments, 577 nodes, and 3 drainage outlets. Sub-catchments were delineated using the Thiessen polygon method, and various deterministic parameters such as area, characteristic width, imperviousness, and average

* Corresponding author: qq3535694239@163.com

slope were calculated. The resulting model is depicted in Figure 1.

2.2 Network Defect Generalization

The pipeline system detected 342 functional defects, such as deposits and obstacles, and 126 structural defects, including misalignments, corrosion and disconnections, as shown in Table 1. These defects were generalized into the model based on detailed analysis from inspection reports and imagery, using the following methodologies:

Sedimentary and obstacle defects alter the shape of cross-section, which can be modeled by setting the pipe to Filled Circular.

Misalignment, disconnection and similar defects mainly change the size and shape of the section at the pipe junctions, which can be modeled by altering change the shape and size of the pipe.

Scale defects reduce the cross-sectional area, which can be simulated by appropriately decreasing the pipe diameter.

Corrosion defects primarily affect the roughness of the inner surface of the drainage pipeline; thus, the Manning coefficient of the pipeline can be adjusted according to the severity of the corrosion.

Defects such as rupture, leakage and concealed connections in straight pipes may cause groundwater or unauthorized sewage to enter the pipe network. These can be simulated by setting up aquifer inflow events based on detection images.

Table 1. Statistics of various defects.

Defect Types	Defect	Number of defects	Percentage
Structural defects	Branch pipe blind connection	6	1.28%
	Misalignment	44	9.40%
	Foreign body penetration	13	2.78%
	Corrosion	39	8.33%
	Rupture	3	0.64%
	Leakage	4	0.85%
	Disconnection	17	3.63%
Functional deficiencies	Deposits	260	55.56%
	Scaling	27	5.77%
	Obstructions	55	11.75%

2.3 Rainfall Data

The rainfall pattern in the study area is predominantly single-peaked, hence the simulation utilized high-precision rainfall data synthesized using the Chicago method, which is particularly effective in calculating short-duration storm intensities. The peak rainfall coefficient for the area is 0.38, and convective storms are predominant, with rainfall durations selected at 2 hours. The storm intensity formula used was developed by the Shaanxi Provincial Climate Center in 2013 using the maximum value method:

$$q = \frac{493.071 \times (1 + 1.693 \cdot lgP)}{(t + 5.368)^{0.540}} \quad (1)$$

In the Formulas:q denotes the intensity of the rainfall, measured in liters per second per hectare, L/(S·ha); P represents the return period, measured in years, a; t signifies the duration of the rainfall event.

3 MODEL CONSTRUCTION

3.1 Hydrologic Modeling

The hydrological model adopted in this study encompasses surface runoff simulation, surface water accumulation, and network flow models.

In the surface runoff simulation, the sub-watershed is divided into permeable surfaces, impermeable surfaces with depression storage, and impermeable surfaces without depression storage. Infiltration losses have the most significant impact on surface runoff, for which the Horton model is utilized to simulate the dynamics of infiltration rates during rainfall events effectively.

The surface water accumulation is modeled using a nonlinear reservoir model. The runoff volume from each sub-watershed is calculated as the sum of the upstream outflow and the local precipitation, minus evaporation and infiltration losses.

The network flow simulation employs a dynamic wave approach to effectively model the complex flow conditions within closed conduits.

3.2 Model calibration

The surface runoff coefficient simulated by StormDesk software was compared with the established comprehensive runoff coefficient to conduct the preliminary calibration of the model. The comprehensive runoff coefficient for the study area was calculated by area-weighted method, results of 0.740.

Calibration under a return period of P=2a involved the trial-and-error method to determine various empirical parameters. Initial values were selected based on measured data and literature, and adjustments were made accordingly. The parameter sets used during the trial-and-error process are shown in Table 2.

Set three was chosen as the final dataset for simulations under different return periods. The variability of the simulated runoff coefficient and the comprehensive runoff coefficient is represented by the coefficient of variation C_v.

$$C_v = \frac{\sigma}{\mu} \times 100\% \quad (2)$$

In the formula, σ represents the difference between the simulated runoff coefficient and the comprehensive runoff coefficient; μ denotes the average value of these coefficients.

Table 2. Range of Model Parameters.

Parameter Name	Initial	Set 1	Set 2	Set 3
N-perv	0.2	0.18	0.2	0.2
N-imperv	0.016	0.018	0.016	0.016

Des-perv(mm)	2.5	2.3	2.3	2.5
Des-imperv(mm)	1.2	1.3	1.2	1.2
Max.Infil.Rate(mm/h)	64	70	72	74
Min.Infil.Rate(mm/h)	2.4	2.5	2.4	2.4
Decay constant(h^{-1})	3.5	3.8	4	4.1
Drying time(d)	6	6	7	7
Zero-imperv(%)	28	27	25	25
Simulated Runoff Coefficient	0.682	0.712	0.729	0.744
Comprehensive Runoff Coefficient	0.740			

The coefficients of variation for different return periods are as outlined in Table 3. Across all return periods, the coefficients of variation do not exceed 10%, indicating that the parameter set performs well.

Table 3. Coefficients of Variation under Different Return Periods.

Return Period	Simulated Runoff Coefficient	Comprehensive Runoff Coefficient	C_v(%)
P=1a	0.677	0.740	8.86
P=2a	0.744		1.87
P=4a	0.798		7.52
P=5a	0.808		8.81

3.3 Model validation

Using the data of 20 heavy rainfall events in the past 3 years, the overflow conditions of simulated nodes was compared with actual situation to verify the simulation's accuracy. As shown in Figure 2, the color intensity represents the frequency of water accumulation in each catchment area during these 20 rainfalls, while the red dots indicate the model's node overflow results.

4 Simulation results and analysis

4.1 Simulation results of pipe network

When all current pipe network defects are present, the total rainfall in the region for P=1a, P=2a, P=4a and P=5a is 26.17 mm, 39.50 mm, 52.84 mm and 57.13 mm. Infiltration loss, surface water storage and surface runoff were analyzed, as shown in Table 4.

Table 4. Runoff analysis in the study area under each recurrence period.

	Infiltration loss(mm)	surface water storage(mm)	surface runoff(mm)
P=1a	4.54	3.91	17.72
P=2a	4.63	5.09	29.78
P=4a	4.70	5.99	42.16
P=5a	4.72	6.24	46.17

As rainfall intensity increases(P=1a to P=5a), the infiltration loss remains stable between 4.54 mm and 4.72 mm, indicating that the local soil reaches saturation state under the design rainstorm intensity of P=1a. Surface water storage increases from 3.91 mm when P=1a to 6.24 mm when P=5a, while surface runoff sharply increases from 17.72 mm to 46.17 mm. This suggests that in the study area, high development and low soil permeability lead to nearly constant infiltration loss, limited surface water storage, and a significant conversion of precipitation into surface runoff and discharge, making the area susceptible to waterlogging during intensity rainfall.

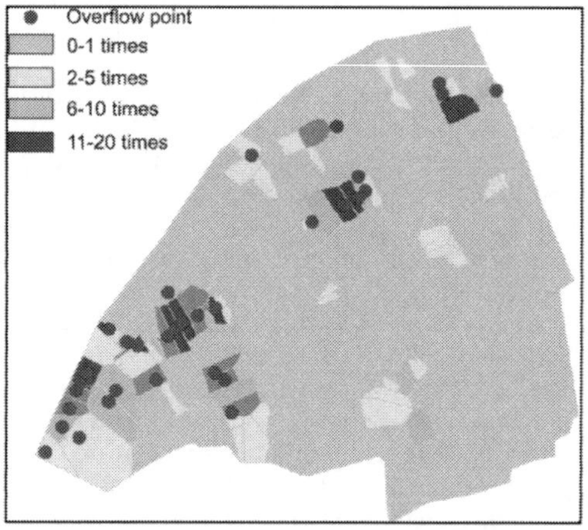

Figure. 2. Fitting of overflow nodes.

The overloading conditions of main pipelines under different design recurrence periods were visualized, as depicted in Figure 3. At P=1a, short-term heavy rainfall resulted in 7.44% of the pipelines in the study area being overloaded. As the rainfall intensity increased, at P=5a, the proportion of overloaded pipelines rose to 19.72%, with approximately half of the overloaded pipelines experiencing overloading durations exceeding 0.5 hours. Integrating road information and pipeline survey data in the study area reveals that under P=1a conditions, the majority of overloaded sections were concentrated on the branch roads. However, under P=5a conditions, only a few trunk pipeline sections remained non-overloaded, placing the entire drainage system under severe high pressure. Therefore, in order to enhance risk mitigation capabilities, it is essential to first improve the drainage capacity of the overloaded trunk pipelines, particularly on Xingan West Road, Jinzhou North Road and Gulou Street, which do not meet P=1a standard.

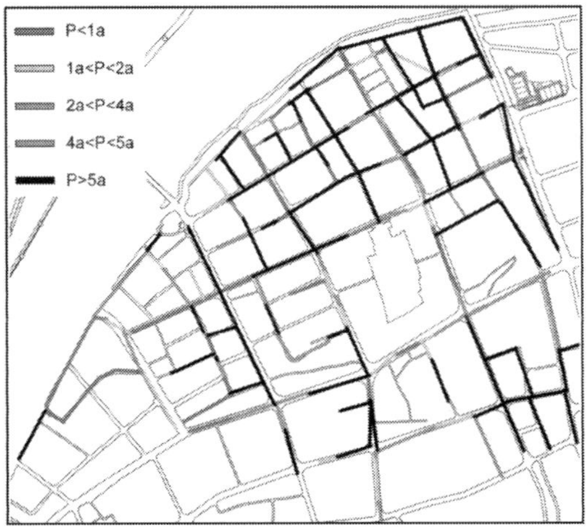

Figure. 3. Distribution map of overloaded pipelines in different recurrence periods.

The study area's pipeline network comprises three drainage outlets, whose service scopes and locations are shown in Figure 1. The peak water depth and peak time for each discharge outlet is detailed in Table 5. Outlet B, serving a larger area (83.5% of the study area), experienced water depths at P=2a, P=4a and P=5a that were 1.14, 1.20 and 1.21 times greater than those P=1a, respectively, and the peak arrival time advancing by 3 minutes, 7 minutes and 9 minutes. Overall, significant changes in water depth occurred between P=1a and P=2a. However, as rainfall intensity increased further, the changes in water depth were minimal, indicating that upstream pipeline were already fully loaded, causing water to accumulate in the upstream pipeline and nodes, or overflow without smoothly discharging through outlets.

Table 5. Statistics of various defects.

Outfall	Outfall A		Outfall B		Outfall C	
	Peak depth	Peak time	Peak depth	Peak time	Peak depth	Pea k time
P=1a	0.52 m	1:03	1.06 m	1:03	0.55 m	0:57
P=2a	0.6 m	0:57	1.20 m	1:00	0.76 m	0:53
P=4a	0.71 m	0:53	1.27 m	0:56	0.84 m	0:50
P=5a	0.74 m	0:52	1.29 m	0:54	0.86 m	0:50

4.2 Overall impact analysis of defects

The defects generalized into the model were entirely repaired, and the resulting simulation outcomes were compared. The specific situation is illustrated in Table 6.

Table 6. Overload condition of pipe network before and after defect repair

	Surcharge conduits		Ponding nodes	
	Before repair	After repair	Before repair	After repair
P=1a	43	34	46	31
P=2a	70	59	86	76
P=4a	108	93	122	112
P=5a	114	105	136	125

In the scenario of P=1a, the repair of defects led to a reduction of surcharge conduits by one-fifth and ponding nodes by one-third. As the rainfall intensity increased, the impact of the repair of defects diminished. At P=5a, the repairs resulted in an 8% reduction in both overloaded pipelines and ponding nodes. It is evident that sedimentation-based defects significantly affect the operational efficiency of the pipe network, with a more pronounced impact during shorter recurrence periods compared to longer ones.

4.3 Impact Analysis of the Defect Area

The main pipe of the west section of Bashan West Road (designated as section A) and the middle section of Xingan West Road (designated as section B), where defects are notably concentrated, are selected for analysis.

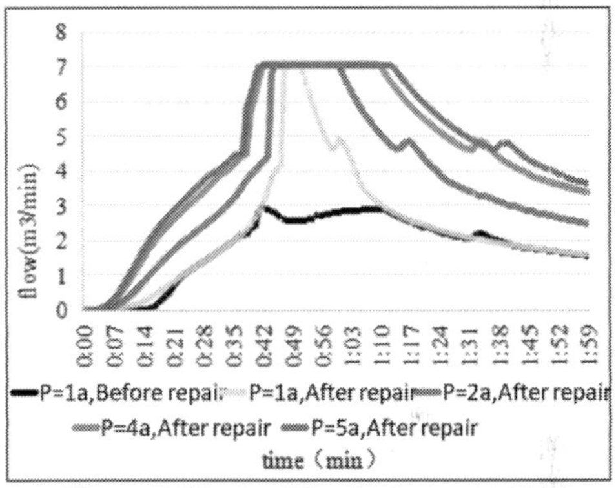

Figure. 4. Change of flow in pipe segment A.

Section A extends 890 m, with diameters ranging from 300-500 mm. It features 27 functional defects and 7 structural defects. After the repair of these defects, when P=1a, the full tube time decreased from 51 min to 12 min. When P=4a, the rainfall intensity increased nearly twice, and the full tube time reached 53 min, which was like that before the repair when P=1a.In addition, the maximum flow rate increased from 2.98 m^3/min to 7.05 m^3/min, and the total flow within a 2-hour rainfall duration escalated from 217 m^3 to 279 m^3. The change of discharge in pipe segment A is shown in Figure 4.

Section B has a total length of 734 m with a diameter of 800 mm. In this segment, 10 sedimentation-related defects and 1 misalignment defect were detected. Following repairs, the maximum flow rate at P=1a increased to 22.76 m^3/min from 16.03 m^3/min. Relative to pipelines of smaller diameter, the influence of sedimentation and other defects appears less pronounced. Distinct from Section A, the peak flow in Section B was delayed by 22 min. Comparatively, while Section A is located at the terminus of the main pipe, Section B benefits from a lengthy upstream main pipe. The extensive upstream inflow is a critical factor contributing to the severe overload of the pipeline in this section. The change of flow rate of pipe section B is shown in Figure 5.

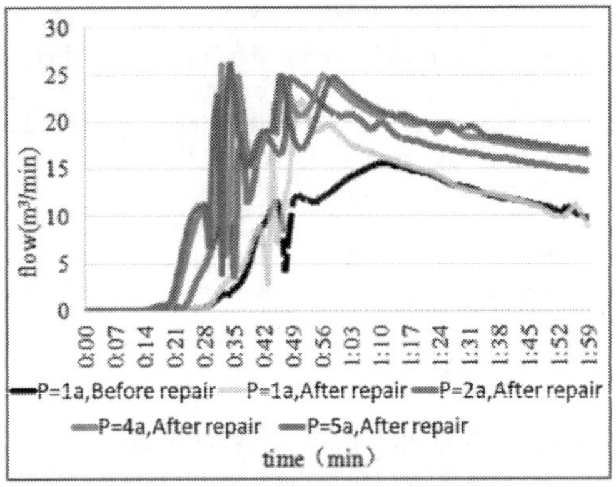

Figure. 5 . Change of flow rate in pipe segment B.

4.4 Impact Analysis of Future Rainfall Variability

Rainfall records from 1956 to 2023 were analyzed. The study area experienced an increasing trend in rainfall from 1964 to 1983 and a decreasing trend from 1984 to 1999. Entering the 21st century, annual rainfall has shown an increasing trend, with peaks surpassing historical highs in 2010 and 2021. The linear trend slope of annual rainfall is 1.121, with significant changes occurring in 1973, 1984, and 2002, indicating a general upward trend for the future.[8].

Table 7. Operational Conditions of the Network under Different Return Periods for Current and Future Scenarios.

	Current		Future	
	P=1a	P=5a	P=1a	P=5a
Surcharge conduits	34	105	47	119
Ponding nodes	31	125	69	138

Using the SWATWeather weather generator, data for the next 20 years was generated, incorporating historical rainfall characteristics. Chicago type rainfall time series for durations of 2 hours under return periods of P=1a and P=5a were generated. The simulation was conducted after repairing defects in the drainage network, and the results of pipeline overloads and node water accumulations are presented in Table 7.

It is observed that the future scenarios of P=1a and P=5a result in higher numbers of overloaded pipelines and waterlogged nodes than the current conditions under the same return periods. If the area's drainage system is retrofitted to meet current standards, it may not suffice under changing rainfall intensities.

5 Conclusions and Suggestions

Based on the drainage network survey and pipeline defect detection data, this paper employed StormDesk software to construct an SWMM model, analyzing both the operational performance and defects of the pipeline network. Conclusions are as follows.

(1) The study area exhibits a high concentration of pipeline defects. Functional defects constitute 74.08% of them, while structural defects make up 25.92%. These predominantly sediment-related defects significantly reduce the operational efficiency of the network under low return periods. However, their impact diminishes as the return period increases.

(2) The low permeability of the underlying surface of the study area results in substantial surface runoff during heavy rainfall. At P=1a, 7.44% of the pipelines are operating beyond capacity, mainly occur in branch pipelines. As rainfall intensity escalates, the operational conditions of the pipe network deteriorates. When P=5a, one-fifth of the pipelines experience prolonged overloads, predominantly in main pipelines. This results in substantial water accumulation in the upstream pipelines and nodes, leading to flooding and potentially severe waterlogging.

(3) The study area's annual rainfall has broken decade-long peak records twice in recent years, with a general upward trend in future rainfall volumes. Subsequent retrofitting processes should adequately consider changes in rainfall intensity, and a revised formula for storm intensity suitable for future scenarios should be compiled.

References

1. Baek, S.S., Ligaray, M., Pyo, J.,et al. (2020) A novel water quality module of the SWMM model for assessing low impact development (LID) in urban watersheds. J. Journal of Hydrology., 586: 124886–124886.
2. Xudong, Z., Kun, Y. (2013) The study of urban rainstorm waterlogging scenario simulation based on GIS and SWMM model — Take the example of Kunming Dongfeng East Road catchment area. In: International Conference on Geoinformatics. Kai Feng. pp. 1-5.
3. Viavattene, C., Ellis, J.B. (2013) The management of urban surface water flood risks: SUDS performance in flood reduction from extreme events. J .Water Science & Technology A Journal of the International Association on Water Pollution Research., 67(1): 99-108.
4. Rossman, L.A. (2009) Storm water management model user's manual. United States Environment Protection Agency, Washington D.C.
5. Sheng, J.G., Dan, Y.D., Liu, C.S. (2012) Study of simulation in storm sewer system of zhenjiang urban by infoworks icm model. J. Applied Mechanics & Materials., 193-194:683-686.
6. Nuhamidah N ,Junaidi A.Performance. (2023) . Evaluation of the Urban Drainage Network Structure Using the SWMM Model. J. International Journal on Advanced Science, Engineering and Information Technology., 13(2):462-468.

7. Yuxiao, Y., Jiake, L. (2024) Simulation of non-point source pollution in the watershed above the Ankang section of the Han River based on the HYPE model. J. Water Resources Protection., 40(03):140-148.
8. Yiwen, L. (2021) Analysis and prediction of rainfall and runoff characteristics in the Ankang section of the Han River Basin. J. Pearl River., 42(06):59-69.

A study of carbon cost pass-through rate differences in the power generation industry

Tiannan Ma[1], Shiying Yao[1], Ruiguang Ma[1], Yun Wang[1], Qian Wang[1], Jinrui Zhang[1], Aimeng Zou[2*]

[1] State Grid Sichuan Economic Research Institute, Chengdu, Sichuan, China
[2] Guiyang Power Supply Bureau, Guizhou Power Grid Co., Ltd., Guiyang, GuiZhou, China

Abstract. In the context of the operation of China's carbon emissions trading market, carbon cost pass-through will profoundly impact the country's power generation industry. However, the core of studying carbon cost pass-through in the power generation industry is determining the carbon cost pass-through rate (CCPTR). Based on this, this paper first takes the Cournot model as the basic principle, which analyzes the impact of different power generation enterprises' numbers and the characteristics of electricity market demand on the carbon pass-through results. Secondly, based on the Cournot model, this paper further CCPTR under marginal cost variability. The research results show that when the demand function is linear, the larger the CCPTR, the larger the proportion of carbon cost pass-through. When the demand function is isoelastic, the smaller the CCPTR, the smaller the proportion of carbon cost pass-through.

1 Introduction

With the power sector accounting for more than 40 percent of China's carbon emissions, the sector still has a long way to go in reducing carbon emissions[1]. China's carbon emissions trading market officially began operations in July 2021, and the power generation industry became one of the first to include objects[2]. The market exhibits flexibility and transparency, optimizing resource allocation and reducing the cost of emission reduction for the entire society[3-4].

As shown in Fig.1, The carbon trading market will have a direct and far-reaching impact on the production and operation of power generation companies through carbon cost pass-through[5]. As the carbon trading market construction continues, countries or regions with more mature carbon trading markets have gradually reduced the proportion of free allocation of carbon allowances and adopted auctions or other paid allocation methods. The goal of power generation enterprises to maximize their profits is: as long as the conduction is enforceable, they will try their best to pass all the carbon cost into the electricity price. Therefore, regardless of the allocation method, the carbon cost of power generation enterprises participating in carbon trading market transactions will eventually be partially or fully passed on to end-users through the electricity price.

Power generation enterprises pass on a percentage of the cost of carbon emissions to consumers by increasing the electricity price, which is known as the carbon cost pass-through rate (CCPTR)[6-7]. Hence, the critical aspect of studying the carbon cost pass-through in power generation enterprises is determining the CCPTR. For the estimation of the CCPTR, the existing literature mainly adopts two research methods: econometric analysis and techno-economic simulation. Due to the maturity of the EU carbon market, most of the econometric studies are based on the panel data of EU countries[8]. Hintermann based on a structural econometric model, evaluated the impacts of price changes on the relationship between supply and demand of electricity in the EU carbon market, and then measured the corresponding the CCPTR[9]. In other national regional carbon markets, Nelson et al. and Nazifi et al. investigated the interrelationship between carbon price and electricity price within the Australian municipal market[10-12].

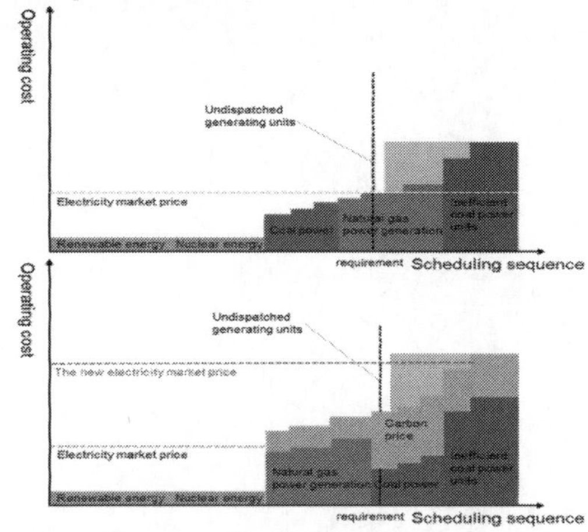

Fig.1 Influence of carbon price on generation cost and scheduling sequence

Considering the unique characteristics of electricity as a commodity and the relatively immature state of China's

* Corresponding author: 2437831572@qq.com

electricity market, it can be seen as an imperfectly competitive market with more oligopolistic traits. This paper utilizes the Cournot model to assess the influence of different manufacturers and the demand characteristics of the electricity market on the results of carbon cost pass-through. Furthermore, we analyze the CCPTR to electricity prices under various electricity market structures based on the Cournot model.

2 Carbon Pass-through model

2.1. Model building

Each power generation enterprise is assumed to develop its generation strategy (power generation) independently in the Cournot model. The total marginal cost MC_f^c of power generation enterprises f under the environmental constraints of the carbon market is shown as follows.

$$MC_f^c = C_f'(q_f) + cc_f \quad (1)$$

where q_f denotes the power generation of power generation enterprises $f(f = 1,2,\cdots,N; N \geq 2)$; $c_f'(q_f)$ represents the marginal generation cost of power generation enterprises f in addition to the carbon emission cost; cc_f denotes the marginal carbon emission cost.

In the Cournot model, the equilibrium condition for profit maximization by power generation enterprises f f is that marginal revenue (MR_f) equals marginal cost (MC_f^c), and thus it can be obtained:

$$q_f p'(q_f + Q_{-f}) + p(q_f + Q_{-f}) - c_f'(q_f) - cc_f = 0 \quad (2)$$

where $Q = q_f + Q_{-f}$ is the total generation of all power in the electricity market; Q_{-f} is the total power generation of power generation enterprises other than the power generation enterprises f ; $p(\cdot)$ is the market electricity price, which is the inverse demand function of the power generation q_f ; and $p'(q_f + Q_{-f})$ is the first-order derivative of power generation q_f concerning the market electricity price p.

The cost of carbon emissions varies across different generating units, so the marginal cost varies across power generation enterprises. Based on this, the equilibrium conditions to satisfy the profit maximization of the whole electricity market under the environmental constraints of the carbon trading market are as follows:

$$Qp'(Q) + Np(Q) - \sum_f c_f'(q_f) - \sum_f cc_f = 0 \quad (3)$$

Since there is an intrinsic connection between the power generation enterprise's power generation and its own marginal cost of carbon emissions trading, i.e., $q_f = q_f(cc_f)$. According to the implicit function theorem, Eq. (3) is obtained by taking the partial derivation of the marginal cost of carbon emission cc_f :

$$(N+1)p'(Q)\frac{dQ}{dcc_f} + Qp''(Q)\frac{dQ}{dcc_f} - \sum_f c_f''(q_f)\frac{dq_f}{dcc_f} - 1 = 0 \quad (4)$$

Assuming that the power generation enterprises are symmetric (i.e., $q_f = \frac{Q}{N}$, then $cc_f''(q_f) = c_f''(\frac{Q}{N})$. In this case it can be obtained:

$$(N+1)p'(Q)\frac{dQ}{dcc_f} + Qp''(Q)\frac{dQ}{dcc_f} - c_f''(q_f)N\frac{dq_f}{dcc_f} - 1 = 0 \quad (5)$$

CCPTR is $\frac{dp}{dcc_f} = p'(Q)\frac{dQ}{dcc_f}$, which is obtained by substituting CCPTR into Eq. (6):

$$CCPTR = \frac{dP}{dcc_f} = \frac{Np'(Q)}{(N+1)p'(Q) + Qp''(Q) - Nc''(\frac{Q}{N})} \quad (6)$$

where $\frac{p''(Q)}{p'(Q)}$ is the slope of the inverse demand function.

The relationship between CCPTR and the number of power generation enterprises, power generation, and the slope of the inverse demand function is apparent. Currently, there is a lack of authoritative research to determine the demand elasticity of electricity. Therefore, this paper further analyses the changes in CCPTR under different price elasticities of power demand, based on the CCPTR obtained by the Cournot model. The results of the CCTPR changes are shown in Fig.2.

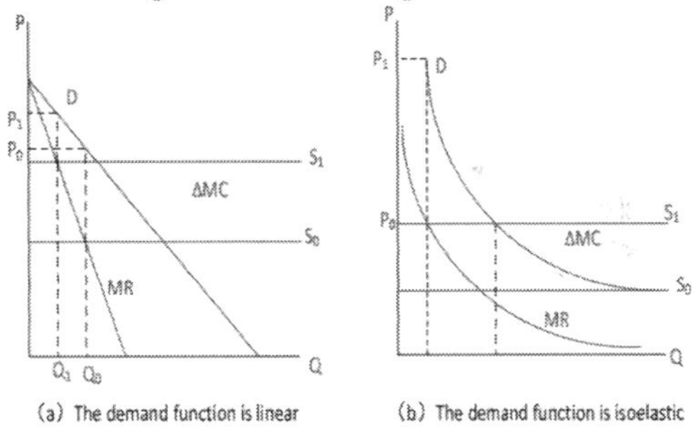

Fig.2 The demand function of under isoelastic and linear

(1) The demand function of electricity is nonlinear

Assume that the nonlinear inverse demand function of electricity is:

$$p(Q) = p_0(\frac{Q}{Q_O})^{-\frac{1}{\zeta}} \quad (7)$$

where ζ is the elasticity coefficient of electricity demand, and Q_o and p_o are the equilibrium electricity quantity and price in the electricity market without considering the carbon trading scenario, respectively.

Substituting the first-order derivative and second-order derivative of the inverse demand function of electricity into Eq. (7), we can get $\frac{Qp''(Q)}{p'(Q)} = -\frac{1}{\zeta} - 1$. Therefore, the CCPTR can be expressed as:

$$CCPTR = \frac{dp}{dcc_f} = \frac{N}{N - \frac{1}{\zeta}} \quad (8)$$

Eq. (8) states that the CCPTR is related to the number of power generation enterprises and the demand elasticity of electricity in the electricity market when the demand elasticity of electricity is isoelastic.

(2) The demand function of electricity is a linear

Assuming that the inverse demand function of electricity is:

$$p(Q) = a - bQ \quad (9)$$

where $a > 0$ and $b > 0$ are reacting to the demand elasticity of electricity. At this time, the slope of the inverse demand function is 0. The CCPTR can be expressed as:

$$CCPTR = \frac{N}{N+1} \quad (10)$$

It can be seen that the CCPTR is inversely proportional to the number of power generation enterprises in the electricity market. As the number of power generation enterprises gradually increases, the CCPTR becomes larger and larger.

2.2 Differential analysis of CCPTR under different electricity market structures

(1) With the variable marginal cost and linear demand function

Fig.3 illustrates the carbon cost pass-through for monopoly and perfect competition markets under the variable marginal cost assumption. In the presence of carbon trading, assuming the per unit of electricity generated is the same, the supply or marginal cost curve increases from S_o to S_1 by the amount of carbon cost c. In a perfectly competitive market, the electricity price at equilibrium equals the marginal cost. If the marginal cost increases due to carbon trading, the electricity price in a perfectly competitive market will also increase. However, if demand responds to the price increase, demand will decrease, leading to a decrease in supply. As a result, the marginal cost will decrease as supply decreases. Because the rise in carbon cost is partly offset by a decline in other parts of the marginal cost, the proportion of carbon cost pass-through to electricity price will also be lower under the variable marginal cost and linear demand assumptions.

Fig. 3(a) illustrates the incremental difference between marginal and carbon cost in a perfectly competitive market. c is the carbon cost of carbon emissions trading; f is the incremental marginal cost due to carbon emissions trading; the difference between the incremental carbon and marginal cost of carbon emissions trading is $g = c - f$. The incremental competitive electricity price ($P_1 - P_0$) is lower than the carbon cost under perfect competition and linear demand.

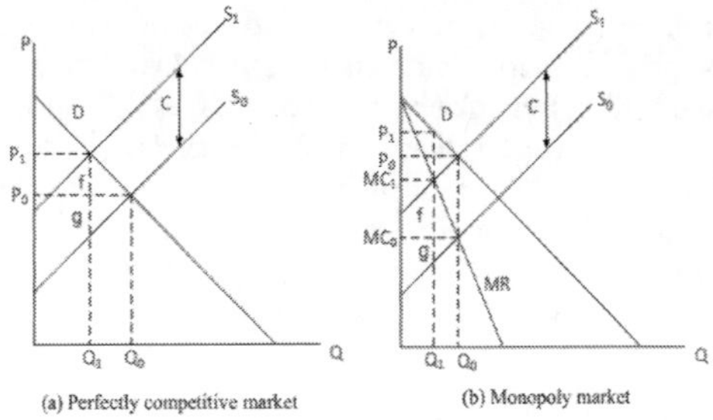

Fig.3 Carbon cost pass-through mechanism under variable marginal cost and linear demand

The difference in CCPTR under variable and constant marginal cost conditions depends on the slopes of the demand and supply curves. In a market structure characterized by linear demand and competition among N power generation enterprises, if CCPTR is defined as $\frac{dP}{dMC}$, it is equal to $\frac{N}{N+1}$ regardless of the change in the slopes of the demand and supply curves as long as the demand curve is linear. If CCPTR is defined as $\frac{dP}{dCC}$ and marginal cost is variable, CCPTR for $\frac{dP}{dCC}$ is always lower than CCPTR defined as $\frac{dP}{dMC}$.

Under linear demand, the inverse demand curve is parameterized as Eq. (11) such that the competitive equilibrium without carbon trading passes through the point (Q_o, P_o), where the elasticity ζ (in absolute value).

$$P = P_o(1 + \frac{1}{\zeta}) - (\frac{P_o}{Q_o \zeta})Q \quad (11)$$

Assuming that there are N symmetric power generation enterprises under equal elastic supply conditions, the marginal power generation cost of a single enterprise except the carbon quota cost is $P_0(\frac{N_{qf}}{Q_o})^b$, according to the condition of profit maximization (MR=MC), Eq. (12) can be obtained as follows:

$$P - (\frac{1}{\zeta N})(\frac{P_o}{Q_o})Q(1+\theta) - P_0(\frac{Q}{Q_o})^b - CC = 0 \quad (12)$$

Derivation of Eq. (11) yields $\frac{dQ}{dP} = -\zeta(\frac{Q_o}{P_o})$, and $CC = 0$ tends to infinity with $\frac{Q}{Q_o} \approx 1$, which yields the derivative at $\frac{dP}{dCC}$.

$$CCPTR = \frac{dp}{dcc} = \frac{1}{1 + \frac{1+\theta}{N} + \zeta b} \quad (13)$$

Because the power generation enterprises' power generation is all positive ($(1+\theta)q_f > 0$), $\theta > 0$. The elasticity of supply $\frac{1}{b}>0$ and the elasticity of demand $\zeta>0$, so the denominator $1+\frac{\theta}{N}+\zeta b>0$. Therefore, the CCPTR is positive. Substituting Eq. (11) into Eq. (12) yields $\frac{\partial CCPTR}{\partial N}$, $\frac{\partial CCPTR}{\partial \zeta}$ and $\frac{\partial CCPTR}{\partial b}$ as functions of N, b, ζ, and CCPTR.

$$\frac{\partial CCPTR}{\partial N}=\frac{(1+\theta)N^{-2}}{(1+\frac{1+\theta}{N}+\zeta b)^2} \tag{14}$$

$$\frac{\partial CCPTR}{\partial \zeta}=\frac{-b}{(1+\frac{1+\theta}{N}+\zeta b)^2} \tag{15}$$

$$\frac{\partial CCPTR}{\partial b}=\frac{-\zeta}{(1+\frac{1+\theta}{N}+\zeta b)^2} \tag{16}$$

Eq. (14) is greater than 0, which indicates that the CCPTR is positively related to the number of power generation enterprises N in the market; Eq. (15) is less than 0, which suggests that CCPTR is negatively related to the demand elasticity. Finally, the assumption that Eq. (16) is less than 0 indicates that the CCPTR positively relates to the supply elasticity $\frac{1}{b}$. Thus, the CCPTR of a market structure containing linear demand and isoelastic supply from power generation enterprises N as follows:

$$CCPTR=\frac{dp}{dcc}=\frac{1}{1+\frac{1}{N}+\zeta b} \tag{17}$$

Where N is the number of active power generation enterprises in the market; ζ is the absolute value of the demand elasticity in the competitive equilibrium (Q_O,P_o) before carbon emissions trading; and b is the supply elasticity function.

(2) With the variable marginal cost and isoelastic demand function

Fig.4 illustrates the carbon cost pass-through for the two extreme cases of monopoly and perfect competition markets under the variable marginal cost and isoelastic demand function. In a market structure with an isoelastic demand function and N power generation enterprises, the CCPTR defined as $\frac{dP}{dcc}$ is always lower than the CCPTR defined as $\frac{dP}{dMC}$ if the marginal cost is upward-sloping.

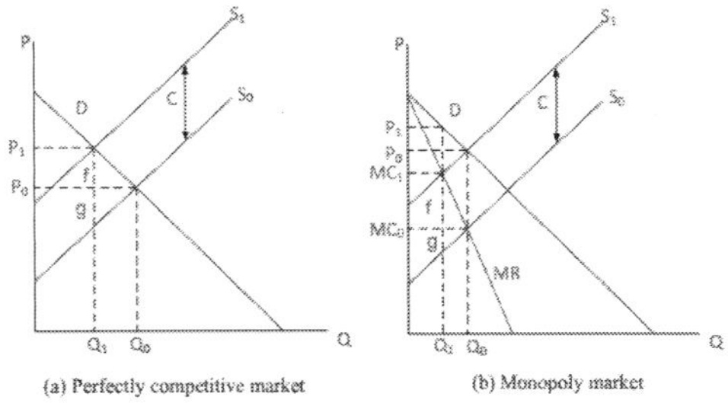

Fig.4 Carbon cost pass-through mechanism under the variable marginal cost and isoelastic demand function

Assuming that the marginal cost is $MC=P_o(\frac{Q}{Q_O})^b$ under the isoelastic supply condition, the marginal cost of the power generation enterprises f without considering the carbon quota trading is $MC_f=P_o(\frac{N_{qf}}{Q_O})^b$, where $\frac{1}{b}$ is the supply elasticity. The isoelastic demand curve is $Q=Q_o(\frac{P}{P_o})^{-\zeta}$, and the inverse demand function can be written as $P=P_o(\frac{Q}{Q_o})^{-\frac{1}{\zeta}}$.

The marginal cost of power generation enterprises f can be expressed as follows:

$$MC_f^c=C_f'(q_f)+cc_f \tag{18}$$

According to the profit maximization condition $MR=MC$, the power generation enterprises f reach equilibrium with the following equation.

$$P(1-\frac{1+\theta}{N\zeta})-P_o(\frac{Q}{Q_o})^b-CC=0 \tag{19}$$

In order to investigate the carbon cost pass-through, Eq. (20) is derived to obtain Eq. (19).

$$CCPTR=\frac{dP}{dCC}=\frac{1}{1-\frac{1+\theta}{N\zeta}+b\zeta P_o^{\zeta b+1}P^{-\zeta b-1}} \tag{20}$$

Substituting $Q=Q_o(\frac{P}{P_o})^{-\zeta}$ and $CC=0$ into Eq. (20), Eq. (21) is obtained:

$$p(1-\frac{1}{N\zeta})-P_0\left(\frac{Q_o(\frac{P}{p_o})^{-\zeta}}{Q_o}\right)^b=0 \tag{21}$$

The simplification gives Eq. (22):

$$p^{-(\zeta b+1)}=(1-\frac{1}{N\zeta})P_o^{-(\zeta b+1)} \tag{22}$$

Finally, Eq. (22) is substituted into Eq. (20) to obtain Eq. (23):

$$CCPTR=\frac{dP}{dCC}=\frac{1}{(1-\frac{1+\theta}{N\zeta})(1+b\zeta)} \tag{23}$$

Therefore, in the case of isoelastic demand and supply, the CCPTR in a market with N power generation enterprises as follows:

$$CCPTR = \frac{dP}{dCC} = \frac{1}{(1-\frac{1}{N\zeta})(1+b\zeta)} \tag{24}$$

If the supply elasticity is much higher than the demand elasticity $1+b\zeta$ close to 1), CCPTR can exceed 1.

To determine the effect of demand elasticity ζ, supply elasticity $\frac{1}{b}$, and the number of power generation N on the CCPTR, the partial derivatives of ζ, b, and N are obtained according to Eq. (24), respectively:

$$\frac{\partial CCPTR}{\partial \zeta} = -(1-\frac{1+\theta}{(N\zeta)(1+b\zeta)})^{-2}((N\zeta)^{-2}N(1+b\zeta) + (1-\frac{1+\theta}{(N\zeta)b})) \tag{25}$$

$$\frac{\partial CCPTR}{\partial b} = -(1-\frac{1+\theta}{(N\zeta)(1+b\zeta)})^{-2}(1-\frac{1+\theta}{(N\zeta)\zeta}) \tag{26}$$

$$\frac{\partial CCPTR}{\partial N} = -(1-\frac{1+\theta}{(N\zeta)(1+b\zeta)})^{-2}((N\zeta)^{-2}\zeta(1+b\zeta)) \tag{27}$$

Eq. (25) is less than 0, indicating that the CCPTR is negatively related to the demand elasticity ζ. Eq. (26) is greater than 0, which means that it suggests that the CCPTR is negatively associated with the supply elasticity $\frac{1}{b}$ (1/b); and Eq. (26) is greater than 0, which means that it indicates that the CCPTR is negatively related to the number of power generation enterprises N.

3. Conclusion

This paper presents a Cournot model of carbon cost pass-through in the electric power industry based on general equilibrium theory. Using the Cournot model, the paper discusses the CCPTR on electricity price under variable marginal cost. The results indicate the following:

(1) When the demand function of electricity is linear, the CCPTR is only positively proportional to the number of power generation enterprises in the electricity market, and as the number of power generation enterprises gradually increases, the CCPTR becomes larger and larger.

(2) When the demand function of electricity is isoelastic, the smaller the number of power generation enterprises or the more inelastic the demand function of electricity in price, the greater the CCPTR. Conversely, the smaller the CCPTR.

(3) When marginal costs vary and demand is linear, the CCPTR is higher with less demand elasticity or more elastic supply.

(4) When marginal costs vary and demand is isoelastic, the CCPTR is lower with more demand elasticity or less elastic supply.

Acknowledgments

This work was financially supported by science and technology project of State Grid Corporation of China. (SGSCJY00NYJS2200027).

References

1. Wang Chenren, Chen Yan, Liu Mneg. Analysis and Suggestions on the Situation of Ensuring Supply and Stabling Price of Fossil Energy in China[J]. Price:Theory & Practice, 2023(03): 31-37+64.\
2. Li Xundong, Tan Qingbo, Zhao Haochen,et al.Carbon Emission Factors and Decoupling Effects of China's Power Industry under the Background of Carbon Peak[J/OL]. Electric Power: 1-14 [2024-04-26]
3. Yu SongMi, Fan Ying, Zhu Lei, et al. Modeling the emission trading scheme from an agent-based perspective: system dynamics emerging from firms' coordination among abatement options[J]. European Journal of Operational Research, 2020, 286 (3): 1113-1128.
4. Wei Y G, Li Y, Wang Z C. Multiple price bubbles in global major emis- sion trading schemes: evidence from European Union, New Zealand, South Korea and China[J]. Energy economics, 2022, 113: 106232.
5. Wei Yigang, Zhu Rongqi, Tan Longyan. Carbon cost pass-through rate and dynamic characteristics of power generation enterprises in China's national carbon market[J]. China Population, Resources and Environment, 2024, (03): 50-59.
6. Dagoumas A S, Polemis M L. Carbon pass-through in the elec- tricity sector: an econometric analysis[J]. Energy economics, 2020, 86: 104621.
7. Feng Yongsheng, Zhou Yamin. Global Hydrogen Energy Development Trend and China's Strategic Choice[J]. Financial Minds,2021, 6(4): 102-123, 143.
8. Kara M , Syri S , Lehtil A ,et al. The impacts of EU CO_2 emissions trading on electricity markets and electricity consumers in Finland[J]. Energy Economics, 2008, 30(2): 193-211.
9. Hintermann B. Pass-through of CO_2 emission costs to hourly electricity prices in Germany[J]. Journal of the Association of Environment and resources an environment, 2022, 20(3): 227-236. DOI:10.1086/688486.
10. Nelson T , Kelley S , Orton F .A literature review of economic studies on carbon pricing and Australian wholesale electricity markets[J]. Energy Policy, 2012, 49: 217-224.
11. Nazifi F. The pass-through rates of carbon costs on to electricity prices within the Australian National Electricity Market[J]. Environmental Economics & Policy Studies, 2016, 18(1): 41-62.
12. Nazifi F, Trück Stefan, Zhu L. Carbon pass-through rates on spot electricity prices in Australia[J]. Energy Economics, 2021, 96: 105178.

AUTHOR INDEX

AUTHOR INDEX

EDP Sciences – Web of Conferences
17, Avenue du Hoggar
Parc d'Activité de Courtabouf
BP 112
F-91944 Les Ulis Cedex A
France

ISBN 979-8-3313-0109-5

CURRAN ASSOCIATES INC.
proceedings.com

EOS Annual Meeting (EOSAM 2021)

EPJ Web of Conferences Volume 255 (2021)

Rome, Italy
13 - 17 September 2021

edp sciences